AF781052

Food Processing and Packaging Technologies - Recent Advances

Edited by Jaya Shankar Tumuluru

Published in London, United Kingdom

Food Processing and Packaging Technologies - Recent Advances
http://dx.doi.org/10.5772/intechopen.100912
Edited by Jaya Shankar Tumuluru

Contributors
Rosana Carvalho Esteves, Abubakar Ibrahim Garba, Duygu Alp, Özcan Bulantekin, Rekha Rani, Bhushan Devidas Meshram, Vaibhav Kisanrao Lule, Shivani Vyawahare, Thomas Amarachukwu Uzombah, Ahmed El Ghorab, Hamdy Shaaban, Khaled El-Massry, Mohamed Abdelgawad, Shaima Mohamed Nabil Moustafa, Ibrahim H. Alsohaimi, Amr Farouk, Pooja Nikhanj, Simran Saini, Mohini Prabha Singh, Gurliin Kaur, Juhi Kumari, Erika Loučanová, Ján Parobek, Martina Nosáľová, Ayman Ibrahim, Tiziana M.P. Cattaneo, Lajos Helyes, Alia Amer, Rubén Vázquez-Medina, Daniel Aguilar-Torres, Omar Jiménez-Ramírez, Gonzalo A. Ramos-López, Juan A. Jimenez-Garcia, Małgorzata Mizielińska, Artur Bartkowiak, Carolina Valenzuela, Valeria Villanueva, Yanelis Ruiz, Fabrizzio Valdés, Marcela Sepúlveda, Ravi Kondle, Kushagra Sharma, Gurpreet Singh, Amit Kotiyal, Jaya Shankar Tumuluru

© The Editor(s) and the Author(s) 2023

The rights of the editor(s) and the author(s) have been asserted in accordance with the Copyright, Designs and Patents Act 1988. All rights to the book as a whole are reserved by INTECHOPEN LIMITED. The book as a whole (compilation) cannot be reproduced, distributed or used for commercial or non-commercial purposes without INTECHOPEN LIMITED's written permission. Enquiries concerning the use of the book should be directed to INTECHOPEN LIMITED rights and permissions department (permissions@intechopen.com).

Violations are liable to prosecution under the governing Copyright Law.

Individual chapters of this publication are distributed under the terms of the Creative Commons Attribution 3.0 Unported License which permits commercial use, distribution and reproduction of the individual chapters, provided the original author(s) and source publication are appropriately acknowledged. If so indicated, certain images may not be included under the Creative Commons license. In such cases users will need to obtain permission from the license holder to reproduce the material. More details and guidelines concerning content reuse and adaptation can be found at http://www.intechopen.com/copyright-policy.html.

Notice

Statements and opinions expressed in the chapters are these of the individual contributors and not necessarily those of the editors or publisher. No responsibility is accepted for the accuracy of information contained in the published chapters. The publisher assumes no responsibility for any damage or injury to persons or property arising out of the use of any materials, instructions, methods or ideas contained in the book.

First published in London, United Kingdom, 2023 by IntechOpen
IntechOpen is the global imprint of INTECHOPEN LIMITED, registered in England and Wales, registration number: 11086078, 5 Princes Gate Court, London, SW7 2QJ, United Kingdom

British Library Cataloguing-in-Publication Data
A catalogue record for this book is available from the British Library

Additional hard and PDF copies can be obtained from orders@intechopen.com

Food Processing and Packaging Technologies - Recent Advances
Edited by Jaya Shankar Tumuluru
p. cm.
Print ISBN 978-1-80356-995-6
Online ISBN 978-1-80356-996-3
eBook (PDF) ISBN 978-1-80356-997-0

We are IntechOpen, the world's leading publisher of Open Access books
Built by scientists, for scientists

6,300+
Open access books available

171,000+
International authors and editors

190M+
Downloads

156
Countries delivered to

Our authors are among the
Top 1%
most cited scientists

12.2%
Contributors from top 500 universities

CLARIVATE ANALYTICS BOOK CITATION INDEX INDEXED

WEB OF SCIENCE™

Selection of our books indexed in the Book Citation Index in Web of Science™ Core Collection (BKCI)

Interested in publishing with us?
Contact book.department@intechopen.com

Numbers displayed above are based on latest data collected.
For more information visit www.intechopen.com

For EU product safety concerns:
IN TECH d.o.o., Prolaz Marije Krucifikse Kozulić 3, 51000 Rijeka, Croatia,
info@intechopen.com or visit our website at intechopen.com.

Meet the editor

Jaya Shankar Tumuluru works as a research agricultural engineer at the Southwestern Cotton Ginning Research Laboratory, the United States Department of Agriculture (USDA), Agricultural Research Service (ARS), Las Cruces, New Mexico, USA. His research focuses on improving the cotton ginning process and using cotton waste for fuels and bioproducts applications. Before joining the USDA, Dr. Tumuluru worked as a distinguished staff engineer at Idaho National Laboratory (INL), one of seventeen national labs in the US Department of Energy complex. He was the principal investigator for multiple US Department of Energy-funded projects on biomass processing for biofuels and bioproduct production. Dr. Tumuluru's research at INL was focused on mechanical preprocessing and thermal pretreatments of biomass. He has published papers on biomass preprocessing and pretreatments, food processing, preservation and storage, and prediction and optimization of the bioprocess using artificial neural networks and evolutionary algorithms. Dr. Tumuluru has published more than 100 papers in journals and conference proceedings; presented more than 85 papers in national and international conferences; edited books on biomass preprocessing, valorization, and volume estimation; edited a monograph on biomass densification; and published more than 15 book chapters. He was named an R&D 100 Award Finalist in 2018 and 2020; received the INL Outstanding Engineering Achievement Award for 2019; the Asian American Engineer of the Year Award for 2018; the INL Outstanding Achievement in Scientific and Technical Publication Award in 2014; the Outstanding Reviewer Award for the American Society of Agricultural and Biological Engineers Food and Process Engineering Division in 2013, 2014, and 2016; the Outstanding Reviewer for the Biomass and Bioenergy Journal in 2015; and the Outstanding Reviewer for the UK Institute of Chemical Engineers in 2011.

Contents

Preface **IX**

Section 1
Food Processing **1**

Chapter 1 **3**
Introductory Chapter: Food Processing, Preservation, and Packaging – A Brief Overview
by Jaya Shankar Tumuluru

Chapter 2 **15**
Acoustic and Thermal Analysis of Food
by Daniel Aguilar-Torres, Omar Jiménez-Ramírez, Juan A. Jimenez-Garcia, Gonzalo A. Ramos-López and Rubén Vázquez-Medina

Chapter 3 **39**
Fresh Cut Fruits and Vegetables Disinfection Pretreatment: A Novel Approach to Extend Fresh Cut's Shelf Life
by Pooja Nikhanj, Mohini Prabha Singh, Simran Saini, Gurliin Kaur and Juhi Kumari

Chapter 4 **63**
Using Nanotechnology for Enhancing the Shelf Life of Fruits
by Ravi Kondle, Kushagra Sharma, Gurpreet Singh and Amit Kotiyal

Chapter 5 **83**
Perspective Chapter: Technological Strategies to Increase Insect Consumption – Transformation of Commodities Meal and Oil into Food/Functional Ingredients
by Valeria Villanueva, Yanelis Ruiz, Fabrizzio Valdés, Marcela Sepúlveda and Carolina Valenzuela

Chapter 6 **115**
Flavoring and Coating Technologies for Processing Methods, Packaging Materials, and Preservation of Food
by Ahmed El Ghorab, Hamdy Shaaban, Ibrahim H. Alsohaimi, Khaled El-Massry, Amr Farouk, Mohamed Abdelgawad and Shaima M.N. Moustafa

Chapter 7 135
Drying Technology Evolution and Global Concerns Related to Food Security and Sustainability
by Ayman Ibrahim, Tiziana M.P. Cattaneo, Alia Amer and Lajos Helyes

Section 2
Food Packaging 183

Chapter 8 185
The Importance of Packaging in Certified Organic Food: A Matter of Conscience
by Rosana Carvalho Esteves

Chapter 9 207
The Perception of Intelligent Packaging Innovation: The Latest Process and Technological Progress
by Erika Loučanová, Ján Parobek and Martina Nosáľová

Chapter 10 229
Overview of Food Antimicrobial Packaging
by Małgorzata Mizielińska and Artur Bartkowiak

Chapter 11 261
Perspective Chapter: Development of Food Packaging Films from Microorganism-Generated Polyhydroxyalkanoates
by Özcan Bulantekin and Duygu Alp

Chapter 12 283
The Place of Packaging System in Advancing Food Preservation for Promoting Food Products' Market Share
by Thomas Amarachukwu Uzombah

Chapter 13 309
Application of Edible Packaging in Dairy and Food Industry
by Bhushan Devidas Meshram, Vaibhav Kisanrao Lule, Shivani Vyawahare and Rekha Rani

Chapter 14 353
Food Preservation Packaging
by Abubakar Ibrahim Garba

Preface

Food processing helps transform raw ingredients into food products with minimal undesired microorganisms, improving storage and stability and making food more suitable for human consumption. Food processing includes physical, chemical, and thermal processes or a combination of these that can affect the physical properties and chemical composition of food while retaining its quality for ease and safety of consumption. Physical, chemical, and thermal methods change food size, shape, texture, appearance, and flavor and improve its sensory characteristics.

Food preservation includes retaining the quality of micro and macronutrients during storage and transportation or shipment nationally and internationally. The preservation technologies that can help retain food quality for long periods are crucial to reduce food loss and increasing sustainability. In addition, food preservation avoids spoilage caused by various chemical and biochemical reactions. Some traditional and modern preservation methods such as drying, chilling, freezing, pasteurization, irradiation, high-pressure, and hurdle technology prevent chemical and biochemical reactions in foods.

Food packaging is the heart of the modern food industry. Most processed foods are sold as packed products. The packaging of foods reduces waste and ensures that the food retains most of the desired qualities during transportation and storage until the consumer consumes it. In addition, a good packing material helps to increase the shelf life of food products during the whole supply chain. Even though food packaging is an integral part of the food industry to store various foods hygienically, there are concerns about using some packaging materials, such as plastic, polyethylene, and styrofoam, as they can release toxins when heated and can be dangerous to consumers. Despite the significant advantages of using packaging materials, there are challenges in using plastics from a food safety point of view. Another major concern about using plastics as a packaging material is environmental pollution. The packaging industry uses a variety of substances, such as dyes for printing labels and glues and adhesives to keep the packages closed; these materials should be non-reactive to the foods. Currently, there is a great emphasis on developing biobased packaging materials that are safe for food and reduce environmental pollution.

The are many benefits of food processing, preservation, and packaging, such as increased food safety, improved nutrition, longer shelf life, and increased economic opportunities. In addition, food processing and preservation help to reduce post-harvest losses. The food processing industry is vital in any country's economy and is about 10% of most developing countries' gross domestic product. Therefore, there should be incentives for developing the food processing industries. In general, investment in the food processing industry is very lucrative because it offers a good return. However, the food processing industry also faces many challenges, such as a lack of infrastructure, skilled workforce, raw materials, energy, storage facilities, and finances.

This book discusses the use of food processing and preservation packaging to tackle the challenges of food safety, nutritional security, and sustainability. Topics covered include edible packaging materials, intelligent packaging materials, nanotechnology for enhancing food shelf life, advanced food packaging systems, green materials for food packaging, antimicrobial packaging materials, food drying technologies, methods of food processing, food analysis using acoustic and thermal methods, food formulations, and functional foods. The research presented herein is useful for students, researchers, and food processing preservation professionals. This volume highlights advances in food processing and packaging systems to help food professionals and engineers increase food quality and preserve food for longer without generating waste.

Jaya Shankar Tumuluru
Southwestern Cotton Ginning Research Laboratory,
United States Department of Agriculture,
Agricultural Research Service,
Las Cruces, New Mexico, USA

Section 1

Food Processing

Chapter 1

Introductory Chapter: Food Processing, Preservation, and Packaging – A Brief Overview

Jaya Shankar Tumuluru

1. Introduction

Food processing, preservation, and packaging are important to increase food availability for human consumption. Food processing includes mechanical, chemical, and thermal methods to process foods to increase their palatability and shelf life. Food processing transforms raw ingredients into food or other intermediate products, and preservation is the process of handling and treating food to control its spoilage by stopping the attack and growth of foodborne diseases causing microbes, avoiding oxidation of fats (rancidity), and maintaining the nutritional value, texture, and flavor of the food [1]. According to Saini et al. [2], the chemicals, microbes, and enzymes present in the food itself result in food spoilage if not processed and preserved. Besides, food and its products must be transported from one place to another. During transit, there are chances to deteriorate the food, loss or decrease in morphological properties, and reduction in the nutritional value of the food. Therefore, it is important to make efforts for food processing and preservation for longer shelf life, stability in quality, maintaining morphological properties, and no change in taste [1].

Various traditional and modern methods were developed, considering the importance of food preservation to reduce post-harvest losses and the chances of food poisoning and other diseases. Commonly used food preservation methods are refrigeration, canning, irradiation, drying, salting, smoking, and fermentation, which help improve the shelf stability of foods such as meats, fruits, vegetables, and fish-based products. Many traditional methods, such as preserving fruits by changing them into jams, had lower energy input and carbon footprint than modern techniques [3]. Some traditional methods involve boiling fruits to kill microbes and decrease the moisture contents, adding enough sugar to prevent the regrowth of microbes, and sealing them in an airtight jar to prevent contamination [3]. Sugaring causes the jam much hypertonic and makes it hard for microbes to survive. Various research areas of food preservation include preventing the growth of bacteria, fungi, or other microorganisms and preventing chemical reactions, such as retarding the oxidation of fats, thus extending the shelf life of food products. In addition, packaging research focuses on improving the shelf life of convenient, ready-to-eat, tasty, and mild processed food products. Currently novel biodegradable packaging materials are gaining importance in addressing the environmental pollution caused by fossil-based packaging materials. This chapter provides an overview of food processing, preservation, and packaging technologies used to develop new products and improve their shelf life.

IntechOpen

2. Food spoilage

By definition, food spoilage is the process that renders food unfit for human consumption [4]. Various factors, such as contamination by microorganisms, insect infestation, or enzyme degradation, cause food spoilage [5]. In addition, physical changes, such as tearing plant or animal tissues, and chemical changes, such as oxidation of certain constituents of food, cause food spoilage. Foods from plant or animal sources degrade after harvest and slaughter due to enzymes in the plant cells and animal tissues and mechanical damage caused during harvesting and post-harvest handling. The enzymes in the foods break down due to chemical reactions catalyzed due to the storage environment causing food spoilage or degradation—the changes or degradation results in off-flavors, textural changes, and loss of nutrients [6]. The common food spoilage bacteria are *Lactobacillus*, yeasts, *Saccharomyces*, and molds (e.g., *Rhizopus*). In addition, bacteria and fungi (yeasts and molds) cause food spoilage and foodborne illnesses. In addition, microorganisms may contaminate foods during harvest, storage, processing, distribution, handling, or preparation. Enzymes that cause the degradation of the food quality are listed in **Table 1** [7].

Enzyme	Food	Type of spoilage
Ascorbic acid oxidase	Vegetables	Vitamin C destruction
Lipase	Cereals	Discoloration
	Milk	Rancidity
	Oils	Rancidity
Lipoxygenase	Vegetables	Vitamin A destruction and off-flavor
Pectic enzyme	Citrus juices	Pectic substances destruction
	Fruits	Softening
Peroxidase	Fruits	Browning reactions
Polyphenol oxidase	Fruits, vegetables	Off-flavor, browning, and vitamin loss
Protease	Eggs	Shelf-life reduction of fresh and dried eggs
	Crab, lobster	Excessive tenderization
	Flour	Gluten formation reduction
Thiaminase	Meats, fish	Thiamine destruction

Table 1.
Enzymes involved in the degradation of food quality.

3. Food storage

Food storage is an important part of food preservation, and many reactions during the storage cause quality degradation. Improper storage can adversely impact nutrient contents. For example, vitamin C and thiamine may be lost from foods during storage, especially at elevated temperatures [8]. Also, during storage, the food changes color, loses texture, and develops off-flavors. Therefore, the proper food storage system should be designed to avoid fresh foods with undesirable changes while retaining the maximum quality. One of the important parameters to consider in designing the food system is the temperature [7]. Lower temperature storage is considered to reduce most of the reactions and results in minimum quality losses. In terms of storage environments, careful control of atmospheric gases such as oxygen, carbon dioxide, and ethylene of storage environments can extend fresh food storage [9]. For example, in North America, the apple industry utilizes controlled-atmosphere storage facilities to preserve the fruit's quality.

4. Food processing and preservation

Food processing by chemical or physical means converts the harvested food ingredients into food or other intermediate products [1]. It is the process of producing raw food ingredients into marketable products that consumers can easily prepare and use. In addition, it is easy to keep processed foods stored for a long time—examples include canned and frozen fruits and vegetables and fortified foods (micro or micronutrient-rich foods).

Food processing must be balanced with food preservation also. Food preservation is to stop or slow down the spoilage of food, and loss of quality, and improve the edibility of food for a longer time. Some technologies that increase food products' shelf life are heating, drying, canning, freezing, and others [10]. These technologies inactivate the microorganisms responsible for food spoilage and foodborne diseases or inhibit their growth. The major advantage of food processing is that it makes it more edible, palatable, and safe to consume and increases the shelf life after harvesting. New technologies were developed to meet the food requirements regarding quality, flavor, taste, and shelf life. Some commonly used food processing methods are chopping or slicing, mincing, liquefaction, fermentation, emulsification, cooking, mixing, and gasification, such as adding gas to bread or soft drinks [11]. In addition, some novel processing technologies were also developed to produce durable products available all year round regardless of their seasonality which, in turn, leads to reduced post-harvest loss.

Most food processes utilize six-unit operations: heat transfer, fluid flow, mass transfer, mixing, size adjustment, and separation [12]. The scope of food processing describes unit operations occurring after the harvest of raw materials until they are processed into food products, packaged, and shipped for retailing. The food preservation methods eliminate harmful pathogens present in the food and minimize or eliminate spoilage microorganisms and enzymes for the shelf-life extension. Food's physical, chemical, and thermal properties help identify the extent of process uniformity during physical, chemical, and thermal processes such as grinding, mixing, chemical modification, pasteurization, and sterilization. Therefore, food scientists and process engineers must adequately characterize the food's thermophysical and chemical properties.

4.1 Food preservation

The goal of food preservation is to prevent the growth of bacteria, fungi, or other microorganisms and retarding the oxidation of fats that cause rancidity, thus promoting longer shelf life and reducing hazards from eating the food [1]. The main goal of preservation is to increase the safety of food products. If the safety of the foods is compromised, it can result in contamination and cause widespread illness. Several food preservation methods are designed specifically to preserve food. Traditional methods are currently used to preserve food and extend its shelf life. Some traditional food preservation methods, such as heating, cooling, pickling, boiling, sugars, and others, are discussed below [3].

4.1.1 Freezing

Freezing is a process used to preserve a wide range of foods. The challenge of frozen foods is the change in texture and structure. For example, rapid freezing can adversely affect the texture of the foods [3].

4.1.2 Pickling

Pickling is a process where the foods are preserved in an edible and antimicrobial liquid, vinegar or vegetable oil, or anaerobic fermentation to increase the shelf life of foods [13]. There are two types of pickling: a) fermentation and b) chemical pickling. In the fermentation process, the bacteria in the liquid produces agents acting as preservatives, and in chemical pickling, the food is preserved in edible liquids that kill the microorganism and bacteria. The pickling method changes the food's texture, flavor, and taste. In Asia, pickling is widely used for many vegetables, including carrots, cauliflower, lemon, and raw mangoes [3]. In North America and other European countries, eggs, fish, and meat are also pickled [3]. During pickling, organic acids such as lactic acid and acetic acid are produced, which act as preservative agents. In addition to acids, brine is also used for food preservation. Both the acids and brine result in inhibiting the bacteria from growing.

4.1.3 Curing

The curing is used for vegetables, meat, and fish, where the moisture content is reduced using the osmosis dehydration process [14]. The osmotic dehydration process helps to reduce the moisture content of the foods such as fruits and vegetables, which helps to reduce microbial damage [14]. Curing also helps to improve the flavoring of the foods. Curing is done by adding salt, nitrates, sugars, and nitrites. Adding salt to the food products slows the oxidation process, which helps reduce the rancidity in the food products.

4.1.4 Fermentation

Food fermentation utilizes microorganisms' growth and metabolic activity to stabilize and transform food materials [15]. Fermentation is used for foods such as beer, wine, and cheese produced using microbes. During fermentation, the storage conditions, such as temperature, salt, oxygen level, and others, should be maintained to produce the microbes to preserve the food products.

4.1.5 Drying

This is one of the oldest technologies used for food preservation. This process exposes the food to sunlight to dry naturally. In addition, various types of commercial dryers are developed for drying foods—drying results in the evaporation of moisture content from food, further reducing the water activity and preventing the microorganisms from deteriorating the foods [1].

4.2 Modern methods for food preservation

4.2.1 Refrigeration

The cooling technique preserves the foods by reducing the growth of microorganisms and enzyme activity [16]. Some food products commonly stored using cooling methods are meat, dairy, and fish, thus increasing the shelf-life of the products. Refrigeration is cooling the environment artificially to bring the temperature below ambient temperature. The refrigerator's temperature ranges from 4 to 10°C. Typically, bacterial growth is slowed at these temperatures, and most bacteria enter a dormant phase. For example, the enzymes which result in the degradation of food products at room temperature, under refrigeration slow the catalytic rate of enzymes and prevent the degradation. Refrigeration is used in many homes today to preserve fresh fruits, vegetables, meat, and milk products.

4.2.2 Sugaring

During the sugar preservation of the food products, the sugar makes the foods hypertonic, and microbes do not survey in hypertonic solutions. The hypertonic solutions suck the water from the microbes and dehydrate them. The foods such as fruits and vegetables are stored in sugar or honey; examples are hams and jellies. Also, many soft drinks and concentrates are prepared based on this principle.

4.3 Modern methods of food preservation

4.3.1 Pasteurization

Most microorganisms and spores could be destroyed by applying sufficient heat to food items. Therefore, high-temperature short time (HTST) and low-temperature short time (LTST) are commonly used to preserve foods [17].

4.3.2 Freeze drying

Freeze drying is also known as lyophilization. During this process, the moisture in the foods is removed under frozen and vacuum conditions at lower temperatures. The principle of this process is that the food's water is evaporated by sublimation at low pressure [3]. The low pressure and temperature during processing help the foods retain most of their quality attributes and shape. The process is used for heat-sensitive foods. The other technology widely used to preserve heat-sensitive foods is vacuum drying.

4.3.3 Vacuum packing

During vacuum packing, the foods are placed in a plastic bag, and a vacuum is created inside the bag, by sucking the air inside the bag [3]. Under no air storage conditions, the microbes do not grow and survive. This method is used for preserving nuts as it avoids oxidation reactions, such as rancidity, and preserves the flavor.

4.3.4 Irradiation

The foods are exposed to β-particles or γ-rays radiation during this process [3]. The radiation helps kill bacteria, molds, pests, and others. The World Health Organization and Food and Agricultural Organization approved this process [3]. This process is used for species, condiments, and fresh fruits.

4.3.5 Chemical preservatives

Antimicrobial chemical agents are added for the long-term preservation of foods [3]. These chemical agents are added in small quantities, where large amounts can be toxic. For example, acid and benzoates are used for food preservation in acidic foods, such as jams, salad dressings, juices, pickles, carbonated drinks, and soy sauce. Sorbic acids and sorbates are used for cheese, wine, baked foods, and others. In the case of meats, nitrates and nitrites are used to prevent the botulism toxin [3]. For fruits and vegetables, sulfur dioxide and sulfites are used, whereas propionic acid and propionates are used for baked foods.

4.3.6 Pascalization

In this process, the foods are exposed to very high pressure to the tune of high pressure, such as 70,000 lb. per square inch [3]. This process helps to retain the food's flavor, freshness, texture, and nutrients and destroys the microbes. After this treatment, the rate of food spoilage decreases significantly. This technique is mostly used for juices and meats.

4.3.7 Biopreservation

This process uses natural microbes or antimicrobials for preservation to improve the shelf life of foods [3]. In this process, beneficial bacterial or fermentation products are used to control and inactive the microorganism's growth [3]. Lactic and acetic acid bacteria are used as biopreservatives. These produce lactic acid, acetic acid, bacteriocins, and hydrogen peroxide, which act as antimicrobials and preserve the foods. D'Amico de Alcântara et al. [18] studied the antibacterial activity of *Lactobacillus rhamnosu* against *Pseudomonas fluorescens* and *Pseudomonas putida*, both isolated from refrigerated raw milk. These authors found that the antibacterial activity is due to organic acids produced.

4.3.8 Hurdle technology

In hurdle technology, more than one approach is used to inactive the microbes. The multiple preservation approaches act as hurdles for microorganisms and prevent their spoilage. Leistner [19] used hurdle technology, combining various hurdles that secure food from spoilage and preserve nutritional quality. Also, the hurdles do not induce smell or change the texture of preserved foods. Some of the hurdle approaches are high temperature, in combination with pressure, acidity, and adding biopreservatives. The selection of hurdles depends on the food's nature and potential pathogens.

4.3.9 Nonthermal plasma

In this process, the food surface is exposed to a flame of ionized gas molecules, such as nitrogen or helium to kill the microbes on the food's surface [3].

4.3.10 Modified atmosphere

In this process, the oxygen is reduced, and carbon dioxide is increased in the storage environment. For example, vegetable and fruit salad bags are stored in reduced oxygen and higher carbon dioxide environments [9, 20, 21]. These storage environments can result in the loss of some of the nutrients in the foods. This method is also used for the preservation of grains. The carbon dioxide used helps to prevent the growth of insects, molds, and oxidation reactions and prevent grain damage. In the sealed room, the oxygen levels are reduced by infusing nitrogen gas [3]. On average, oxygen is 21% in the air, whereas, in such facilities, the oxygen level is reduced to 1 and 2%. Hermetic storage is an airtight storage widely used for storing grains. This method reduces the grain's respiration and prevents the growth of insects, fungi, and pests [22].

4.4 Benefits of food processing and preservation

The key reason is to maintain the quality and stability of a product. Fresh juice, for example, will easily phase-separate after extraction, and enzymes start to degrade valuable components, such as antioxidants. If these enzymes are inactivated by pasteurizing (heat-treating) the juice, spoilage can be slowed down [23]. This also applies to products that include a large amount of fat, as they easily become rancid when enzymes are still active and oxygen is present. Processing also makes some foodstuffs more digestible by softening tissue or breaking it down [1].

Furthermore, processing increases the variety of food products available to the consumer and makes them more convenient, meeting the demands of our on-the-go lifestyle. Lastly, food processing is important in ensuring that food is affordable. Furthermore, proper storage of foods, such as fruits, vegetables, and grains, can help reduce the spoilage and waste of foods and solve the food shortage in developing and underdeveloped countries.

Some of the major benefits and drawbacks of food processing and preservation are as follows:

- Reduces the bacteria growth, which can cause illness or diseases. For example, drying and pickling dehydrate the food product and alter the pH, preventing harmful microorganisms' growth.
- Improves the food product's shelf life.
- Processed food contains artificial ingredients.
- Over-processing of foods makes the food pleasant and leads to overconsumption.
- Most processed foods have high amounts of added sugar, which is very unhealthy.

5. Food packaging

Food packaging controls the storage environment and creates conditions extending food storage and shelf life. Some commonly used packaging materials are a) flexible paper, thin laminates, or plastic film; b) semi-rigid material such as aluminum foil, laminates, paperboard, and thermoformed plastic; and c) rigid material such as metal, glass, or thick plastic [24]. The most commonly used packaging material is plastic, which is cheap, lightweight, and easy to form in different shapes and sizes. In addition, plastic films with selective permeability properties to gases, carbon dioxide, and oxygen can be selected for storage [24, 25]. These films have led to the development of modified atmospheric packaging systems. By selecting the right barrier properties of the packaging materials, the shelf life of food products can be extended.

For example, intermediate moisture foods need to be protected from moisture during storage. Therefore, low moisture permeability packaging materials, such as polyvinyl chloride, polyvinylidene chloride, and polypropylene, can be used [26]. In the case of foods with high fatty acids, low gas permeability films can be used to reduce oxidation reactions. In the case of fresh fruits and vegetables, as they respire, packaging material such as polyethylene, which has high gas permeability, can be used. Currently, smart package systems offer properties that meet the special needs of certain foods [27]. For example, packages made with oxygen-absorbing materials remove oxygen from the inside, thus protecting oxygen-sensitive products from oxidation [28]. However, temperature-sensitive films exhibit an abrupt change in gas permeability when they are subjected to a temperature above or below a set constant. These films change from crystalline to amorphous at a given temperature, causing the gas permeability to change substantially. There are significant health and environmental concerns regarding using petroleum-based packaging materials such as plastics, polyethylene, and styrofoam, as they can release toxins into the foods when heated and can be dangerous to consumers, also they do not decompose for a long time, creating environmental issues. Packaging films made from biodegradable are gaining much importance in overcoming environmental issues. The packaging materials are made from polyhydroxyalkanoates (PHAs), biodegradable, and synthesized by various microorganisms, replacing petroleum-based plastics. These materials are environmentally friendly and, simultaneously, can be completely degraded by various microorganisms in a short time (less than a year).

6. Conclusions

Food processing and preservation are important to prevent food loss, improve storage stability, and retain most nutrients during storage. Improperly stored food can cause foodborne diseases if consumed and result in revenue loss. It is estimated that about 15% of the food produced is lost after post-harvest. Many traditional and modern preservation techniques have been developed for food storage. Traditional methods commonly used are pickling, curing, drying, and fermentation. The modern methods used for food preservation are pasteurization, freeze drying, vacuum packing, irradiation, chemical preservatives, pascalization, biopreservation, hurdle technology, nonthermal plasma, and a modified and controlled atmosphere. Both the traditional and modern methods can help improve the shelf stability of foods, such as meats, fruits, vegetables, and fish-based products. Although many existing techniques are used for food preservation, the selection of the preservation technique is based on technical and economic feasibility. In addition, novel and environmental friendly packaging methods are gaining much importance in retaining food quality during storage.

U.S. department of agriculture disclaimer

The findings and conclusions in this [publication/presentation/blog/report] are those of the author(s) and should not be construed to represent any official USDA or U.S. Government determination or policy. Mention of trade names or commercial products in this publication is solely for the purpose of providing specific information and does not imply recommendation or endorsement by the U.S. Department of Agriculture. USDA is an equal opportunity provider and employer.

Author details

Jaya Shankar Tumuluru
Southwestern Cotton Ginning Research Laboratory, United States Department of Agriculture-Agricultural Research Service, Las Cruces, New Mexico, USA

*Address all correspondence to: jayashankar.tumuluru@usda.gov

IntechOpen

© 2023 The Author(s). Licensee IntechOpen. This chapter is distributed under the terms of the Creative Commons Attribution License (http://creativecommons.org/licenses/by/3.0), which permits unrestricted use, distribution, and reproduction in any medium, provided the original work is properly cited.

References

[1] Amit SK, Uddin MM, Rahman R, Islam SMR, Khan MS. A review on mechanisms and commercial aspects of food preservation and processing. Agriculture & Food Security. 2017;**6**:51. DOI: 10.1186/s40066-017-0130-8

[2] Saini RV, Vaid P, Saini NK, Siwal SS, Gupta VK, Thakur VK, et al. Recent advancements in the technologies detecting food spoiling agents. Journal of Functional Biomaterials. 2021;**12**(4):67. DOI: 10.3390/jfb12040067

[3] Kumar A. Food preservation: Traditional and modern techniques. Acta Scientific Nutritional Health. 2019;**3**(12):45-49

[4] Petruzzi L, Corbo MR, Sinigaglia M, Bevilacqua A. Chapter 1: Microbial spoilage of foods: Fundamentals. In: Bevilacqua A, Corbo MR, Sinigaglia M, editors. Woodhead Publishing Series in Food Science, Technology and Nutrition, The Microbiological Quality of Food. UK: Woodhead Publishing; 2017. pp. 1-21

[5] Lorenzo JM, Munekata PE, Dominguez R, Pateiro M, Saraiva JA, Franco D. Main groups of microorganisms of relevance for food safety and stability: General aspects and overall description. In: Barba FJ, Sant'Ana AS, Orlien V, Koubaa M, editors. Innovative Technologies for Food Preservation. Cambridge, MA, USA: Academic Press; 2017. pp. 53-107

[6] Erkmen O, Bozoglu TF. Enzymatic and nonenzymatic food spoilage. In: Erkmen O, Bozoglu TF, editors. Food Microbiology: Principles into Practice. UK: John Wiley & Sons, Ltd.; 2016. pp. 401-406

[7] Singh RP, Wilfred DN. Food Preservation. Encyclopedia Britannica. 5 January 2023. Available from: https://www.britannica.com/topic/food-preservation [Accessed: 21 January 2023]

[8] Giannakourou MC, Taoukis PS. Effect of alternative preservation steps and storage on vitamin C stability in fruit and vegetable products: Critical review and kinetic modelling approaches. Foods. 2021;**10**(11):2630

[9] Caleb OJ, Mahajan PV, Al-Julanda Al-Said F, Opara UL. Modified atmosphere packaging technology of fresh and fresh-cut produce and the microbial consequences—A review. Food and Bioprocess Technology. 2012;**6**(2):303-329

[10] Blackburn CdeW. Managing microbial food spoilage: An overview. Food Spoilage Microorganisms. In: Blackburn CdeW, editor. Woodhead Publishing Series in Food Science, Technology and Nutrition, Food Spoilage Microorganisms. UK: Woodhead Publishing; 2006. pp. 147-170

[11] Jain A. The future of food processing technology in India. International Journal of Innovative Research in Science, Engineering and Technology (IJIRSET). 2022;**11**(2):1453-1458

[12] Park SH, Lamsal BP, Balasubramaniam VM. Principles of food processing. In: Clark S, Jung S, Lamsal B, editors. Food Processing: Principles and Applications. 2nd ed. UK: John Wiley & Sons, Ltd.; 2014. pp. 1-15

[13] Behera SS, El Sheikha AF, Hammami R, Kumar A. Traditionally fermented pickles: How the microbial diversity associated with their nutritional and health benefits? Journal

of Functional Foods. 2020;**70**:103971. DOI: 10.1016/j.jff.2020.103971

[14] Yadav AK, Singh SV. Osmotic dehydration of fruits and vegetables: A review. Journal of Food Science and Technology. 2014;**51**(9):1654-1673. DOI: 10.1007/s13197-012-0659-2

[15] Terefe NS. Food fermentation. Reference Module in Food Science. 2016:1-2. DOI: 10.1016/B978-0-08-100596-5.03420-X

[16] Berk Z. Refrigeration: Chilling and Freezing. Food Process Engineering and Technology. 2nd ed. 2013. pp. 439-460. DOI: 10.1016/B978-0-12-415923-5.00019-8

[17] Tadini CC, Gut JA. The importance of heating unit operations in the food industry to obtain safe and high-quality products. Frontiers in Nutrition. 2022;**9**:853638. DOI: 10.3389/fnut.2022.853638

[18] D'Amico de Alcântara AL, Bruzaroski SR, Luiz LL, Batista de Souza CH, Poli-Frederico RC, Fagnani R, et al. Antimicrobial activity of Lactobacillus rhamnosus against Pseudomonas fluorescens and Pseudomonas putida from raw milk. Journal of Food Processing and Preservation. 2019;**2019**(43):e14082. DOI: 10.1111/jfpp.14082

[19] Leistner I. Basic aspects of food preservation by hurdle technology. International Journal of Food Microbiology. 2000;**55**:181-186

[20] Annis PC, Dowsett HA. Low oxygen disinfestation of grain: Exposure periods needed for high mortality. In: Proc. International Conference on Controlled Atmosphere and Fumigation, Winnipeg, June 1992. Jerusalem: Caspit Press; 1993. pp. 71-83

[21] Annis PC, Morton R. The acute mortality effects of carbon dioxide on various life stages of Sitophilus oryzae. Journal of Stored Products Research. 1997;**33**:115-124

[22] Annis PC, Banks HJ. Is hermetic storage of grains feasible in modern agricultural systems? In: Corey SA, Dall DJ, Milne WM, editors. Pest Control and Sustainable Agriculture. Australia: CSIRO; 1993. pp. 479-482

[23] Umair M, Jabeen S, Ke Z, Jabbar S, Javed F, Abid M, et al. Thermal treatment alternatives for enzymes inactivation in fruit juices: Recent breakthroughs and advancements. Ultrasonics Sonochemistry. 2022;**86**:105999. DOI: 10.1016/j.ultsonch.2022.105999

[24] Marsh K, Bugusu B. Food packaging—Roles, materials, and environmental issues. Journal of Food Science. 2007;**72**(3):R39-R55

[25] Hopewell J, Dvorak R, Kosior E. Plastics recycling: Challenges and opportunities. Philosophical Transactions of the Royal Society B: Biological Sciences. 2009;**364**(1526):2115-2126

[26] Siracusa V. Food packaging permeability behaviour: A report. International Journal of Polymer Science. 2012;**2012**:302029. DOI: 10.1155/2012/302029

[27] Biji KB, Ravishankar CN, Mohan CO, Srinivasa Gopal TK. Smart packaging systems for food applications: A review. Journal of Food Science and Technology. 2015;**52**(10):6125-6135

[28] Cichello SA. Oxygen absorbers in food preservation: A review. Journal of Food Science and Technology. 2015;**52**(4):1889-1895

Chapter 2

Acoustic and Thermal Analysis of Food

Daniel Aguilar-Torres, Omar Jiménez-Ramírez, Juan A. Jimenez-Garcia, Gonzalo A. Ramos-López and Rubén Vázquez-Medina

Abstract

Exploring the food acoustic features can help to understand and effectively apply some preservation treatments that extend their expiration date. The food composition and properties are crucial issues in their acoustic behavior when stimulated with acoustic waves. If these waves are varied in frequency and intensity, the temperature of food could be affected facilitating the moisture removal or degrading its nutritional condition. Therefore, we presented a guide to determine and apply the most influential spectral component of ultrasound waves on apple and tomato when dehydrated in an ultrasound-assisted dehydration system. In this guide, applying the finite element method, we study, simulate, and analyze the acoustic and thermic behavior of apple and tomato inside a chamber when radiated with acoustic waves at (1 Hz, 1 MHz) by using up to three piezoelectric transducers. From the physical parameters defined in the simulation environment for apple and tomato, we find the relevant spectral components that can produce temperature changes in each food sample considering the radiation time and the food sample location. This work represents an analysis guide that allows for determining the best conditions for the acoustic radiation of foods, avoiding their structural and nutritional damage, and seeking the design of energy-efficient processes.

Keywords: food acoustics, food drying, ultrasound waves, food acoustic behavior, food thermal behavior

1. Introduction

Commercially available food must have an expiration date. After that date, we should consider that food has lost a significant nutrient content, or it has been degraded by contamination to detrimental levels to human health. Different methods have been developed for food preservation. For example, dehydration, refrigeration/freezing, fermentation, canning, pasteurization, and incorporation of chemical additives. Food dehydration is a reliable method commonly used for food preservation, based on food moisture removal, it must be performed by using energy-, cost-, and time-efficient technologies, and it must use practical methods that do not interfere with time spent on personal and professional daily activities. In this sense, it should be considered that, according to Kumar *et al.* [1] and Merone *et al.* [2], food dehydration implies energy-intensive processes approximately using 25% of the total energy consumed by the food industry. In addition, food dehydration may be a practical process with hygiene and temperature conditions. Therefore, under the demands of today's life, food dehydration must be an automatically controlled process ensuring that the food structure, content, and quality are preserved on time. In general, food dehydration systems have low efficiency, and consequently, the development of efficient systems is a relevant scientific and technological challenge [3–5]. Hence, when developing or selecting a food dehydration system, its efficiency must be evaluated considering removed moisture content and the moisture removal rate. On the other hand, energy consumption and dehydrated food quality are two critical requirements for the design, implementation, or selection of a dehydration system [6]. In this sense, ultrasound-assisted dehydration systems can be a promising alternative [7], because they can help to reduce energy consumption [8–10]. However, we must keep in mind that the internal microstructure of food could change drastically when it is immersed in ultrasound waves, reducing its resistance to water diffusion, and increasing its temperature. But, an ultrasound-assisted dehydration system could have higher energy, cost, and time efficiency than simple convection dehydration systems [11–13]. Additionally, they have been demonstrated to be reproducible processes that avoid further wastewater treatment and additional energy use [14, 15].

An overview of ultrasound waves application in other technologies for food processing can be revised in the works published in 2021 by Singla and Sit [16] and Khadhraoui *et al.* [17]. In conventional dehydration technologies combined with ultrasound systems, it is worth noting that the ultrasound-assisted dehydration systems can increase the dehydration rates or decrease the dehydration temperature since the ultrasound waves strongly accelerate mass transfer maintaining food quality. For example, in 2017, Fei *et al.* showed that ultrasound osmotic dehydration produces samples with reduced sugar, ascorbic acid, and soluble protein content at significantly higher rates than osmotic dehydration, and the food samples showed a better texture and microstructure. The ultrasound osmotic dehydration process not only retained the nutrient composition and flavor material more effectively but also improved the texture and efficiency of osmosis-treated mushrooms [18]. Complementary to the application of ultrasound waves in food dehydration systems, we can mention that, in 2016, Başlar *et al.* [19] reviewed different ultrasound-assisted dehydration processes, including convective, osmotic, vacuum, and freeze dehydration applications, as well as the various types of ultrasonic equipment used. They summarized the mechanisms, applications, advantages, disadvantages, and recent investigations of ultrasound-assisted dehydration concluding that ultrasound treatments can potentially provide significant improvement in food dehydration, as the dehydration process simultaneously accelerates heat and mass transfers in the system. In general, they showed that the dehydration methods combined with ultrasound-assisted dehydration resulted in less dehydration temperature and duration. Finally, they suggested that these two advantages in ultrasound-assisted dehydration systems may avoid the reduction effect of food quality in comparison with other dehydration techniques.

Therefore, considering the work of Başlar *et al.* [19], we have compared different dehydration systems against ultrasound-assisted versions of the same systems. In **Table 1**, we have included convective, osmotic, vacuum, and freeze dehydration systems considering that the ultrasound-assisted dehydration systems offer higher dehydration rates or lower food dehydration temperatures than conventional systems.

System	Advantages	Disadvantages
Convective dehydration	1. Easy operation. 2. It prevents microbial growth. 3. It reduces water activity in food. 4. It prevents enzymatic activity. 5. It allows easy packaging and transport of food.	1. High energy consumption. 2. Low dehydration rate. 3. Low product quality. 4. High dehydration temperature. 5. Long dehydration time.
Ultrasound-assisted convective dehydration	1. It reduces dehydration time. 2. It reduces dehydration temperature operation. 3. It increases the moisture transfer rate. 4. It minimizes quality losses in the final products. 5. Low-energy consumption	1. The operating frequency must be selected and adjusted. 2. An acoustic decoupling is produced between the media on which the ultrasound signals are propagated.
Osmotic dehydration	1. It minimizes the flavoring compound losses. 2. It minimizes the aromatic substance's losses. 3. It minimizes color losses. 4. The dehydrated products have a better appearance and nutritional properties.	1. Mass transfer is limited. 2. Low-mass transfer. 3. Mass transfer completely stops after a certain time.
Ultrasound-assisted osmotic dehydration	1. It enhances mass transfer during osmotic dehydration. 2. It improves the fluxes of osmotic solute to the intercellular of the dewatered food. 3. It reduces the dehydration time and cost.	1. The choice of appropriate parameters in the ultrasound generator is a considerable issue. 2. The ultrasound signal effects are reduced at high temperatures.
Vacuum dehydration	1. Low temperature. 2. It prevents food oxidation. 3. It is an energy-efficient dehydration process. 4. Low dehydration time. 5. It retains the organoleptic and nutritional properties of food. 6. Risk of occupational accidents.	1. It does not perform a continuous dehydration process. 2. It is an expensive process due to high maintenance costs. 3. It does not remove sufficient moisture from the center to the surface in food samples. 4. Low-heat transfer. 5. It requires pretreatment or combination with other methods to improve the dehydration process efficiency.
Ultrasound-assisted vacuum dehydration	1. It has a near-perfect rehydration rate. 2. Fast food dehydration. 3. It enhances food quality and appearance. 4. It accelerates mass transfer from the food center. 5. Dehydration rate increases without increasing the temperature in the dehydration chamber. 6. It improves diffusion and raises the convective mass transfer rate. 7. It has a reduced overall processing time. 8. Low ultrasound power.	1. Specific pressure and temperature parameters are required for the dehydration process of each food.

System	Advantages	Disadvantages
Freeze dehydration	1. Preservation of the activity of food nutraceuticals and pharmaceuticals. 2. Preservation of food flavor and aroma. 3. Fast water penetration and recovery of the original characteristics during a rehydration process.	1. Long dehydration times (hours or days). 2. High operating costs. 3. Limited industrial application. 4. To reduce its dehydration time, the dehydration system must be assisted by other processes.
Ultrasound-assisted freeze dehydration	1. It increases the freeze-dehydration efficiency. 2. It increases the dehydration rate. 3. It reduces operating costs. 4. Dehydration times less than 1 day. 5. It does not deteriorate food quality.	1. The food sample under dehydration must be small. 2. The process requires specific values of ultrasound signal frequency and power, as well as temperature and airflow in the dehydration chamber for each food.

Table 1.
Advantages and disadvantages of conventional and conventional ultrasound-assisted dehydration systems.

Regarding the applicability status of dehydration technology, we must say that ultrasound-assisted dehydration is mainly done at the laboratory level, which is used to study the dehydration kinetics and physicochemical, microbiological, structural, and rehydration characteristics of some foods [20]. Ultrasound-assisted dehydration systems are based on conventional dehydration techniques that are assisted by ultrasound waves resulting in different methods, such as ultrasound-assisted convective dehydration [21], ultrasound-assisted osmotic dehydration [22], ultrasound-assisted vacuum dehydration [23], and ultrasound-assisted freeze dehydration [24]. In ultrasound-assisted dehydration systems, ultrasound waves are used to increase the dehydration rates or decrease the dehydration temperature, since these waves strongly accelerate mass transfer maintaining food quality.

On the other hand, with respect to commercial alternatives, it can be stated that some ultrasound-assisted systems are beginning to emerge, such as the one used for dehydration of heat-sensitive biological materials and food processing. Some examples of these commercial systems include the following:

1. The Center for Advanced Research in Drying (CARD) is devoted to research on the drying of moist and porous materials, such as food, agricultural products, chemical products, and biopharmaceuticals. It is investigating the use of high-power ultrasound as a nonthermal, faster, and less energy-intensive method for protein dehydration [25].

2. Dehydration Assisted by Ultrasound by PUSONICS SL is a technology that offers power ultrasonic plate-transducers capable of transmitting sonic and ultrasonic energy through the air for drying heat-sensitive materials, where micro-vibrations could prevent spoilage due to damaging high-temperature levels [26].

3. Ultrasound enhanced osmotic dehydrating method and apparatus is an infiltration dehydrating method with ultrasonic strengthen and correlated device applicable to ripe fresh fruits with smooth epidermises, such as apples and pears [27].

4. Ultrasound applied to food processing as yogurt fermentation, extraction of flavors and bioactive compounds, milk homogenization, sugar crystallization for confectionery, edible oils hydrogenation, honey liquefaction, juices stabilization, wine and liquor aging, accelerated ice cream freezing, batter aeration, chocolate crystallization and conching, and meat tenderization [28, 29].

Therefore, in this work, we focus on sound–assisted convection dehydration systems, which could reduce dehydration time and cost by at least 30% compared to conventional methods [30] for food dehydration. Furthermore, it should be noted that these dehydration systems are not harmful to humans as microwave, gamma radiation, or electromagnetic field pulses systems; therefore, they can be applied in the food industry for safety reasons [31, 32]. Thereby, it is important to study, analyze, and understand the food acoustic-thermal behavior when dehydrated by using an ultrasound-assisted convection dehydration system. In this regard, we presented a guide to determine and apply the most influential spectral components of ultrasound waves on food under dehydration in an ultrasound-assisted dehydration system for increasing the temperature of food under dehydration. For this purpose, we have defined two case studies; in the first one, we study and analyze the thermo-acoustic behavior of apple samples, and in the second study case, we consider tomato samples. In both case studies, food samples are immersed in ultrasound waves and they are radiated using one and three transducers. With these case studies, after identifying the intensity and frequency of the ultrasound waves that have the greatest influence on the dehydration process of each food, we intended to observe the temperature changes experienced by food samples while they are being dehydrated. In this way, this work allows us to find the best conditions for food acoustic radiation avoiding structural and nutritional damage to the food. Also, in this guide, we show how to apply the finite element method to simulate and analyze the thermal and acoustic behavior of foods under dehydration inside the dehydration chamber of an ultrasound-assisted system.

The rest of the chapter is organized as follows. Section 2 describes some scientific works related to the analysis of ultrasound-assisted food dehydration systems. Section 3 describes the two case studies for which the thermo-acoustic analysis was performed, the used configurations for the dehydration chamber, the food composition and properties, a brief overview of transducers and acoustic waves, and the selected simulation environment. Section 4 provides the acoustic and thermal mathematical model used for the thermo-acoustic analysis by the finite element method, the spatial behavior of the ultrasound waves inside the dehydration chamber, and the temporal behavior of the temperature change in the food samples. Section 5 shows the opportunities for future work when ultrasound–assisted convection systems are used in food dehydration. Finally, Section 6 is devoted to conclusions and work perspectives.

2. Related works

There are some works reporting analysis of ultrasound-assisted food dehydration systems. For example, in 2006, García-Pérez *et al.* proposed an ultrasound-assisted convection dehydration system consisted of an aluminum–cylinder drying chamber able to create a high-intensity ultrasound field at 21.8 kHz [8]. In 2011, Khmelev *et al.* [9] also proposed an ultrasound-assisted dehydration system consisted of a resonant drying chamber to amplify the ultrasound waves. They determined that the efficiency of their dehydration system was 20% higher than the efficiency of a pure convection system. In 2015, Fernandes *et al.* examined the influence of ultrasound waves on apple dehydration and estimated the effective moisture diffusivity [33]. Lastly, Sabarez *et al.* developed and tested a high-intensity ultrasound system to assist a conventional dehydration system, which was more efficient between 46 and 57% than a conventional dehydration system [10]. Also, there are other works related to the energy and environmental analysis of ultrasound-assisted food dehydration. For example, in 2022 Chavan *et al.* simulated dehydration at an industrial scale for some vegetables [34]. On the other hand, there are works related to parametric studies about ultrasound-assisted food dehydration systems. For example, in 2020, Huang *et al.* considered the air flow rate, ultrasonic power, and mass loading in the analysis of a food ultrasound-assisted dehydration system [35].

Considering this variety of scientific works, we decided to include in this chapter the study and simulation analysis of the thermo-acoustic behavior of apple and tomato samples when they are dehydrated in an ultrasound-assisted convection system. We performed this simulation analysis by using the finite element method (FEM) implemented in COMSOL Multiphysics™. In this way, we analyzed the temporal and spectral behavior of the dehydration system, which we estimated, as a function of the radiation time and the ultrasound waves frequency, and the temperature changes of the 5 mm-size food plates (apple or tomato). This simulation analysis also identifies changes in food temperature based on the location of the samples within the dehydration chamber.

3. Case studies

In order to study the ultrasound waves behavior and their thermal effects on apple and tomato samples, we proposed an ultrasound-assisted dehydration system based on the forced convection dehydration system shown in **Figure 1** considering two case studies. In the first case study, represented by **Figure 2a**, we used one piezoelectric

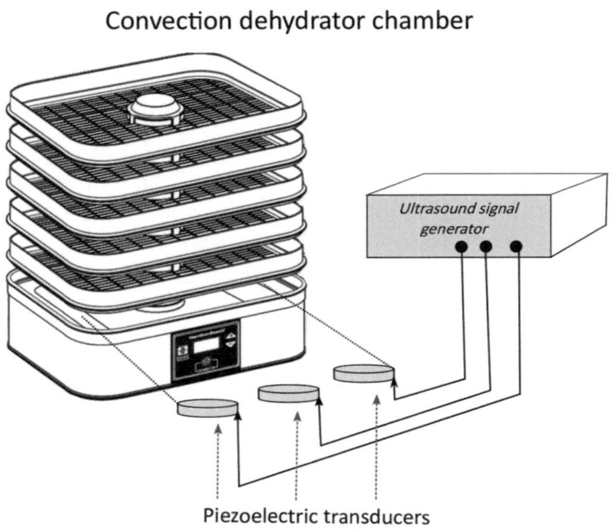

Figure 1.
Ultrasonic dehydration chamber.

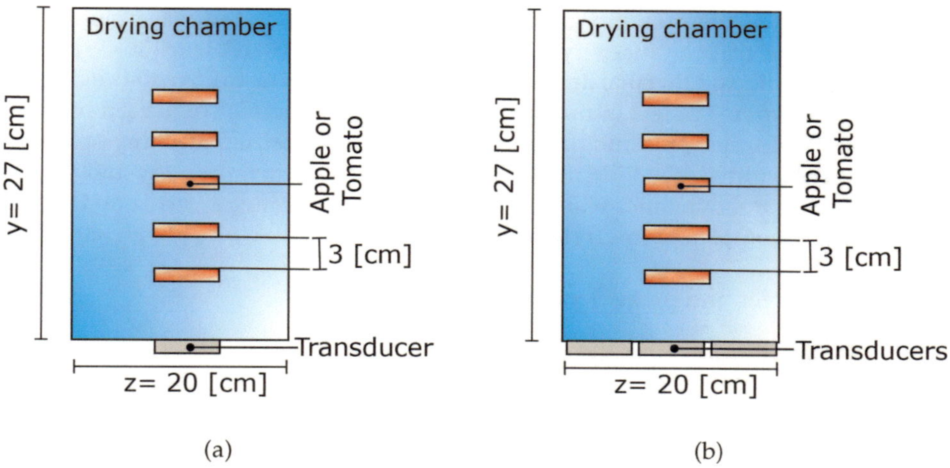

Figure 2.
Proposed geometry for the thermo-acoustic simulation when apples and tomatoes were dehydrated using: (a) one PZT, and (b) three PZT's.

transducer of size (0.5 cm × 4 cm). In the second case study, represented by **Figure 2b**, we use three piezoelectric transducers as the first case study. Considering a transversal cut at (x, z)-plane, note in **Figure 2** that the dehydration system has a (27 cm × 20 cm × 20 cm)-chamber composed of five gridded trays spaced 3 cm from each other, an electronic module to regulate the dehydration temperature, an ultrasound waves generator, and up to three piezoelectric transducers placed at the dehydration chamber base.

In both case studies, we consider 0.5 cm-thick apple and tomato plates placed in the five gridded trays and assumed that the dehydration chamber is in equilibrium conditions at a temperature of 60°C. From the geometries proposed in **Figure 2**, and considering that the simplified geometry of the dehydration chamber derived by the transversal cut at (x, z)-plane can help to simplify the thermo-acoustic analysis of the proposed system, we show an example of the FEM simulation results for both case studies considering the effects of the ultrasound waves on apple and tomato when a conventional convection dehydration system is used.

3.1 Food composition and properties

In order to perform a thermal and acoustic FEM analysis for apple and tomato inside an ultrasonic-assisted convection dehydrator, in **Table 2**, we summarize the

Parameter	Apple	Tomato	Units
Density (ρ)	840	560	[kg/m^3]
Sound speed (c)	49.89	231.45	[m/s]
Thermal capacity (C_p)	3810	4080	[J/(kgK)]
Thermal conductivity (k)	0.418	0.66	[W/(mK)]
Acoustic absorption coefficient (α)	0.22	0.35	[dB/m]

Table 2.
Thermo–acoustic parameters of apple and tomato.

physical parameters of apple and tomato, such as density, sound propagation velocity, heat capacity, thermal conductivity, and sound absorption coefficient [36–39].

For the acoustic and thermal analysis for the apple and tomato, we defined the parameters shown in **Table 2** in order to determine the dynamics produced by the drying chamber in the apple and tomato. We should emphasize that if any of these parameters are omitted, the FEM analysis cannot be performed properly and the spatial and temporal dynamics will have a significant error since acoustic and thermal phenomena are completely interrelated.

3.2 Acoustic waves and piezoelectric transducers

Firstly, a wave can be defined as the energy and momentum transfer from one point in a medium to another point in the same medium without net matter transport between the two points [40]. When the waves require a medium for their propagation are called mechanical or elastic waves [41]. In this case, the medium particles perform a periodic motion around a mean position as the wave propagates through the medium. A mechanical wave is produced when a particle is perturbed in the propagating medium and interacts with the neighboring particle and its energy is transmitted to the next particle (due to the inertia of the medium) [42]. The perturbed particles return to equilibrium due to the medium elasticity after a finite time. Thus, when the mechanical wave motion is produced the following parameters must be considered: frequency, propagation velocity, period, phase, and wavelength.

On the other hand, the piezoelectric transducer is an important component of ultrasound instrumentation systems [43]. Piezoelectric transducers convert electrical waves into mechanical vibrations and mechanical vibrations into electrical waves [44]. These devices are mainly used to generate waves in the ultrasound range (frequencies higher than 20 kHz) at low, medium, and high intensity. Piezoelectric transducers can be produced by using ceramics, quartz, Rochelle salts, and metal alloys to be used in ultrasound wave generators applicable in multiple industrial areas [45]. For example, drying, ultrasonic cleaning, fuel oil injection into burners, and medical treatments, among others.

3.3 Simulation environment

To describe the acoustic and thermal behavior of apple and tomato in the proposed dehydrator, COMSOL Multiphysics™ has been chosen as the simulation environment, which is a software tool for finite element analysis useful in various physics and engineering applications, especially for couple or multiphysics phenomena [46]. The possibility of analyzing different physical phenomena integrated into COMSOL Multiphysics™ allows the user to model and analyze scenarios involving multiple interacting physical phenomena. The phenomena that can be modeled in COMSOL Multiphysics™ are related to acoustics, electromagnetism, micro-electromechanical systems (MEMS), microwaves, radio frequency components, semiconductor devices, and wave propagation, among many others. In this work, the acoustics and heat transfer modules have been used to perform the acoustic and thermal analysis of the proposed system.

The finite element method is based on dividing the body, structure, or domain over which the equations characterizing the phenomena physical behavior are defined into subdomains called finite elements [47]. The finite elements set forms a domain partition also called discretization or mesh. Due to the subdivisions generated in the geometry, the mathematical equations that govern the phenomena physical behavior could not be solved in an exact way, but in an approximate way, since the solution that results in the simulation environment depend on the nodes and elements number, as well as of the elements size and type defined in the mesh [48]. From the approximation provided by the solution of the equations describing the desired physical phenomenon from the generated finite elements, it is possible to describe the desired system behavior.

Using a COMSOL Multiphysics™ simulation environment and the FEM, the acoustic and thermal analysis of the proposed system for apple and tomato has been performed from the geometry shown in **Figure 2**. Additionally, we have used the mathematical models developed in Subsection 4.1, and we have shown the obtained results in Subsections 4.2 and 4.3.

4. Analysis methodology

In order to perform the thermo-acoustic analysis when apple and tomato are dehydrated in the ultrasound-assisted convection dehydrator, we define the mathematical models that describe the behavior of the ultrasound waves inside the dehydration chamber and within the food samples, as well as the temperature change of the food samples. Subsection 4.1 shows the model describing the ultrasound waves propagation in the system and the thermodynamic model that describes the temperature change inside the food. Subsequently, Subsection 4.2 shows the system acoustic analysis and Subsection 4.3 shows the thermic analysis of food samples.

4.1 Mathematical model

The mathematical model used to describe the ultrasound waves propagation is derived from the reduction of the mass, momentum, energy, and state balance equations. According to Blackstock and Everest in [49, 50], mass, momentum, energy, and states are defined by Eqs. (1)–(4), respectively.

$$\frac{D\rho}{Dt}+\rho\frac{\partial u}{\partial x}=0, \tag{1}$$

$$\rho\frac{Du}{Dt}+\frac{\partial P}{\partial x}=0, \tag{2}$$

$$\rho\frac{D\varepsilon}{Dt}+P\frac{\partial u}{\partial x}=-\frac{\partial q}{\partial x}, \tag{3}$$

$$P=c_0^2\delta\rho\left[1+\frac{B}{2!A}\frac{\delta\rho}{\rho_0}+\frac{C}{3!A}\left(\frac{\delta\rho}{\rho_0}\right)^2+\dots\right], \tag{4}$$

where $\frac{D}{Dt}$ is the Stokes derivative of the variable studied, ρ is the medium's density, P is the medium's pressure, ε is the medium's internal energy, q is the heat flow inside the dehydration chamber, c_0 is the sound speed, δ_ρ is the density excess ($\delta\rho=\rho-\rho_0$), and A, B, and C are the coefficients of the Taylor series for P.

Also, according to Kinsler [51], the wave equation defined by Eq. (5) is used considering that it idealizes many types of wave motion produced in an isolated medium that does not exchange energy, momentum, or mass with its environment.

$$c^2\nabla^2 u - \frac{\partial^2 u}{\partial^2 t} = 0, \tag{5}$$

where u represents acoustic waves, ∇^2 represents the Laplacian applied to u, c is the wave speed, and t is time.

It should be noted that the mathematical model given by Eq. (5) is not enough when describing the waves behavior in a fluid with losses. Then, in order to include those losses, it is considered that u depends on the energy dissipation in a three-dimensional viscous medium [52]. Now, considering that the complex wave number $k = \beta + j\alpha$ is used to calculate a solution by a harmonic time, from Eq. (5) the expression shown in Eq. (6) is obtained,

$$u = u_0 e^{-\alpha x} e^{j(wt-\beta x)}, \tag{6}$$

where u_0 is the wave amplitude in $t = 0$, α is the absorption coefficient, and β is the wave cycles number per distance unit.

In a similar way, solving Eq. (5) for P, Eq. (7) is obtained.

$$P = P_0 e^{-\alpha x} e^{j(wt-\beta x)}, \tag{7}$$

where P_0 is the wave pressure in $t = 0$.

Based on the geometry proposed for the FEM simulation, an approximation of the thermodynamic model for food has been made. For this purpose, the model has been developed considering the equilibrium equation that describes the thermal system considering as a particular case the first law of thermodynamics, where it is considered that the heat transfer system does not generate work, and the system dynamics is a function of heat flow and temperature. The system variables description to be considered taking into account the heat flows, temperatures, thermal capacitances, and thermal resistances are shown in **Figure 3**.

where, Q_u is the heat flux generated by the ultrasound wave, Q_1, Q_2, and Q_3 are the heat fluxes leaving the system, R_{f1}, R_{f2}, and R_{f3} are the thermal resistances of the sample walls, C_f is the food thermal capacitance, T_f is the food temperature, and T_{dc} is the drying chamber temperature.

Considering heat fluxes, thermal resistances, food temperature, and environment temperature, equations for each thermal element that composes the food (walls of the food describing the heat flow through them) are written according to Eqs. (8)–(10).

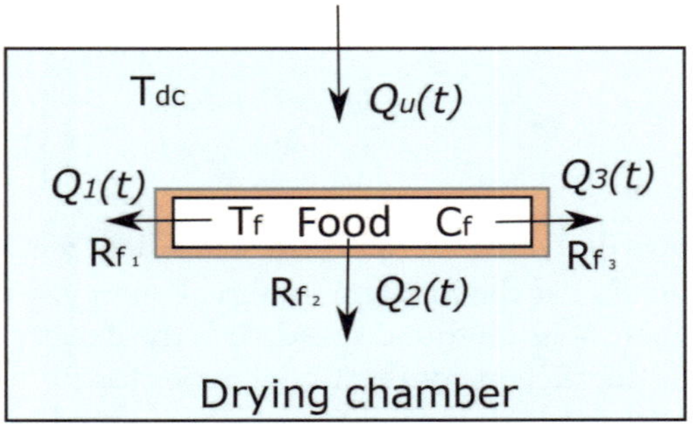

Figure 3.
System variables for thermodynamic modeling.

$$Q_1 R_{f1} = T_f - T_{dc}, \tag{8}$$

$$Q_2 R_{f2} = T_f - T_{dc}, \tag{9}$$

$$Q_3 R_{f3} = T_f - T_{dc}. \tag{10}$$

Once the element equations are defined, the equilibrium equation is established, starting from the first law of thermodynamics, which will define the system behavior from the incoming and outgoing system flows considering that there is only heat transfer, and no work is generated. This equilibrium equation can be written by Eq. (11).

$$C_f \frac{\mathrm{d}T_f}{\mathrm{d}t} = \sum Q_{input} - \sum Q_{output}, \tag{11}$$

where Q_{input} are the incoming heat fluxes, and Q_{output} are the outgoing heat fluxes.

Since the equilibrium equation is a function of the incoming and outgoing heat fluxes, each of these must be defined considering the system elements. In the system, the inflow will only be given by Q_u and the outflows can be calculated from Eqs. (8)–(10). Substituting the heat fluxes in the equilibrium equation and taking as output the rate of food temperature change as a time function, the system dynamics are described by Eq. (12).

$$\frac{\mathrm{d}T_f}{\mathrm{d}t} = \frac{Q_u}{C_f} - \frac{3}{C_f R_{fT}} T_f - \frac{3}{C_f R_{fT}} T_{dc}, \tag{12}$$

where T_f is the food temperature, t is the time, Q_u is the heat flux produced by the ultrasound, C_f is the food heat capacity, R_{fT} is the total thermal resistance generated by the apple walls, and T_{dc} is the drying chamber temperature.

In this case, it is contemplated that $Q_u = 2\alpha I$, where α is the local acoustic absorption coefficient of the food, and I is the local sound intensity. In the same way, $C_f = \rho_f C_{pf} A$ where ρ_f is the food density, C_{pf} is the food-specific heat, and A is the food transverse area. Thus, Eq. (12) is also expressed by Eq. (13).

$$\frac{\mathrm{d}T_f}{\mathrm{d}t} = \frac{2\alpha I}{\rho_f C_{pf} A} - \frac{3}{\rho_f C_{pf} A R_{fT}} T_f - \frac{3}{\rho_f C_{pf} A R_{fT}} T_{dc} \tag{13}$$

If Eq. (13) is lumped and considering that $R_{fT} = \frac{\Delta x}{k_f}$, where Δx is the thickness of the sample, and k_f is the food thermal conductivity, then, it would be written as:

$$\frac{\mathrm{d}T_f}{\mathrm{d}t} = \frac{2\alpha I}{\rho_f C_{pf} A} - \frac{3k_f}{\rho_f C_{pf} A} \frac{\Delta T_f}{\Delta x} \tag{14}$$

where $\Delta T_f = T_f - T_{dc}$, and $I = \mathbb{R}\frac{1}{2}Pv$, where P is the pressure generated by the sound wave and v is the particle velocity.

Therefore, v can be written as shown in Eq. (15).

$$v = \frac{\partial \varepsilon}{\partial t} \tag{15}$$

where ε is expressed as is shown by Eq. (16).

$$\overline{\varepsilon} = \frac{\partial \overline{\varepsilon}}{\partial x}\hat{x} + \frac{\partial \overline{\varepsilon}}{\partial y}\hat{y} + \frac{\partial \overline{\varepsilon}}{\partial z}\hat{z} \tag{16}$$

By substituting the Eqs. (15) and (16) in we obtain that the rate of change of temperature with respect to time of the food sample is described by Eq. (17).

$$\frac{\partial T_f}{\partial t} = \frac{\alpha P}{\rho_f C_{pf} A}\frac{\partial \overline{\varepsilon}}{\partial x} - \frac{3k_f}{\rho_f C_{pf} A}\nabla T \tag{17}$$

In order to describe the acoustic and thermal behavior of the apple and the tomato in the spatial and temporal domain, the FEM will be used to solve the solutions to Eqs. (6) and (17). The description of the analysis performed for each of the cases is shown in Subsections 4.2 and 4.3.

4.2 Acoustic behavior analysis

To perform the acoustic analysis in the spatial domain by means of the FEM in COMSOL Multiphysics™, the following programming sequence is proposed:

1. *Programming environment startup:* the programming of the simulation environment in COMSOL Multiphysics^{TM} should be as follows: open COMSOL Multiphysics^{TM} → new document → model wizard → space dimension selection → 2D → physics selection → acoustics → acoustic pressure → frequency (acpr) → click in add →study → frequency domain → done. This programming sequence will open the simulation environment already configured for the acoustic analysis.

2. *Geometry construction:* to generate the geometry according to the dimensions given in **Figure 2**, follow the following path: geometry → units → cm → geometry → primitives → rectangle → width (enter value in cm of the rectangle to be constructed) → height (enter value in cm of the rectangle to be constructed) → location (enter coordinates according to **Figure 2**) → construct all. This step must be performed for each of the rectangles of the geometry proposed in **Figure 2**.

3. *Assignment of materials:* for the selection of materials corresponding to each geometry the following path must be followed: materials → add material → fluids → gases → air → add material → materials → blank material → material properties → select properties (Parameters of **Table 2**) → add properties → name material (tomato or apple as the case may be) → select air material → select geometric entities that compose it → select food material → select geometric entities that compose it. In this way, you will have assigned the corresponding materials to each of the geometries that represent the dehydration chamber and the food samples.

4. *Physics configuration:* to configure the physics (acoustics), the following sequence must be followed: acoustic pressure, frequency (acpr) → domain → acoustic

pressure → select geometries corresponding to air → domain → acoustic pressure → select geometries corresponding to food → contour → pressure → select geometries corresponding to piezoelectric transducers → enter test pressure value (2 Pa)→ initial values → pressure value and temperature (2 Pa at 60°C). From this sequence, the physics will be configured to perform the acoustic analysis of the food samples.

5. *Mesh construction:* to build the mesh follow the following path: mesh → select the geometries that will be interacting with the acoustic field → sequence type → physics controlled → element size → extra fine → build mesh. With these steps, a mesh will be generated with enough elements to have a good approximation of the interaction of the acoustic field in the dehydration chamber.

6. *Study configuration:* the following sequence is used to set up the study: study → frequency → frequency units → Hertz → study frequencies → enter desired frequencies (1 Hz - 1 MHz) → physics selection → acoustic pressure → frequency → mesh → mesh 1 → value of dependent variables → defined by physics → compute. From this sequence, the simulation environment will perform the calculation of the approximation of the acoustic field behavior and the frequency spectrum of the food samples.

7. *Results:* for the visualization of the obtained results, just select the results option and the frequency spectrum, sound pressure, and sound pressure level graphs will be displayed. In this way, the acoustic analysis will be finished.

From these steps, it is possible to analyze the acoustic and spatial behaviors of the ultrasound waves at (1 Hz, 1 MHz) inside the dehydration chamber. Then, we can determine the optimal operating frequencies at which the dehydration system can perform the most efficient dehydration on each test food (apple and tomato) considering the average system pressure. Thus, in **Figure 4**, we can identify the spectral component the most influential spectral component for each food under consideration; that is, we selected the frequency band with the highest average pressure. This frequency band is centered around 34 kHz (see **Figure 4a**) when apple samples are considered and around 70 kHz (see **Figure 4b**) for tomato samples. Consequently,

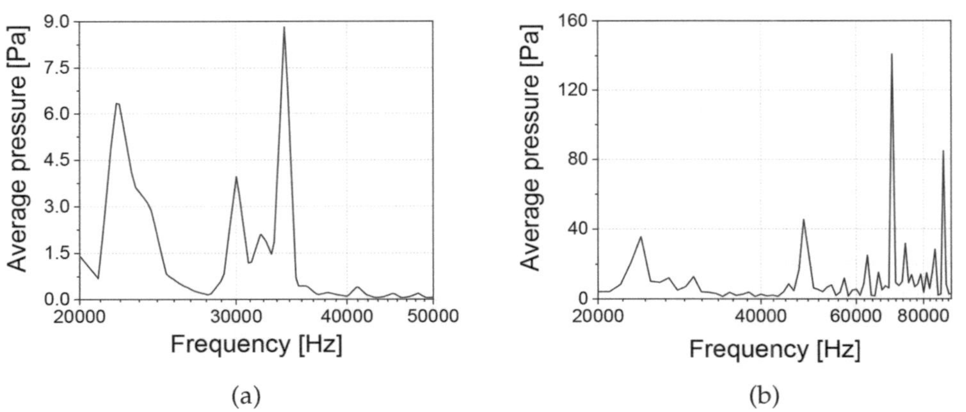

Figure 4.
Apple and tomato frequency spectrum: (a) apple and (b) tomato.

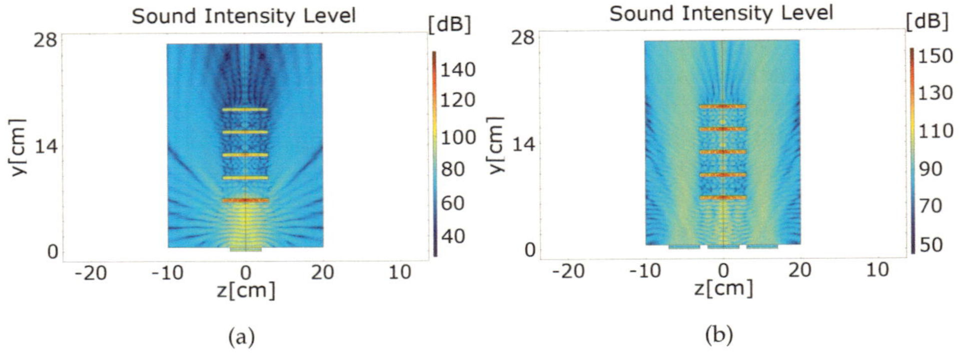

Figure 5.
Acoustic field when apple samples are radiated by ultrasound waves with fundamental frequency at 34 kHz: (a) one piezoelectric transducer and (b) three piezoelectric transducers.

Figure 5 shows the spatial behavior of the ultrasound waves when apple samples are radiated by a single piezoelectric transducer (see **Figure 5a**) and three piezoelectric transducers (see **Figure 5b**) at 34 kHz and 2 Pa of pressure.

Also, **Figure 5** shows that the moisture removal at the surface level or internally in the food is a function of the intensity and frequency of the ultrasound waves applied. It should be considered that in this study a sweep of frequencies between 1 Hz and 1 MHz was performed and it was noticed that there are frequencies with greater influence than others. On the other hand, also trying not to exceed a temperature of 70°C in the foods under dehydration to avoid their structural and nutritional damage, we determined the most appropriate pressure that should be exerted on the food. Note that when three piezoelectric transducers are used, the ultrasound waves have a more homogeneous spatial distribution than a single piezoelectric transducer is considered (see **Figure 5b**). In addition, the apple sample closest to the piezoelectric transducer has the greatest influence by ultrasound waves. Therefore, the acoustic field distribution also influences the apple samples, and it depends on the transducer number.

In a similar way to apple samples, **Figure 6** shows the spatial behavior of the acoustic field when the tomato samples are radiated by ultrasound waves by a single transducer (see **Figure 6a**) and three piezoelectric transducers (**Figure 6b**) at 70 kHz

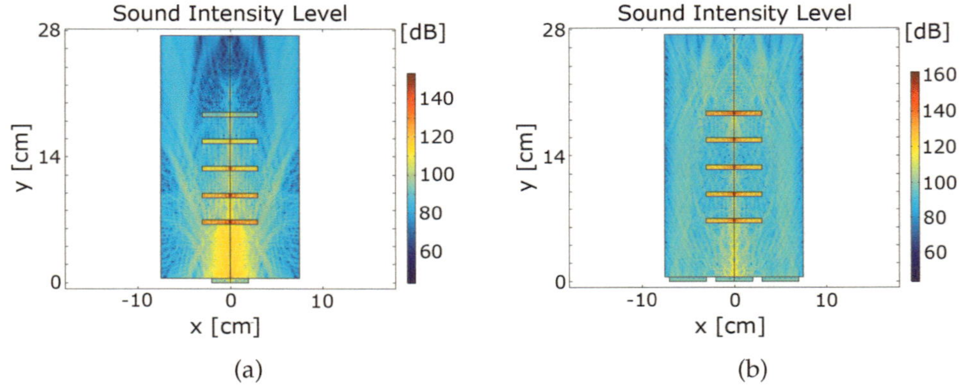

Figure 6.
Acoustic field when tomato samples are radiated by ultrasound waves with fundamental frequency at 70 kHz: (a) one piezoelectric transducer, and (b) three piezoelectric transducers.

and 2 Pa. Note that the acoustic field generated by the ultrasound waves is distributed more homogeneously when the tomato samples are radiated by three piezoelectric transducers, while when they are radiated with only one piezoelectric transducer the acoustic field has more influence on the tomato samples that are closer to the piezoelectric transducer.

It should be noted that in **Figures 5** and **6** the sound intensity level is not uniform in food samples, which implies that the moisture removal is different at the surface level than inside the food. In this way, the moisture removal is a function of the intensity and frequency of the ultrasound waves applied. In addition, it should be remembered that in this study we made a sweep of frequencies between 1 Hz and 1 MHz and we noticed that there are frequencies with greater influence than others. On the other hand, also trying not to exceed a temperature of 70°C in the foods to be dehydrated, to avoid their structural and nutritional damage.

Now, from these results, Subsection 4.3 shows the ultrasound waves influence at 34 kHz and 70 kHz on the temperature change for apple and tomato samples, respectively.

4.3 Thermic behavior analysis

In order to perform the thermal analysis on the apple and tomato samples, a multiphysical analysis is performed in which the energy produced in the food samples by the ultrasound waves is used to obtain the temperature change in each of them during a dehydration time of 40,000 seconds. To perform the thermal analysis of the ultrasound-assisted convection dehydration system, the following steps are established:

1. *Programming simulation environment:* the construction of the geometry and the integration of the materials will be maintained, as the thermal analysis will be carried out only as an add-on segment to the acoustic analysis. To configure the simulation environment for the thermal analysis we must follow the following sequence within the same file used for the thermal analysis: physics → add physics → heat transfer → study → add study → time domain. With this sequence, the simulation environment for thermal analysis is now configured.

2. *Physics configuration:* to configure the heat transfer physics, follow the following path: heat transfer → domain → heat source → select the geometries corresponding to the chamber and the food samples → contour → temperature → enter initial temperature (60°C) → initial conditions (initial temperature 60°C) → heat source → enter the command acpr.Qpw (indicates the heat produced in the system by the acoustic source). With these steps, the physics will already be configured.

3. *Mesh construction:* for the thermal analysis, a mesh must be constructed independently of the one generated for the acoustic analysis, for this the following sequence is performed: mesh → new mesh → add mesh → sequence size → controlled by physics → element size → fine. This mesh will be the one that will help to give the approximation of the thermal analysis in the function of the acoustic analysis.

4. *Study configuration:* to set up the study, follow the following sequence: study → time domain → time units → seconds → times → select time range (0–40,000 seconds) → select physics and variables → heat transfer → dependent

variable values → settings → user-controlled → method → solution → study → study in frequency → frequency parameter value → selected according to the optimum operating frequency for each food → compute. In this way, the simulation environment will be able to give the approximation of the temperature change in the food samples as a function of the frequency that is radiating them.

5. *Results:* to visualize the results obtained, simply select the results option and the temperature contour graphs of the dehydration system and the temperature rise curve for each food sample will be displayed. Thus, the thermal analysis according to the optimum operating frequency of the dehydration chamber is finished.

Using the above steps, we can determine the temperature change in the dehydration system. Note that the food mass should not exceed 70°C. This temperature limit is given according to Michelice and Ohaco [53] in their guide to food dehydration and drying, and as experimentally proven by Tao *et al.* [54] in their tests carried out in the dehydration of blackberries. This temperature has been established since the use of higher temperatures may cause structural and nutritional damage to foods. For this purpose, simulation runs must be performed by varying the ultrasound waves pressure until the food samples temperature do not exceed 70°C.

For the first case study, **Figure 7** shows the thermal analysis performed on the five levels of apple samples at 34 kHz, 80 kPa, and an initial temperature of 60°C. **Figure 7a** shows the isothermal curves generated inside the dehydration system after 40,000 seconds. **Figure 7a** shows the effect of ultrasound waves radiation on food samples. Note that the food samples closest to the piezoelectric transducer reaches a maximum temperature of 70.8°C, while the other samples have minimal increases in temperature. The temperature change in the apple samples is shown in **Figure 7b**. Note that the first food samples have the greatest influence on their temperature change. It should be noted that in **Figure 7b** each food sample is denoted with its Cartesian coordinates (x,z) within the system, being so for the first sample (0, 6.7), the second (0, 9.7), the third (0, 12.7), the fourth (0, 15.7), and the fifth (0, 18.7).

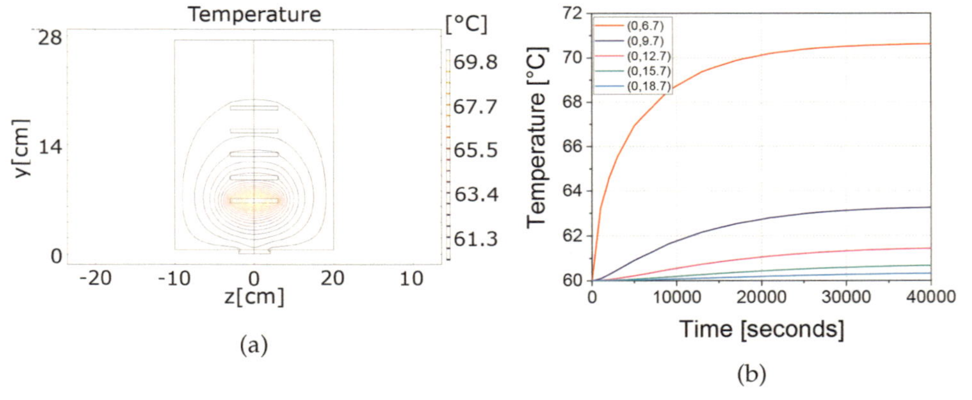

Figure 7.
Temperature of the apple samples radiated with a piezoelectric transducer.

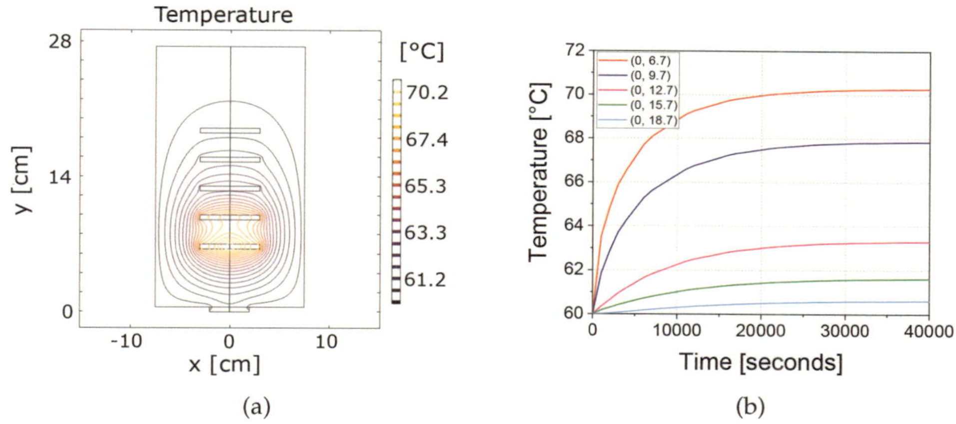

Figure 8.
Temperature of the tomato samples radiated with a piezoelectric transducer.

Similarly, **Figure 8** shows the temperature changes of five tomato samples radiated with ultrasound waves using a piezoelectric transducer at 70 kHz and 22 kPa for 40,000 seconds. **Figure 8a** shows the isothermal curves inside the dehydration system, in which it is observed that the first and second samples have a significant increase in temperature reaching a maximum of 70.1°C, while the other samples have smaller temperature changes. In **Figure 8b**, these temperature changes can be better observed from the curves plotted as a function of time.

However, **Figure 9** shows the temperature change of the apple samples radiated with ultrasound waves using three piezoelectric transducers at 34 kHz, 52 kPa, for 40,000 seconds. Note that the maximum temperature reached was 70.2°C. **Figure 9a** shows the isothermal curves generated after 400,000 seconds in apple samples located at the five levels are more homogeneous compared to the isothermal curves in **Figure 7a**. These temperature changes are more visible in **Figure 9**. These curves show that all the apple samples have a significant temperature change due to the ultrasound waves radiation and that the difference between each of them is not so large compared to those obtained for a single transducer in **Figure** 7.

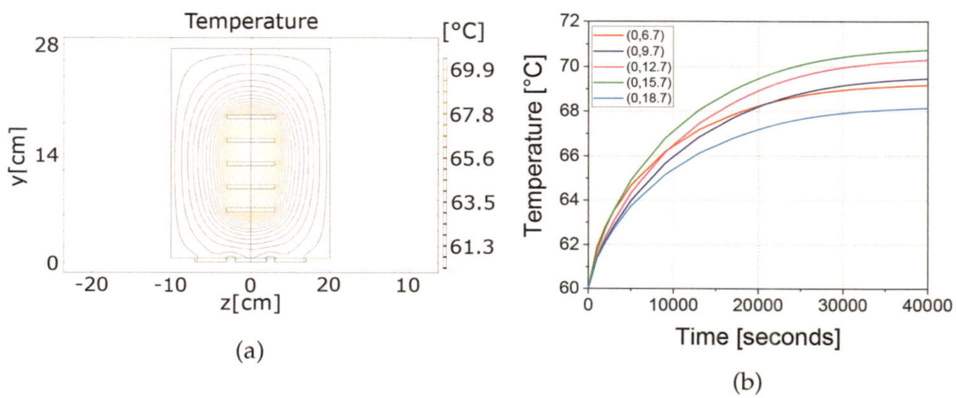

Figure 9.
Temperature of the apple samples radiated with three piezoelectric transducers.

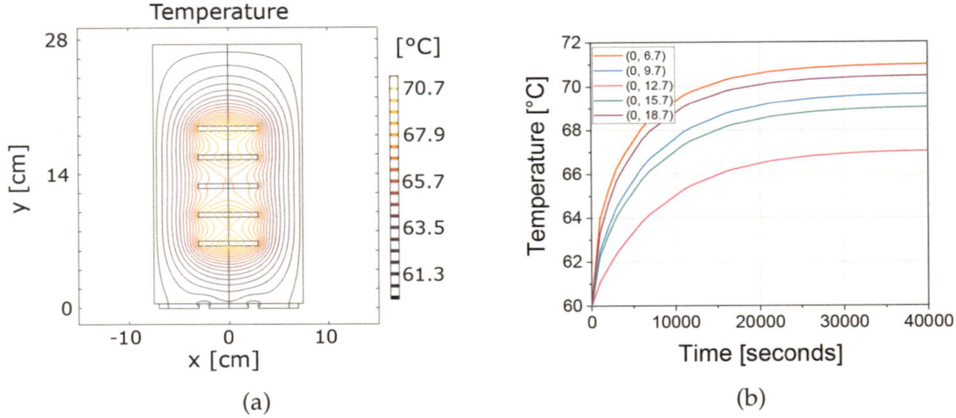

Figure 10.
Temperature change of tomato when radiated with three piezoelectric transducers.

Now, the temperature change in the tomato samples radiated by ultrasound waves using three piezoelectric transducers at 70 kHz and 5 kPa during 40,000 seconds is shown in **Figure 10**. Note in **Figure 10a** that the isothermal curves for the tomato samples show homogeneous temperature changes reaching a maximum temperature of 70.8°C. These temperature changes in the tomato samples are more evident in **Figure 10b**. Note the temperature increase on five levels of food samples occurs to a greater extent when radiated with three piezoelectric transducers than if the samples are radiated with a single transducer as shown in **Figure 9b**.

From the thermal-acoustic analysis performed, it is possible to highlight different aspects that are relevant at the moment of designing a convection dehydration system assisted by ultrasound waves: (a) the distribution of the acoustic field produced by the ultrasound waves depends on the frequency, the number of transducers and the dimensions of the dehydration chamber, (b) the increase in the temperature of the food samples is a function of the frequency and pressure with which the samples are radiated, and (c) it is possible to radiate more than one sample simultaneously within the same dehydration chamber placed at different distances from the radiation source. Thus, in this way, it is possible to give an approximation of the thermo-acoustic behavior of different foods within the processes of dehydration by convection assisted by ultrasound waves.

5. Discussion and work perspectives

The performed analysis describes the thermo-acoustic behavior of ultrasound waves and food samples in an ultrasound-assisted convection dehydration system considering two case studies. In the first one, we use separately apple and tomato samples placed in five gridded trays spaced 3 cm from each other and ultrasound–radiated inside the dehydration chamber by using one piezoelectric transducer. In the second case study, we use the same conditions of the spatial distribution of food samples but using three piezoelectric transducers. From the thermo-acoustic analysis, the optimal operating frequencies were identified for apple (around 34 kHz) and tomato (around 70 kHz) samples. Also, from a temperature control on food samples, we identified the operating pressures that avoid the structural and nutritional damage of the food samples under consideration in the dehydration process. By obtaining the acoustic and thermal behavior of apple and tomato samples in an ultrasound-assisted convection dehydration system, we can design an ultrasound-assisted dehydration system that improves the quality, time, cost, and energy of small– and large-scale dehydration processes. The development of new technologies focused on ultrasound-assisted convection dehydrator systems for food dehydration remains a challenge for the scientific community when considering acoustic, thermal, mechanical, and biochemical approaches in the design, purchase, or implementation of dehydration systems applied to human consumption foods. Moreover, from these new technologies, it is possible to obtain dehydrated products with high nutritional content and little structural damage that, without leaving behind a good taste and texture, may allow the human being to store food and its subsequent consumption in a long-term time. Finally, it is necessary to emphasize that although there are several commercially available ultrasound–assisted dehydration systems, there is still a gap to be filled so that ultrasound-assisted systems can be applied in multiple food fields at an industrial level, and at the same time, that these systems tend to be used on a daily basis at affordable prices in households around the world.

6. Conclusions

From the thermo-acoustic analysis for 0.5 cm-plate apple and tomato samples inside an ultrasound-assisted convection dehydrator assisted by ultrasound waves, and considering the physical dimensioning of the dehydration chamber, the physical parameters of the apple and tomato, and the wave and heat equations, we can determine the spectral behavior of ultrasound waves inside food and the temperature changes on the apple and tomato samples. Based on this analysis, we identify the optimum operating frequencies from the average pressure applied in the dehydration system for apple and tomato samples using ultrasound waves at 34 kHz and 70 kHz, respectively. Using these operating frequencies, we performed a spatial analysis when the ultrasound-assisted convection dehydrator was implemented with one and three piezoelectric transducers radiating a test acoustic field at 2 Pa. However, the spectral analysis showed that using ultrasound waves at 34 kHz for apples, the acoustic field was more uniformly distributed on apple samples located at five levels inside the dehydration chamber using three piezoelectric transducers. The same phenomenon occurred when the tomato was radiated at 70 kHz with three piezoelectric transducers. In addition, considering the distribution of the acoustic field in the two case studies, the temporal analysis to obtain the temperature changes inside the apple and tomato samples at the test frequencies was performed. The results obtained showed that when the five racks were used to place the apple and tomato samples and they were radiated using one piezoelectric transducer, only the temperature of food sample closest to the transducer was increased, while the rest of the food samples remained without significant temperature changes. Note that for the case of a single piezoelectric transducer, the pressure levels ranged between 80 kPa for apples and 22 kPa for the tomato to reach a temperature close to 70°C, which avoids structural and nutritional damage to the foods. When three piezoelectric transducers were used, uniform temperature changes were observed, and a similar temperature increase was observed in food samples of the five racks. Under these conditions, the apple and tomato samples reached temperatures between 67 and 70°C. When apple and tomato samples were radiated with three piezoelectric transducers using pressure between 52 kPa and 5 kPa respectively, less energy was required than when the food samples were radiated with one transducer. Thus, from the FEM analysis, it was possible to determine the optimal operating conditions at which an ultrasound-assisted convection dehydrator for apple and tomato samples can operate most efficiently based on its spectral and thermodynamic behavior.

Acknowledgements

The authors thank Instituto Politécnico Nacional (IPN-México) for financial support under grant numbers SIP–20220531 (Rubén Vázquez–Medina), SIP–20220572 (Omar Jiménez-Ramírez), and SIP-20220933 (Gonzalo Alonso Ramos-López). Daniel Aguilar-Torres (CVU-829790) thanks for the scholarship provided by Consejo Nacional de Ciencia y Tecnología (Mexico).

Conflict of interest

"The authors declare no conflict of interest."

Author details

Daniel Aguilar-Torres[1†], Omar Jiménez-Ramírez[1†], Juan A. Jimenez-Garcia[2†], Gonzalo A. Ramos-López[3†] and Rubén Vázquez-Medina[3*†]

1 Instituto Politécnico Nacional, Escuela Superior de Ingeniería Mecánica y Eléctrica Unidad Culhuacan, Mexico City, Mexico

2 Universidad Autónoma del Estado de México, Centro Universitario Nezahualcóyotl, Estado de México, Mexico

3 Instituto Politécnico Nacional, Centro de Investigación en Ciencia Aplicada y Tecnología Avanzada Unidad Querétaro, Queretaro, Mexico

*Address all correspondence to: ruvazquez@ipn.mx

† These authors contributed equally.

IntechOpen

© 2022 The Author(s). Licensee IntechOpen. This chapter is distributed under the terms of the Creative Commons Attribution License (http://creativecommons.org/licenses/by/3.0), which permits unrestricted use, distribution, and reproduction in any medium, provided the original work is properly cited.

References

[1] Kumar C, Karim MA, Joardder MU. Intermittent drying of food products: A critical review. Journal of Food Engineering. 2014;**121**:48-57. DOI: 10.1016/j.jfoodeng.2013.08.014

[2] Merone D, Colucci D, Fissore D, Sanjuan N, Carcel JA. Energy and environmental analysis of ultrasound-assisted atmospheric freeze-drying of food. Journal of Food Engineering. 2020; **283**:110031. DOI: 10.1016/j.jfoodeng.2020.110031

[3] Ozcan-Sinir G, Ozkan-Karabacak A, Tamer CE, Copur OU. The effect of hot air, vacuum and microwave drying on drying characteristics, rehydration capacity, color, total phenolic content and antioxidant capacity of Kumquat (Citrus japonica). Food Science and Technology. 2018;**39**:475-484. DOI: 10.1590/fst.34417

[4] Mar V, Riera E, Pérez G, García-Pérez JV. The use of ultrasound for drying, degassing and defoaming of foods. Innovative Food Processing Technologies: A Comprehensive Review. 2021:415-438. DOI: 10.1016/B978-0-08-100596-5.22957-0

[5] Nowacka M, Dadan M. Ultrasound-assisted drying of food. In: Emerging Food Processing Technologies. New York, NY: Humana; 2022. pp. 93-112. DOI: 10.1007/978-1-0716-2136-37

[6] Zhang M, Chen H, Mujumdar AS, Tang J, Miao S, Wang Y. Recent developments in high-quality drying of vegetables, fruits, and aquatic products. Critical Reviews in Food Science and Nutrition. 2017;**57**(6):1239-1255

[7] Wang X, Xu S, Wu Z, Li Y, Wang Y, Wu Z, et al. A novel ultrasound-assisted vacuum drying technique for improving drying efficiency and physicochemical properties of Schisandra chinensis extract powder. Food Science & Nutrition. 2022; **10**:49-59. DOI: 10.1002/fsn3.2645

[8] García-Pérez J, Cárcel J, de la Fuente-Blanco S, de Sarabia ER-F. Ultrasonic drying of foodstuff in a fluidized bed: Parametric study. Ultrasonics. 2006;**44**: e539-e543

[9] Khmelev VN, Shalunov AV, Barsukov RV, Abramenko DS, Lebedev AN. Studies of ultrasonic dehydration efficiency. Journal of Zhejiang University-SCIENCE A. 2011;**12**:247-254

[10] Kowalski SJ, Mierzwa D, Stasiak M. Ultrasound-assisted convective drying of apples at different process conditions. Drying Technology. 2017;**35**(8):939-947

[11] Magalhães ML, Cartaxo SJ, Gallão MI, García-Pérez JV, Cárcel JA, Rodrigues S, et al. Drying intensification combining ultrasound pre-treatment and ultrasound-assisted air drying. Journal of Food Engineering. 2017;**215**:72-77

[12] Santacatalina JV, Soriano JR, Cárcel JA, Garcia-Perez JV. Influence of air velocity and temperature on ultrasonically assisted low temperature drying of eggplant. Food and Bioproducts Processing. 2016;**100**: 282-291

[13] Cao X, Zhang M, Mujumdar AS, Zhong Q, Wang Z. Effects of ultrasonic pretreatments on quality, energy consumption and sterilization of barley grass in freeze drying. Ultrasonics Sonochemistry. 2018;**40**:333-340

[14] Sabarez HT, Gallego-Juarez JA, Riera E. Ultrasonic-assisted convective drying of apple slices. Drying Technology. 2012;**30**:989-997

[15] Szadzińska J, Mierzwa D, Musielak G. Ultrasound-assisted convective drying of white mushrooms (Agaricus bisporus). Chemical Engineering and Processing-Process Intensification. 2022;**172**:108803

[16] Singla M, Sit N. Application of ultrasound in combination with other technologies in food processing: A review. Ultrasonics Sonochemistry. 2021;**73**:105506. DOI: 10.1016/j.ultsonch.2021.105506

[17] Khadhraoui B, Ummat V, Tiwari BK, Fabiano-Tixier AS, Chemat F. Review of ultrasound combinations with hybrid and innovative techniques for extraction and processing of food and natural products. Ultrasonics Sonochemistry. 2021;**76**:105625. DOI: 10.1016/j.ultsonch.2021.105625

[18] Fei P, Lifu C, Wenjian Y, Liyan Z, Yong F, Ning M, et al. Comparison of osmotic dehydration and ultrasound-assisted osmotic dehydration on the state of water, texture, and nutrition of Agaricus bisporus. CyTA-Journal of Food. 2018;**16**(1):181-189. DOI: 10.1080/19476337.2017.1365774

[19] Başlar M, Toker ÖS, Karasu S, Tekin ZH, Yildirim HB. Ultrasonic applications for food dehydration. In: Handbook of Ultrasonics and Sonochemistry. Singapore: Springer; 2016. pp. 1247-1270. DOI: 10.1007/978-981-287-278-464

[20] Ojha S, Schlüter OK. Other ultrasound-assisted processes. In: Innovative and Emerging Technologies in the Bio-Marine Food Sector. Vol. 11. Academic Press; 2022. pp. 129-147. DOI: 10.3390/foods11010122

[21] Zhu R, Jiang S, Li D, Law CL, Han Y, Tao Y, et al. Dehydration of apple slices by sequential drying pretreatments and airborne ultrasound-assisted air drying: Study on mass transfer, profiles of phenolics and organic acids and PPO activity. Innovative Food Science & Emerging Technologies. 2022;**75**:102871. DOI: 10.1016/j.ifset.2021.102871

[22] Li L, Yu Y, Xu Y, Wu J, Yu Y, Peng J, et al. Effect of ultrasound-assisted osmotic dehydration pretreatment on the drying characteristics and quality properties of Sanhua plum (Prunus salicina L.). Lwt. 2021;**138**:110653. DOI: 10.1016/j.lwt.2020.110653

[23] Li Y, Wang X, Wu Z, Wan N, Yang M. Dehydration of hawthorn fruit juices using ultrasound-assisted vacuum drying. Ultrasonics Sonochemistry. 2020;**68**:105219. DOI: 10.1016/j.ultsonch.2020.105219

[24] Gong Y, Li J, Li J, Fan L, Wang L. Effect of ultrasound-assisted freeze-dried on microstructure, bioactive substances, and antioxidant activity of Flos Sophorae Immaturus. Food Bioscience. 2022;**49**:101913. DOI: 10.1016/j.fbio.2022.101913

[25] The Global Home of Chemical Engineers (AIChE). Catalyzing Commercialization: Ultrasonic Dehydration of Protein Suspensions. New York, USA. 2020. Available from: https://www.aiche.org/resources/publications/cep/2020/june/catalyzingcommercializationultrasonic-dehydration-protein-suspensions [Accessed: September 6, 2021]

[26] Advanced Power Ultrasonic Technologies (Pulsonics). Dehydration Assisted by Ultrasound. Madrid, Spain. 2021. Available from: https://www.pusonics.es/dehydration-assisted-by-ultrasound [Accessed: September 6, 2021]

[27] Huai X, Jiang R, Liu D, Sun B. Ultrasonic Enhanced Osmotic

Dehydrating Method and Apparatus. China. 2004. Available from: https://patents.google.com/patent/CN1641300A/en [Accessed: September 6, 2021]

[28] Hielscher Ultrasonics. Ultrasonication and Its Manifold Applications in Food Processing. Germany. 2004. Available from: https://www.hielscher.com/ultrasonication-and-its-manifold-applications-in-food-processing.htm [Accessed: September 6, 2021]

[29] SharperTek. Ultrasonic Honey Processing. USA. 2019. Available from: https://www.sharpertek.com/ho.html [Accessed: September 6, 2021]

[30] Mulet A, Cárcel JA, Sanjuan N, Bon J. New food drying technologies-use of ultrasound. Food Science and Technology International. 2003;**9**(3): 215-221. DOI: 10.1177/1082013203034641

[31] Musielak G, Mierzwa D, Kroehnke J. Food drying enhancement by ultrasound–A review. Trends in Food Science & Technology. 2016;**56**:126-141. DOI: 10.1016/j.tifs.2016.08.003

[32] Kentish S, Ashokkumar M. The physical and chemical effects of ultrasound. In: Ultrasound Technologies for Food and Bioprocessing. New York, NY: Springer; 2011. pp. 1-12. DOI: 10.1007/978-1-4419-7472-3

[33] Fernandes FA, Rodrigues S, Cárcel JA, García-Pérez JV. Ultrasound-assisted air-drying of apple (Malus domestica L.) and its effects on the vitamin of the dried product. Food and Bioprocess Technology. 2015;**8**(7):1503-1511

[34] Chavan P, Sharma P, Sharma SR, Mittal TC, Jaiswal AK. Application of high-intensity ultrasound to improve food processing efficiency: A review. Food. 2022;**11**(1):122

[35] Huang D, Men K, Li D, Wen T, Gong Z, Sunden B, et al. Application of ultrasound technology in the drying of food products. Ultrasonics Sonochemistry. 2020;**63**:104950

[36] Manjarrez LM. Alimentos Ciencia e Ingeniería 7-1. Ecuador: UTA; 1999

[37] Çengel YA, Ghajar AJ. Tablas y diagramas de propiedades (Sistema Internacional) de Transferencia de calor y masa. México DF, México: McGraw-Hill; 2011. pp. 865-891

[38] Sweat VE. Thermal Properties of Foods. Engineering Properties of Foods. Vol. 49. New York, USA: Marcel Dekker; 1986. pp. 99-139. ISBN: 0-8247-8943-1

[39] Anon. Thermal Properties, Food Resource. Corvallis, Oregon: Oregon State University; 2007b. Available from: http://food.oregonstate.edu/energy/t10.html

[40] Billingham J, King AC. Wave Motion. Vol. 24. Cambridge, United Kingdom: Cambridge University Press; 2000. ISBN: 0-521-63257-9

[41] Hansen, CH. Fundamentals of acoustics. Occupational Exposure to Noise: Evaluation, Prevention and Control. World Health Organization, 2001;**1**(3):23-52

[42] Bruneau M. Fundamentals of Acoustics. London, United Kingdom: John Wiley & Sons; 2013. ISBN: 978-1-905209-25-5

[43] Safari A, Akdogan EK, editors. Piezoelectric and Acoustic Materials for Transducer Applications. New York, USA: Springer Science & Business Media; 2008. ISBN: 978-0-387-76538-9

[44] Arnau VA. Piezoelectric Transducers and Applications. Vol. 72. Berlin,

Germany: Springer Book Archive; 2004. pp. 1-37. DOI: 10.1007/978-3-662-05361-4

[45] Macrelli E, Romani A, Paganelli RP, Sangiorgi E, Tartagni M. Piezoelectric transducers for real-time evaluation of fruit firmness. Part I: Theory and development of acoustic techniques. Sensors and Actuators A: Physical. 2013; **201**:487-496

[46] Comsol C, Comsol DMUSG. Multiphysics^{TM} v. 5.5. COMSOL. COMSOL Multiphysics. 2020;**5**

[47] Reddy JN. Introduction to the Finite Element Method. New York, USA: McGraw-Hill Education; 2019. ISBN: 9781259861901

[48] Szabó B, Babuška I. Finite Element Analysis: Method. Verification and Validation. New Jersey, USA: John Wiley and Sons; 2021. ISBN: 9-7811-11942-638-7

[49] Blackstock D. Fundamentals of Physical Acoustics. Vol. 794. New York: Wiley; 2000

[50] Everest FA, Pohlmann KC. Master Handbook of Acoustics. McGraw-Hill Education; 2022

[51] Kinsler LE, Frey AR, Coppens AB, Sanders JV. Fundamentals of Acoustics. Hoboken, NJ, USA: John Wiley & Sons; 2000

[52] Nédélec JC. The helmholtz equation. In: Acoustic and Electromagnetic Equations. Applied Mathematical Sciences. Vol. 144. New York, NY: Springer; 2001. DOI: 10.1007/978-1-4757-4393-72

[53] De Michelis A, Ohaco E. Deshidratación y desecado de frutas, hortalizas y hongos. In: Procedimientos hogareños y Comerciales de Pequeña escala. 2012

[54] Tao Y, Li D, Chai WS, Show PL, Yang X, Manickam S, et al. Comparison between airborne ultrasound and contact ultrasound to intensify air drying of blackberry: Heat and mass transfer simulation, energy consumption and quality evaluation. Ultrasonics Sonochemistry. 2021;**72**:105410. DOI: 10.1016/j.ultsonch.2020.105410

Chapter 3

Fresh Cut Fruits and Vegetables Disinfection Pretreatment: A Novel Approach to Extend Fresh Cut's Shelf Life

Pooja Nikhanj, Mohini Prabha Singh, Simran Saini, Gurliin Kaur and Juhi Kumari

Abstract

Fresh cut fruits and vegetable have gained penetration and popularity since last few years. These fresh cut commodities are in great demand among the consumers as these are ready-to-eat fresh and provide all essential nutrients. The increasing trend in fresh cut produce tends to increase the investment in research and development to address various issues regarding the product supply, refrigeration, packaging technology, processing and shelf life extension. Cutting and peeling causes physical damage to the raw fruit and vegetable that make them more perishable. In these review latest developments that plays the key role in extending the shelf life of the fresh cut are discussed. These technologies help in reducing the microbial load over the fresh cut without much altering the physicochemical properties. Future researches should consider various combined technologies which allow better preservation as well as supplemented with nutritional factors.

Keywords: fresh cut, sodium hypochlorite, shelf life, microbial inhibition, physico-chemical

1. Introduction

India ranks second in fruit production after China with production of 98,579 ('000MT) under 6648 ('000Ha) production area (NHB, 2018–2019). There is marginal rise of India's horticulture produce in 2019–2020. According to the estimation of ministry of Agriculture and farmer's welfare, total horticulture production in the country stood 310.74 million Tonnes in 2018–2019. This is marginal higher than the production in 2017–2018. 97.97 million Tonnes of fruit production are estimated compared to 96.45 million Tonnes in 2017–2018. A major contribution to total fruits production of India is contributed by Punjab with 94.80 ('000MT) under production area of 2001.69 ('000Ha). Presently Kinnow, Guava, Mango, Pear, Sweet Orange, Litchi, Peach and Ber are major fruits; while Limes/Lemons, Amla, Grapes, Plum, Banana, Pomegranate, Phalsa, Sapota, Papaya etc. are the minor fruits grown in the

Punjab region. Fruits are a great source of nutrients, dietary fiber, minerals, vitamins and energy and its consumption is linked to reduce the risk of cardiovascular diseases and flabbiness. Fruits also are the rich source of phytochemicals and other phenolic compounds that function as anti-inflammatory agents, antioxidants and phytoestrogen [1]. Intake of vegetables and fruits has been increased as today's society is becoming more concerned about health and role of raw fruits and vegetables for improving and maintaining consumer and human health. Increasing demand of nutritious and healthy products, fresh produce is always at the priority of the consumer list because of natural nutritious quality, flavors and freshness. The International Fresh-cut produce association (IFPA) defines as fruits and vegetables that have been trimmed and/or peeled and/or cut into 100% useable product that is bagged pre-packaged to offer consumers high nutrition, convenience and flavor while still maintaining its freshness. The fresh-cut vegetables and fruits are in great demand among the consumers as these are ready-to-eat, fresh and nutritious. Today, people are replacing unhealthy junk food with fresh cut vegetables and fruits. The increasing trend in fresh cut produce tends to increase the investment in research and development to address various issues regarding the product supply, refrigeration, packaging technology and processing equipment. Fresh cut produce is gaining popularity among the consumers which lead to the availability at the retail level also. This results in the expansion of industries favoring production of fresh cut fruits which accelerates the quick services at restaurants and in retail shops. There are many produce of fresh vegetable and fruits in the market like fresh cut salads, fruits and vegetables in the market. Many industries are committed in developing the products so as to continue delivering the reliable product in the market. Fresh cut vegetable salad dominate among the minimally processed fresh cut produce but as the popularity of fresh cut fruits among the baby boomers and young generation of the society will probably overtake the sale of fresh cut salad in coming years. There are certain cons related to the productionof fresh cut produces. The process of damage can be defined in two major patterns which directly or indirectly influence each other in their production. Physiological spoilage is caused due to metabolic and enzymatic activity of the plant tissue and microbial spoilage is caused due to proliferation of microorganisms. Cutting and peeling of fresh raw fruit causes physical damage to the fruit and them more perishable than the intact fruit [2]. The quality factor of fresh cut vegetables and fruit product is analyzed by the consumer is the combination of properties or characteristics that decide their value to the consumer. It is often assumed the "if it looks good, it tastes good." Quality of fresh cut fruits is usually determined by the various properties which include nutritive value, flavor, appearance and texture. On accessing the product quality consumer assess product appearance and probably the color of the product [3]. Quality of fresh intact fruits depends upon the cultivator, pre harvest and harvesting methods, handling methods while when the factor of "fresh-cut" fruits come into play it renders the shelf life of the produce because of highly perishable nature. Fresh cut salads are gaining popularity in the market as various negative effects are addressed during processing while on the other hand fresh cut fruits do not show exceptional growth due to its biochemical and physiological phenomena which accelerates the perishable nature of the fruits and vegetables. Major factor that affects the quality of the fresh cut methods of preparation which includes types of tools used, surface area and size of the cut slices, type of water used for washing and handling as well as storage conditions which includes humidity, packaging, maintaining optimum temperature and sanitation conditions.

Proper washing/cleaning of fresh-cut product right after cutting is the most important step in fresh cut production to reduce the pathogenic micro flora from the commodities [4]. Food safety guidelines (FDA) for the fresh-cut product industry usually stipulate a washing or sanitizing step to eradicate pesticide residue, dirt and microorganisms responsible for decay and quality loss. The effectiveness of the washing pretreatment depends upon the quality of the water, pH, temperature of the operation, type and concentration of the disinfectant and contact time with the commodity. The product should rinse and washed with the water visually free from dirt, dust and other debris [5]. The U.S. Food and Drug Administration (FDA) guide to reduce the microbial count on fresh-cut vegetables and fruits to safety level point out that water quality is the very important step on fresh cut processing (FDA/CFSAN 2008). It is also considered that using sanitized water for washing purpose reduces the initial microbial load on the fresh cut produce is accountable to extend the shelf life of the produce [6]. So, there is a need to give main focus on the disinfectant washing to decrease microbial count on raw vegetables and fruits, being an economic technique to enhance the fresh cut's shelf life. Additionally, use of chemical disinfectants along with cold refrigeration storage is an effective way to extend the shelf-life of fresh-cut fruits up to a reasonable level. Further, there is a need to test for various disinfectants to evaluate their efficacy in reducing microbial load over the fresh cut fruits and the best packaging material to be discovered to enhance the shelf life and quality of fresh-cut fruits. Being the most economic and easy method to conduct, disinfectant pretreatment washing of the cut fruits with an optimum disinfectant following an optimized process conditions provides the farmers to opt for the new technology to establish the Fresh cut market in Punjab. This book chapter reviews the various techniques to disinfect the fresh cut fruits and vegetables with novelty lies in the process parameter optimization including type of disinfectant, concentration of disinfectant, pH and dipping time for disinfection will help the fresh cut industries to opt for the process technology to enhance the shelf life of the fresh cut fruits and vegetables.

2. Global trends of fresh cut produce

Healthy and nutritious foods are of critical importance among the young population of the country. Vegetables and fruits are imported round the year to meet the demands of the country. With increased dining away from home, fresh cut products are taking an ever-increasing role because of time saving in cooking of other food materials. Fresh-cut vegetable and fruits are the prime rapid budding food categories in United States superstores. The requirement of fresh cut commodities is also increasing being a healthier preference over the fast food sectors. The freshcut manufacturing industries are flourishing in various European nations with the United Kingdom, France, and Italy as share pioneers. In the Eastern nations, Europe is among the nations with highest advancement in fresh cut industries. In the above stated developed countries, fresh cut vegetables and fruits are developed at the industrial scale but in India and other developing countries, the production is yet at a small scale. Now with the rising awareness and concern in this category of commodities, there is a massive opportunity of fresh cut market in the emerging countries [7]. Small scale industry also plays a significant share in supplying packages for fresh cut commodities to the developing countries as per the demand by the consumers. Small scale businesses and small vendors are the fundamental distributors of fresh cut commodities in most emerging nations. In 2006, 27% of freshcut produce in the United States was sold in the food service sector, while 73% was sold in retail. Freshcut produce sales increased in value from US$3.3 billion in 1999 to US$15.5 billion in 2007 [8]. Fresh cut packed salads and vegetables showed a growth trend in 2008, while sales in fresh cut fruits declined. Fresh cut organic salads are now being mainstreamed across the United States and feeding consumer desire for healthy food. Fresh cut commodities are widely available in restaurants and retail outlets in US. Fresh cut fruits and vegetables are sold in parts of the Asian countries as well. Fresh cut vegetables are in greater demand as compared to fresh cut fruits in Thailand [9]. With this growing demand of fresh cut commodities, Thailand is likely to show a continued growth trend. Recently in India, German Food Company which deals with fresh cut fruits and vegetables has formed a joint venture with exotic fruits importer IG interventions and launched the sliced fruits packs. In 2017, the fresh cut and ready-to-eat product trend has grown by 4% in both volume and value compared to 2016 industrial brand have increased their share by 8% despite their market share already being 40%,"explains AIIPA fresh-cut president [10].

3. Quality of fresh cut fruits

Quality is a combination of characteristics that determines the value of produce to the consumer. Fresh produce is expected by the consumer without any defects having optimum maturity, physicochemical characteristic and must be in fresh condition. The condition of fresh vegetable and fruits relates to their sensory quality, general appearance, nutritional quality and flavor [11]. Overall quality characteristics impacting the fresh cut fruits and vegetables are compiled in **Figure 1**. Consumers judge the quality of vegetables and fruits on the basis of their appearance and firmness (external attributes). Important aspect to be considered during the preservation of fresh cut produce is tissue color and control of surface browning and discoloration. Polyphenol oxidase (PPO) is the enzyme is responsible for the oxidative browning of the surfaces

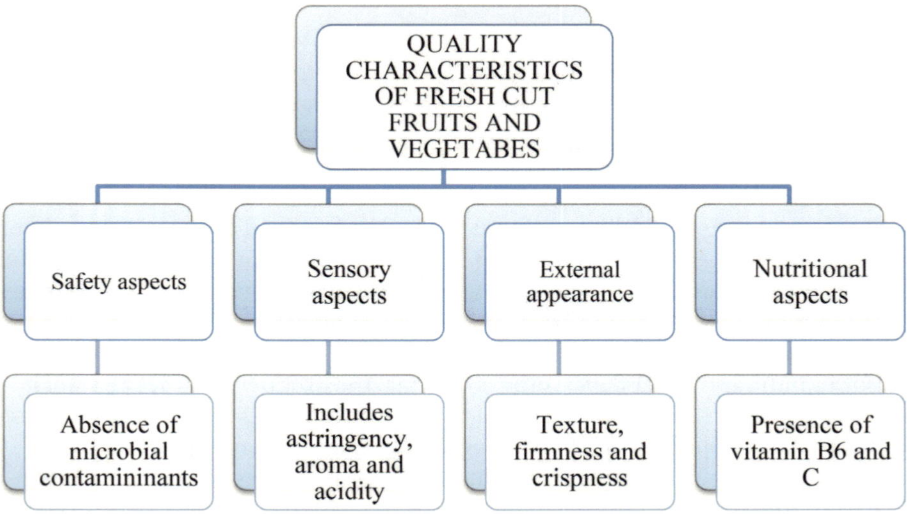

Figure 1.
Quality characteristics of fresh cut fruits and vegetables.

of fresh cut produces. In this reaction, phenolic compounds present in the fruits and vegetables get converted into dark colored compound in the presence of oxygen after coming in contact with environmental conditions. Extent of browning varies by growing conditions, cultivator and commodity characteristics. Several researches have been done to reduce the PPO-mediated discoloration. Usually a simple visual, appraisal of appearance and quality, hopefully objective is desired. A detailed scale to analyze the color quality of both fresh cut and whole fruit is also followed [12]. Consumer often buys the product for the first time based on appearance or impulse. Apart from this, other intrinsic factors such as texture and flavor also plays role in determining the quality. Flavor is comprised of aroma and taste which mainly relates to sugar, acids, salts, bitter compounds and volatile components. Flavor in fruits and vegetables arises from numerous biosynthetic pathway and wide range of aldehydes, alcohols, esters, ketones, lactones, sulfur and nitrogen containing compounds.

Volatile compounds play major contribution to impact the aroma in fruits such as banana, apple, pear, guava, papaya, strawberry and melon. During the storage of fresh cut produces, flavor losses that act as the direct consequence of senescence. Flavor changes occur during the storage of fresh cut fruits and vegetables may affect the consumerpreferability towards buying of fresh cuts. Consumer generally considers flavor as the most important attribute for fruits and vegetables, but textural defect in interaction of flavor also responsible for the consumer rejection. Texture comprises of both mechanical and structural properties of the food, as well as sensory component in mouth and hand, all of which can be measured by several nondestructive and destructive instruments or objective method [13]. Subsequent purchases by consumers are, however, depends upon their eating experience (taste, aroma experience attribute or texture intrinsic).

Other quality parameters like nutritional and safety attributes also influence consumer decisions on making repeated purchases of the commodity. Fruits with highest quality should be used for the processing of fresh cut commodities. Fruits and vegetables act as the major source of vitamins, fibers, minerals and carbohydrate. There are certain metabolites such as anthocyanins, phenolic acids and flavonoids are obtained from fruits and vegetables which plays integral part in human metabolism and also assumed to beneficial for human health due to its biological properties including antiviral, anti-inflammatory, antibacterial, antioxidant function [14]. These above mentioned components are considered to be highly beneficial in curing certain diseases such as coronary heart diseases, osteoporosis, stroke, and hypertension. Daily intake of 400 g of fruits and vegetables is recommended by the world Health Organization in 2003 [15]. Processing of raw fruits and vegetables results in alteration of nutrients. After cutting of the fresh fruits and vegetable, the antioxidant capacity of the produce may increase (celery, carrot, parsnip, potato, white cabbage, sweet potato) [16] or decrease (potato, melon, zucchini, cabbage) during storage. Carotene and Vitamin C content decrease very little during short term storage of fresh cut fruits [17]. However, Vitamin C and ascorbic acid content usually decreases after cutting, especially during long term storage [18]. Safety is also the major component for maintaining the quality aspect of fresh produces. Safe food should be free from physical, chemical and biological hazards. Physical hazard include glass, stones, plastic, hair, jewelry and metals which may be intentionally or unintentionally get into the produce during the production process. Chemical hazards include natural substances for, e.g., mycotoxins, alkaloids, allergens and enzyme inhibitors, chemicals like pesticides and toxic elements like lead, arsenic, zinc, and cadmium. These elements affect the human health more rapidly as compared to biological hazards. This chemical hazard enters into the fruits and vegetable during production process and postharvest handling. Biological hazards are caused by the pathogenic microorganisms. These microorganisms cause intoxication and food borne infections when consumed. Microbial population which may be insufficient to cause food spoilage but may be sufficient to cause illness in human beings on consumption of fresh produces. Thus, the fresh cut vegetable and fruits which seems perfect in appearance is not guaranteed to be microbiologically safe. These pathogenic microorganisms may contaminate the fruits and vegetables through irrigated water, soil, badly treated manure, sewage and poor worker hygiene or through poor postharvest handlings. There are many factors that impact on the quality of fresh cut produce is described below:

4. Factors impacting quality of fresh produce

Factors impacting fresh produce may be categorized as physical, chemical or microbiological factors (**Figure 2**). The quality of the fruits is influenced at various stages from growing to harvesting and even during storage. The quality of the produce before its harvesting is influenced by the genotype, road stock, climate, cultivars, cultural practices, maturity and ripening [19]. So, it is essential to choose the appropriate cultivars to assure the quality characteristics of the fruit destined for fresh cut processing. Climatic conditions and cultural practices also affect the nutritional quality of the fresh produce. The growing season and location influence

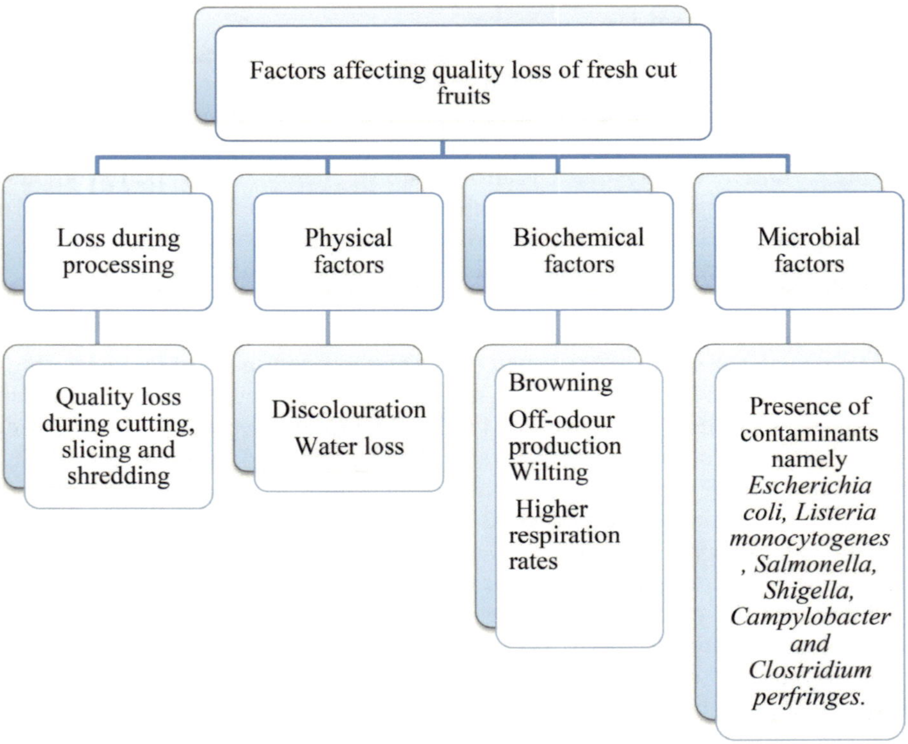

Figure 2.
Factors involved in quality loss of fresh cut fruits.

the level of carotene, riboflavin, ascorbic acid and other nutrients. Low light intensity results in the low levels of ascorbic acid in the plant tissues. Heavy rainfall results in mechanical damage to fruits which make them prone to be attacked by the microorganisms. Agricultural practices like trimming and pruning increases the crop load and size of the fruit. Nutrient composition of the soil has an immediate impact on the fruit texture [20], taste and appearance. Gayán et al. [21] studied that deficiency of calcium in soil will result in tissue softening after harvest. Contamination of the produce starts right from the fields which results in outbreak of food borne illness [22]. Use of raw manure to edible crops and contaminated water can transfer pathogens to the crop resulting in diseases. Harvest techniques also affects the fresh produce quality. Maturity of fresh fruits at the time of harvest directly effect on texture. Immature fruit contain insoluble pectic substances of high molecular weight known as protopectin. On ripening of fruit, these pectin polymer decreases and water soluble pectin are formed. This pectin imparts the characteristic textural changes in the fruit leading to a soft and mushy consistency [23]. Over ripened fruits are generally susceptible to damage during cutting and thus, are not suitable for fresh cut processing. For the fresh cut produce, fruits are harvested at the optimum maturity stage in order to ensure the best eating quality produce. Fruits like banana, guava and papaya continue to ripen even after harvest by exposing them to ethylene. Fruits like pineapple, litches, oranges and muskmelon are harvested mature because no ripening and flavor development occurs after harvest. Fresh produce quality and safety is affected by the post harvestmanagement factors. Management of storage conditions like relative humidity and temperature and handling of the produce influence the quality after harvesting. Temperature, relative humidity, the composition of the gaseous environment influence respiratory and physiological processes. Chemical and microbial contamination compromises the safety of fresh produce. Microbial contamination can be transmitted through improper cultural practices, through contact with unclean surfaces and soil, unhygienic working conditions composition of gases in environment and physical or mechanical damage to the produce. Loss of vitamin C accelerates by the mechanical injury and can and can increase the susceptibility to spoil by the microorganisms. Proper handling is required after harvesting to avoid physical and mechanical injury and the avoidance of microbiological and chemical contamination. Vegetables and fruits must be stored under optimal relative humidity and temperature conditions. Ethylene sensitive green leafy vegetables, watermelons, herbs must be stored separated from high ethylene producers like peaches, tomatoes.

Apart from all the risk factors microbiological safety is the major concern in the fresh cut industries. Many factors may be involved in the epidemiology of the produce associated diseases. In case of fresh cut produce risk can be divided into two categories. First one deals with the conditions or factors contaminating fresh produce with the microorganism during cultivation or harvesting [24]. These include use of contaminated water for irrigation, poor agricultural practices, and use of chemical sprays in irrigation water, application of improperly composted manure and lack of training and good personal hygiene among the workers. The second category of microbial risk is at the cutting and slicing operations in the industry. Internal tissue of the fruit is generally free from the comtamination due to waxy and outer peel. However, cutting and slicing breaks the physical barrier and allows the juices to come out from the internal tissue on the surface of the fruit. This juice contains nutrients which accelerate the growth of microorganism. Surface exposure results in the growth of many potential contaminants and pathogens on cut surfaces [25]. Microbiological risk factors have been determined by the researchers and include the following:

several pathogens (e.g., *Listeria monocytogenes* and *Aeromonas hydrophila*) are psychotropic and can grow at temperatures used to store the fresh cut produce, there is no kill step(like cooking) in the process to eliminate the potential microorganism, the longer shelf life (0–14 days) which is common, due to sophisticated packaging and good temperature control may provide sufficient time for pathogen to grow. Modified atmosphere may inhibit the growth of spoilage organism, but certain organism like *Listeria monocytogenes*, survive and thrive under such conditions [26].

Quality of fresh cut can be maintained between the harvesting and processing. Fruits are harvested when they reach maturity. Methods of harvesting, extent of handling, temperature, storage time influence the quality of fresh produce. Trained labour should be employed so as to prevent the produce from the damage. Fruit and vegetables which are damaged by the insects, animals or by any other physical damage are not fit for the production of fresh cut produce as these are more susceptible to microbial contamination. Fresh produce should be handled properly to assure the safety and quality. Pre cooling of harvested produce is done to reduce the field heat, reduces the impact of ethylene on ethylene sensitive produce, prevent wilting, prevent the quality loss by suppressing respiration and enzymatic degradation, and slow down the microbial activity.

In the preparation of the fresh cut produce highest priority is given to the safety the product. Washing of raw fruits and vegetables eliminate number of pathogens but fail to eliminate human pathogen. So, it is essential to evict the pathogen on produce with various physical and chemical treatments supported by food safety programmes including Hazard Analysis Critical Control Point (HACCP), Good Manufacturing Practices (GMPs), Good Agricultural Practices (GAPs).

5. Latest technologies to ensure safety and to increase shelf life of the fresh cut fruits and vegetables

5.1 Chemical treatment

5.1.1 Dipping in disinfectants

Washing of the fresh produce is done to remove dirt and to reduce the microbial contamination during processing. Sanitation and post harvest handling effects the microbial population on the quality of fresh cut produces. Washing of vegetables and fruits results in insignificant decrease in microbial population. Use of disinfectants and sanitizers like peroxyacetic acid, chlorine, hydrogen peroxide, sodium chloride or ozone can do reduction to 1–2 log units in the initial population of the microorganism on the fresh produce. Chlorine is used as the most commonly used sanitizer for washing purposes. Chlorine concentration from 50 to 200 ppm is used for washing fruits and vegetables [27]. Despite chlorine is used as the disinfectant for decontamination, it has been in several European countries due to chemical hazards associated with it [28]. There are many alternatives used to chlorine nowadays. Chlorine dioxide (ClO_2) is approved to use in flumes water for washing of fresh produce. It is highly effective at neutral pH. Chlorine dioxide is also responsible for formation hazardous by product such as chlorate and chlorite. 200 ppm of chlorine dioxide is used for sanitization of equipments while 3 ppm is used for the washing of uncut produce. Another chlorine based sanitizer acidified sodium chlorite has strong oxidizing capacity. This chemical is approved by United States Environment Protection Agency (USEPA) and United State Food and Drug Administration (USFDA).500-1200 ppm of concentration is used in dipping and spraying process on fresh fruits and vegetable, including fresh-cut. Sodium hypochlorite (NaOCl) is widely known as liquid bleach due to its bleaching property. It has various properties and globally used at household and industrial level in different concentration. NaOCl exihibits broad spectrum anti microbial activity and is used as disinfectant for various purposes. 0.5% solution of NaOCl is known as strong chlorine solution is used for areas disinfecting with body fluids. Weak chlorine solution is a 0.05% of NaOCl is used for washing hands. Study was done to compare the effectiveness of 100 mg\l of sodium hypochlorite (SH) and 500 mg\l acidified sodium chlorite (ACH) on the prevention of enzymatic browning and growth of microbial population on fresh cut produce. ACH washing reduce the microbial contamination and prevent the browning of fresh cut produce during storage. 500 mg/l ACH is effective against ant browning and anti microbial treatment of fresh cut produce [29].

5.1.2 Acid electrolyzed water

Acid electrolyzed water technique was initially developed in Japan. In this technique the diluted salt solution is passed through the containing anode and cathode separated by the membrane. Inanode side of the cell AEW is produced and results in the production of certain ions like $HOCl^-$, OCl^-, Cl_2 gas [30]. AWE is used as the disinfectant alternative to chlorine to its antimicrobial property and minimum effect on the nutritional quality of the food and health of the person. AWE provides the ensurance to the shelf life extension and safety of the fresh cut fruits and vegetables [31]. AWE act as effective technique for the inactivation of microorganism. Recently, slightly acid electrolyzed water is gaining popularity due to its less adverse effects on human health and strong bactericidal efficacy than the acid electrolyzed water [32]. Another application of AEW is NEW (neutral electrolyzed water). Function of NEW is same as AEW except that the product produced at anode is redirected to cathode and resulted into the formation of neutral solution [33]. Free Cl- and OH- ions create the high oxidation and reduction potential which result in bactericidal property [34].

5.1.3 Nanoparticles

Nanotechnology is another emerging field in science. This technique is now being used in preservation of fresh cut fruits and vegetables. In nanotechnology nanosized particles that are in the range of 1–100 nm in dimension are used to alter the physical and chemical properties of the specific material [35]. Nanopartices of various materials have been used. Silver nanoparticles has been used as antimicrobial agent in the preservation. Due this antimicrobial property these nanoparticles damage broad spectrum of microorganisms by various mechanism such as activation of antioxidant enzymes, DNA damage, depletion of antioxidant molecules, structural changes in nuclear membrane and cell wall [36]. Coating of silver nanoparticles-PVP on green asparagus leads to extended shelf life of 25 days at 2°C and 20 days at 10°C respectively [37]. Hybrid of cellulose-silver nanoparticle in combination with MAP is beneficial for the preservation of melon stored at 4°C for 10 days by retarding the senescence of melon [38]. Apart from silver nanoparticle, nanoparticles of TiO_2 and ZnO are also used in the preservation of fruits and vegetable. Nanoemulsions is one of application came under the category of nanoscience. Nanoemulsionresults in higher stability in terms of coalescence of oil droplet, gravitational separation, enhanced activity of emulsified oils, flocculation and higher surface area to droplet volume ratio

[39]. Nanoemulsions of lemongrass and oregano oil results in the reduction of microbial population to several units [9]. Likewise, an edible coating of lemongrass oil is responsible for the inactivation of *E. coli* on fresh cut apple during storage [40]. Food industries are benefitted a lot from the antimicrobial nanomaterials but safety issues to use them at commercial level are of more concern. These nanoparticles penetrate into product tissue and alter the chemical composition. Therefore, further study is required to before commercializing this technology.

5.1.4 Ozone

Fresh cut fruits and vegetables are treated with ozone in gaseous or aqueous form to extend their shelf life. Ozone came under the category of generally regarded as safe and has been approved by US FDA as direct contact food sanitizers [41]. Ozone reacts with the intracellular enzymes, nucleic acids, spore coat, components of envelop or viral capsid of microorganism. Gaseous treatment of ozone (950 μl/20 min) on sliced lettuce and spinach results in 1.0–1.5 log reduction in *E. coli* and *Listeria innocua* [42]. 9 ppm exposure of gaseous ozone results in the reduction of 2.89, 3.06 and 2.56 log in *E. coli* O_{157}, *Listeria monocytogenes, Salmonella Typhimurium*in fresh cut bell pepper. Fresh cut melon [43], apple, papaya [44] were treated with aqueous ozone (1.4 mg/5–10 min) shows significant reduction in the bacterial count while certain biochemical properties such as total phenols, PPO and POD activites, ethylene production and MDA content were reduced. Although ozone is being used in the preservation of fresh cut fruits and vegetables, special care must be given to the after effects caused by the long exposure and high concentration of it. Long exposure and increase concentration causes various health related problems so in this regard federal occupational safety and health administration in united states specified the limits on working environment (0.1 ppm for long term exposure and 0.3 ppm for short term exposure) [45]. All chemical approached for fresh cut fruits and vegetables are presented alongwith their advantages and disadvantages in **Table 1**.

Technology	Pros and Cons	Reference
Dipping in disinfectants	Use of disinfectants and sanitizers like peroxyacetic acid, chlorine, hydrogen peroxide, sodium chloride or ozone can do reduction to 1–2 log units in the initial population of the microorganism on the fresh produce.	[27]
Acid electrolyzed water	It provides the ensurance to the shelf life extension and safety of the fresh cut fruits and vegetables. Free Cl- and OH- ions create the high oxidation and reduction potential which result in bactericidal property.	[46]
Nanoparticles	The antimicrobial property these nanoparticles damage broad spectrum of microorganisms by various mechanism such as activation of antioxidant enzymes, DNA damage, depletion of antioxidant molecules, structural changes in nuclear membrane and cell wall.	[45]
Ozone	Fresh cut fruits and vegetables are treated with ozone in gaseous or aqueous form to extend their shelf life. Gaseous treatment of ozone (950 μl/20 min) on sliced lettuce and spinach results in 1.0–1.5 log reduction in *E. coli* and *Listeria innocua*.	[35]

Table 1.
Latest chemical technologies to ensure safety and to increase shelf life of the fresh cut fruits and vegetables.

5.2 Physical treatment

5.2.1 Modified atmosphere packaging

MAP is the effective technique used for the preservation and extending shelf life of the fresh cut produce. In this technique, gas composition around the product is replaced with the inert gases like CO_2, O_2 and N_2. MAP inhibits the growth of microorganisms by creating unfavorable conditions. Higher O_2 concentration is used for the preservation of aroma producing volatile substances in fresh cut honeydew melon and cantaloupe stored for 12 days at 5°C. Meng et al. [47] reported that the concentration of O_2 greater than 70% is proved to effective against the microbial growth and enzymatic browning in mushroom slices, shredded chicory endives and grated celeriac. The color evaluation under high O_2 concentration of these three fresh cut produce donot exceed through 6–7 days while the control sample (5% O_2 compensated with N_2) was rejected after 3–4 days. Higher O_2 concentration 50 or above 90% is effective in inhibiting the microbial growth, enzymatic browning in fresh cut lettuce stored for 6 days at 7°C [48]. Monnin et al. [49] through research concluded that the combination of high O_2 and ascorbic acid are highly conducive to the inhibition of peroxidase and polyphenol oxidase activites and maintenance of biochemical properties of fresh cut eggplant result in extended in shelf life for 12 days at 4°C.The sample taken as control was also microbiologically accepted but the sensory score was less than the treated sample. 50% O_2 and 50% CO_2 shows the strong inhibition on the growth of yeast and the production of volatile compounds in fresh cut pineapple. This method was also reported to be responsible for the extension of shelf life. Apart from above mentioned gases, CO at low levels was also approved as GRAS as a gas used in MAP. CO is generally used in preservation of meat through MAP as it reduces the metmyoglobin and maintaining the cherry red color of meat. Very less literature is present which confirm the use of CO in fresh cut produce. Niemira [50] studied the use of CO (<175 ml/l) treatment for 20 min in fresh cut lotus root slices. This reduces the activites of PPO and POD. They also reported that the phenylalaineammonialyase (PAL) and malonaldehyde (MDA) content of treated sample were 17% lower and 40% higher than the non treated sample stored at 5°C for 8 days respectively but the controversy regarding the use of CO is still there. There are many other gases which are recently being used in MAP such as argon, helium and nitrous oxide. These noble gases indirectly affect the metabolism of the plant tissue. These gases increase the diffusivity of the O_2 from the plant tissue there by affecting certain biochemical pathways and making unfavorable conditions for the growth of microorganisms. Study was done by [51] on browning in sliced apples. N_2O and Ar (90% N_2O, 5% O_2, 5% CO_2 and 5% CO_2, 5%, 25% Ar) was used under high pressure in MAP. Results shows 15% and 25% browning in N_2O and Ar treated respectively whereas 60% in control sample stored at 5°C for 12 days. On the other hand there is increase in firmness and total soluble solid content in the treated sample. The effect of noble gases (89.3% N_2O, 89.9% Ar, 90.1% He) was compared with the air packaging in fresh cut watercress. Result shows that the C_2H_4 emission and the rate of respiration were low as compared to control but there was no effect on the growth of psychotropic and *Enterobactericeae* was observed. They suggest that the combination of other technologies with MAP to ensure the microbial safety in case of watercress [52].

5.2.2 Electron beam irradiation

Food irradiation is done for the preservation of fresh cut fruits and vegetable. Irradiation results in the negligible modification in the nutrients, taste, flavor, color and other qualities. Cobalt-60 is most frequently used radioisotope for the purpose of quality and safety of fresh produce. Lower doses of are proved to be effective in the preservation of food [46]. However these radiationshave the potential to cause cancer if used above permissible limits. EBI donot require any radioactive isotope for ionization. Electron beams are generated with the help of the machine capable of accelerating the electron close to the speed of light at high energy level in the range of 0.15–10 MeV in a vacuum environment [53]. Energy source is the commercial electricity and generator can be easily switched off and on. This technique helps to eliminate the micro flora present over the food by the destructing their DNA structure, membrane proteins and enzymes resulting in death of the organism. The effectiveness of the EBI depends the dosage of irradiation and type of food. Irradiation of blueberries with 2.3 and 3.13 kGy results in the 8.9–28 log CFU/g and 6 CFU/g decrease in the *E. coli* population. Fresh cut cabbage was treated with 2.3 and 4.0 kGy resulting in 4.0 and 7.0 log reduction in *E. coli* respectively. A research was conducted to study the correlation between the shelf life and irradiation dosage on blueberries. The result shows that the untreated blueberries decay to 39% while those treated with 2 and 3 kGy reports 8% and 3% decay respectively [38]. Mushrooms generally have shorter shelf life due to weight loss, enzymatic browning and texture changes. To overcome this mushrooms were treated with EBI at 2 kGy and then they result in the highest total antioxidant capacity, higher whiteness and lowest electrolyte leakage in mushroom. With increase in certain biochemical properties, there is decrease in vitamin C content. Despite of the successful applications discussed above, irradiation level is restricted by Food and Drug Administration (FDA) on fruits and vegetables. The maximum level at which is recommended is 1.0 kGy with the two exceptions, i.e., fresh lettuce and spinach which can be irriatedupto 4.5 kGy. Many trials of irradiation using EBI technique on fresh cut fruits is under progress, promising and encouraging results can be the hope for future research.

5.2.3 Pressurized inert gases

Inert gases such as neon, argon, nitrogen, krypton, xenon are used in the preservation of fresh cut fruits and vegetables as these gases form the ice-like crystal structure called clathrate hydrate. The gas molecules get intraped into the cage like structure by water with the help of vander Waals forces and are stable at lower temperature [54]. Various studies have been done to study the role of inert gases in the preservation of fresh cut fruits and vegetables. Shelf life of fresh cut asparagus spears was extended from 3 to 5 days to 12 days by treating them with argo (Ar), xenon (Xe) under 1.1 MPa (Ar and Xe at 2.9(v:v) in partial pressure) for 24 h at 4°C. Ramos-Villarroel et al. [55] used the pressurized Argon (4 MPa) for 1 h on green peppers which results in decreased water mobility as well as loss of water, ascorbic acid growth of yeast and mold and maintaining the cell integrity by inhibiting the production of MDA and activites of POD and CAT (catalase). Use of argon on green peppers also results in the extension of shelf life to 12 days stored at 4°C as compared to the untreated ones having shelf life of 8 days. Shelf life of the fresh cut apples and pineapples can be extended from 9–7 to 15–12 days by the application of high pressure treatment of Ar at 150 MPa for 10 min and 1.8 MPa for 60 min respectively. However these treated samples possess lower scores in firmness than that of the untreated [56]. Further study was conducted by modifying this experiment by using combination of gases. Combination of argon and xenon as well as argon and nitrogen give promising results [57]. Lower growth of *Saccharomyces cerevisiae* and *E. coli* was observed in fresh cut apple and pineapple when treated with the mixture of xenone and argon under

1.8 MPa (Xe and Ar at 2:9(v:v) in partial pressure as compared to the untreated sample. On the other hand loss of ascorbic acid and total phenols, lower browning were reported under high pressure (10 MPa) nitrogen and argon on fresh cut pineapples.

5.2.4 Ultraviolent light (UV)

UV radiations are the non ionizing radiations having the wavelength in the range of 100 nm to 400 nm. These rays are classified into three different types: UV-A (315–400), UV-B (280–305), UV-C (100–280) [25]. UV-C at 254 nm is most prominently used due to its germicidal properties. UV rays causes the DNA damage in the living organism by inducing the formation of DNA photoproducts such as purimidine 6–4 pyrimidone and cyclobutane pyrimidine dimers which hinders replication and transcription and eventually leading to mutations and cell death [58]. The major advantage of UV rays is its broad spectrum action over microorganism, convenient manipulation and lower cost. Poor penetration of UV rays limits its use in food field. However, UV-C is frequently being used for the surface decontamination of fresh cut fruits and vegetables as enzymatic deterioration and surface spoilage mainly occur on surface [59]. Many studies on the use of UV on fresh cut produce gave the satisfying results. Salvia-Trujillo et al. [60] reports that by treating the fresh cut apples with 1.2 kJ/m^2 UV-C lowers the microbial population to 2 log units as compared to the untreated ones stored at 6°C for 8 days. 12.5 kJ/m^{2}of UV-C shows the significant results in inhibiting PPO activity and browning in fresh cut carambola which remain fresh even after 21 days of storage. Similarly, significant results were observed in case of fresh cut apples [61]. Fresh cut peppers were treated with UV-C which eventually leads to 50% higher firmness as compared to the control sample stored at 12 day. Similarly, maintenance of antioxidant activity (DPPH activity), vitamin C content and total phenolic compounds of fresh cut paprika [62] and mandarin.

Unfortunately, due to the negative effects of UV-C on nutritional and sensory characteristic, it is in limited in use. Extended exposure of UV-C on fresh cut pineapples accelerates browning and significant decrease in Vitamin C content. High doses of UV-C causes weight loss and excess electrolyte leakage in fresh cut green onion [63]. Shelf life of watermelon is increased to 11 days at 5°C by treating it with low UV-C (1.6 and 2.8 kJ/m^2) but when treated it with high UV-C (4.8 and 702 kJ/m^2), shelf life of 8 days. Therefore the use of combination of UV-C with citric acid [64], malic acid, electrolyzed water, modified atmosphere packaging [65], gaseous ozone [66] have been developed. Recently, UV- light emitting diodes are being used due to their long life expectancy, energy efficient, low cost, convenient manipulation, no harm to human eyes and skin and no liberation of mercury waste. UVA-LED shows significant results in disinfecting food by reduction of *E. coli* on fresh cut cabbage and lettuce whilemaking no loss of vitamin C. Although, antibrowning effect is associated with irradiance, the fruit cultivar and exposure time, effective in inhibiting browning in fresh cut pear and apple [67].

5.2.5 Pulsed light (PL)

PL is thenother technique used for decontamination of packaging material and food by inactivating the microorganisms. Short duration and high power pulses are generated with inert gas (generally xenon) lamp and involve broad spectrum white light. PL results in photochemical effect and results in structural changes in DNA of viruses, bacteria and other pathogen and interferes in replication and transcription resulting mutation and eventually death of the organism. The main advantages of this technique areits low energy cost, its great flexibility, and significant reduction in very short time and lack of residual compounds. PL is recently being used for treating fresh cut fruits and vegetables. Treating fruits and vegetables with PL not only reduces the microflora, but also in the maintaience of sensory and nutritional properties of fresh cut produce. The efficiency of PL depends upon the number of pulses and intensity of the pulses. Low intensity may be ineffective while higher intensity may be toxic and cause undesirable damages. Various studies have been done on the preservation of fresh cut with the help of PL technique. Fresh cut mangoes was treated with pulsed light reports the conducive effect on the firmness, carotenoid and color of the fruit stored at 6°C for 7 days where as loss of color, firmnesswas observed in the control sample after 3 days of storage [68]. Carotenoid content oftreated sample was 9 mg/g dry matter as compared to 2 mg/g dry matter of untreated ones. Reduction in yeast and mold and maintenance of chlorophyll a and b in fresh cut avocado has been reported [69]. Exposure to high pulses (12 J/cm) results in significant reduction in *E. coli* and *Listeria innocua* in fresh cut mushroom [70]. Significant reduction was seen in *L. innocua*, *Escherichia coli, S. cerevisiae* in fresh cut apples when exposed to high pulses. However, browning action on the cut surfaces was promoted. It occurs mainly due to increase in temperature or thermal damage during the treatment which accelerates non- enzymatic and enzymatic browning. Moreover some negative effects on color, texture and sensory attributes have been reported [71]. To overcome this disadvantage combined technique was employed by [72]. PL treatment was combined with anti- browning pretreatment by dipping into mixed solution of 1%(w/v) ascorbic acid and 0.1%(w/v) calcium chloride with 71.6 J/cm^2 PL dose shows the effective results in minimizing browning on fresh cut apples. Further research must be done to find improved application of PL combine with other techniques.

5.2.6 Cold plasma (CP)

Cold plasma is anothernon-thermal technique used nowadays for the food decontamination and preservation. Plasma is defined as the fourth state of matter after solid, liquid, gas. It is the quasi neutral ionized gas which consists of photons, negative ions, free electrons, excites or non excited atoms and molecules. Various techniques are being employed for the production of plasma such as lasers, microwaves, magnetic field, electricity, direct and alternating current. Mixture of nitrogen, oxygen or mixture of other nobel gases are used in CP. Recombination process takes place between the active particles with the release of energy as visible and UV light. These active particles in the plasma react with the food substrate releasing the energy into the viruses and bacteria. Although, the exact mechanism of the inactivation is still not known but the primary mechanism of inactivation attributes to direct chemical interaction with the charged species, destruction of cellular components by UV and denaturation of DNA strand [73]. The proportion of the gas mixture and the specific energy source depend upon the chemical composition, temperature and density of the plasma. Apart from these protein, fat content, texture, pH and texture of the food also depends. This treatment is successfully applied for microbial decontamination on strawberry, potato, cherry, cabbage and milk representing significant results. Yeoh et al. [74] reported 1.76, 2.72, 0.94 log reduction of *Salmonella typhimurium* on strawberry, lettuce and potato by the use of CP technique. Another study was done on strawberries by [75] which reports 44–95% reduction of yeast and mold count and 12–85% reduction of mesophilic count. Beside this no significant change occur in color, texture and firmness of the treated product. CP treatment was given for 10,

60 and 120 s to tomato resulted in 3.1, 6.3 and 6.7 log 10 CFU/sample reductions of *Salmonella, E.coli and L.monocytogenes* from the initial sample. Similar but extended treatment was given to strawberries due to its complicated surface [76] reports that 0, 15, 30, 45, 60, 90 and 120 s treatment with CP to blueberries result in the reduction of yeast/molds and total aerobic plate count to 1.5–2.0 CFU/g and 0.8–1.6 log CFU/g as compared to the controlled sample after 1 and 7 days respectively. Above mentioned studies was mainly concentrated on the fresh produce. Very few literatures is present on the application of CP on fresh cut produce due to its early stage of development. Browning area and PPO activity in fresh cut apples is reduced to 65% and 12–58% respectively by treating it with CP for 30 min compared to control after 4 h of storage. In fresh cut melon 17% POD and 7% PME activity inhibition of fresh cut melon is achieved by treating them with CP. Study done by few scientists claimed that the treatment with CP results in improved color retention and reduced browning in fresh cut kiwi fruit during storage. In addition no significant change occur in antioxidant content and antioxidant activity but slight change(up to 10%) was seen in fresh cut apples after treatment [77]. CP has gained much attention during last decade due to its promising results. However, much information about its effects on food quality and mechanism involved is unknown [78]. Further study is needed regarding physiochemical reaction kinetics, sensorial and nutritional properties of foods. Safety issued are also not verified in case of CP. Therefore integrated risk assessment is required for its application at its commercial use.

6. Impact of different techniques on shelf life and sensory aspect

The shelf life of FCF pretreated with chemical preservatives revealed that inspite of the chemical pretreated fruits showed an increase in microbial load; yeast and mold being most prevalent over the FCF during the storage time. However, slow increase in microflora was observed over FCF pretreated with NaOCl. Antimicrobial action of NaOCl leads to reduction in pH and an increase in acidity that adds to the hurdles in the proliferation of the microorganisms during the storage period. Food processing techniques stabilize the product and lengthen their storage and shelf life, production of fresh cut fruits increases their perishability. Enzymatic browning due to oxidation of phenolic compounds lowers the product quality. Sun et al. [79] evaluated the effect of washing of fresh cut potatoes and sweet potatoes with SH and ASC. Results shows that the 500 mg/l concentration is effective alternative to SH at 100 mg/l to inhibit the browning and PPO activity hence allowing longer shelf life.

Quality of fresh cut apples was determined by dipping in organic acids and acidic electrolyzed water. Plesoianu et al. [80] conducted the to investigate the effect of citric acid (2%), benzoic acid (0.2%), sorbic acid (0.2%) and ascorbic acid (0.5%) in acid electrolyzed water. Samples were placed at 8°C for 14 days. The results indicate that the acid electrolyzed water showed less browning on fresh cut sample as compared to ascorbic acid and citric acid. The samples treated with 2% citric acid and acid electrolyzed water significantly maintained the firmness, phenol content and antioxidant activity after 14 days of storage. Nanotechnology is the advance technology for the preservation of fresh cut fruits.

A new example of this technology applied on the minimally processed fresh cut melon. Fresh cut melon (*Cucumis melo* L.) were coated with the alginate-based coating having silver nanoparticles (Ag-MMT) to study the its effect on shelf life of the product. Results shows that the treated sample was effective from microbial and sensory point of view as compared to the controlled sample hence promoting more acceptance. Prolonged shelf life was also recorded 11 days in case of treated sample as compared to 3 days in control [81]. Active packing of nano-ZnO was studied on fresh cut fuji apple by Li et al. in 2011 [82]. Cutting of fresh cut induce ethylene production which was suppressed by nanopackaging. Polyphenoloxidase and pyrogallol peroxidase acitivites were also reduced. Initial browning index was maintained 23.9 which is much lower than the control sample having 31.7 on 12th day. The research concludes that the naonpackaging could be used to increase the shelf life of the fresh cut fuji apples.

Ozone treatments were also studied for preserving the vegetable quality. Carrot color was found to be insignificantly changed after treatment at 450 ppb for 48 h, 7.6 mg l^{-1} for 15 min, and between 1 and 5 mg l^{-1} for 9.5–110.5 min. However, some dry white blotches were observed at 60 μl l^{-1} and some scattered slightly brown discolored blotches of periderm at 50 nl l^{-1} were found. Ozone behaves as postharvest stress condition that results in respiration and ethanol production due to an abnormal metabolism. An ozone supply of 15 μl l^{-1} for 8 h a day for 28 days provides some disease protection with a minimum of physical and physiological damage [83].

Modified atmospheric packaging is very common method used for the preservation of fresh cut fruits and vegetables. Study was conducted to check the effect of MAP used in combination with 2% Natureseal and evaluating the physicochemical and microbial parameters during 21 days at 4°C. Sensory quality was evaluated after 10 days of treatment. The headspace CO_2 level in MAP samples amplified significantly up to 35.3% at the end of storage; while O_2 decreased significantly. Color values were also affected with no changes in Hunter *L* and *a* values, which decreased and increased, respectively. MAP preserved the sensory eminence of fresh-cut pears up to 10 days of storage [84].

Effect of electric beam irradiation was studied on shelf life of summer truffles (tuber aestivum) packed under modified packaging. Effect of two doses of electron beam irradiation (1.5 and 2.5 kGy) on microbial population and sensory characteristics. Samples were analyzed weekly for 42 days stored at 4°C. Results showed the sample treated with 2.5 kGy e-beam has prolonged the shelf life to 42 days as compared to 21 days for the control samples [85].

Shelf life of the fresh-cut green peppers was extended using pressurized argon treatment. Fresh-cut green peppers were treated with pressurized (2–6 MPa) argon for 1 h. Control and argon-treated samples were placed in polystyrene packaging with 5% O_2 and 8% CO_2 and then stored at 4°C and 90% RH for 12 days. A range of quality parameters like chlorophyll content, Water loss, Water mobility, sensory quality, ascorbic acid loss, malondialdehyde (MDA), cell membrane permeability, cell protective enzyme activity and microbial quality were determined after every 2 days. Loss of water and water mobility was greatly reduced by the pressurized argon treatment in fresh-cut green peppers. Likewise, ascorbic acid loss, chlorophyll content and hue angle were also reduced during storage. The pressurized argon treatments were found to retain the cell integrity by inhibiting an increase in MDA and membrane permeability compared to the control samples. The treatment also reduced proliferation of coliforms, yeasts and molds. The fresh-cut pressurized argon treated green peppers can be kept in a fresh-like condition for 12 d at 4°C [47].

Latest technologies like pulsed light and cold plasma discussed in this chapter are of great importance for the preservation of fresh cut. Various researches and papers are published which shows their efficacy. Research was conducted to record fresh-cut 'Golden Delicious' apples' quality attributes after treatment with pulsed light treatment (12 J/cm^2) and a gellan-gum based (0.5% w/v) edible apple fiber enriched coating. Various physicochemical and sensory aspects were analyzed during 14 days storage at 4°C. The combined application of coating and PL treatment retarded the microbiological deterioration of fresh-cut apples and maintained the sensory attribute scores above the rejection limits after prolonged storage [86]. Same technique was applied to maintain the physicochemical and nutritional profile of fresh cut mangoes. Pulsed light treatments were carried out using an automatic flash lamp system (Mulieribus, Claranor) composed of eight lamps situated all around the sample with a total fluence of 8 J^{cm-2}. Pulsed light treatment maintained the firmness, color and the carotenoids of fresh-cut mangoes.

Cold plasma technique is used for the preservation of minimally processed products. Misra et al. [87] reported the control and CP treated tomatoes showed decrease in respiration rate during the storage. However, the respiration rates were similar for the control and CP-treated tomatoes at the end of storage. The results of Tappi et al. [88] showed that the plasma treatment can cause an alteration of the cellular respiratory pathway. Misra et al. [89] utilized cold plasma treatment for strawberries in modified atmosphere packaging that revealed no significant increase in respiration rate. Tappi et al. [88] reported that cold plasma treatments results in an increase of firmness in fresh-cut apples. The amount of firmness was 18.9 N for the control and 21.8 N for plasma-treated samples at 15 kV for 10 min. The highest firmness was found in cold plasma treated mushrooms which indicates the impending application of this inventive technology in escalating the shelf-life and quality of mushroom after harvesting [90].

7. Future trend in processing of fresh cut fruits and vegetables

In last few years there is revolution in fresh cut industry. Stepping of women into the jobs cause radical change in the lifestyle and very less time is left for the preparation of meal for the whole family. Industrial kitchen have to prepare food for large number of people with limited number of labour. Moreover, consumers are becoming more health conscious as a result there is change in their food choices and prefer fresh and convenient product. This scenario created a challenges as well as opportunity for products into the market like fresh cut fruits and vegetables as a way to increase consumption of vegetables and fruits. Fresh cut product results in greater demand among the consumer for quality appearance, convenience and healthy nutrition. Various fresh cut products are already present in the market boosting and attracting consumers. Minimally processed food occupy special place in the market and is one of the major growing segment in the food industry. Being organic, these products stand out from rest of the products. Fresh cut salads mixes are fastest growing categories growing at the rate of 200% over the past 3 years.

Due to the raw materials diversity, processing conditions and packaging systems used in the production of fresh-cut products, it is impossible to institute a one-size fits-all approach to attain microbial safety. Rather, the producer has to vigilantly consider an extensive variety of factors and hurdles—quality of raw material, hygienic conditions, storage temperature, water content, acidity, modified atmosphere conditions—in formative ways to control microbial growth. Through the potential selection and combination of these factors, the producer is able to concluding the optimum shelf life of the product and ensures safe products for end consumers. Hence, safety and sanitation are the top priority parameters for fresh-cut processors. New design essentials of processing equipment are a critical part of this uninterrupted evolution of food safety and sanitation. These elements aid the plant managers to implement cleaning practices more effectively.

The high influence placed on hygienic design of fresh-cut processing facility and equipments is a new trend now days. Since microorganisms are ever-present and mutate incessantly by adapting to different types of disinfectants and sanitizers, it is extremely important to develop vibrant sanitary protocols to control microbial contamination efficiently. To all stakeholders in the fresh-cuts industry (from regulators to processors to equipment manufacturers), equipment designed with hygienic goals in mind is fast becoming an area of primary concern. For this reason, sanitation design and protocols will continue to develop that organizations need to apply collective knowledge and advanced engineering to create safer and more proficient processes. Organizations such as UFPA (United Fresh Produce Association), PMA (Produce Marketing Association) and the USDA offer plant owners and managers information on what to look for in processing equipment that will best suit their needs within a plant. These organizations also offer important guidance to manufacturers to continually grow and develop their easy-to-clean designs. Maximizing use of the knowledge base is the best way to maintain the advance in safety and sanitation process in the fresh-cut industry. Additionally, the industry and its organizations are operational with equipment manufacturers to establish sanitary equipment design guidance in a proactive effort to offer basic flowcharts, and checklists to help in the evaluation of effective sanitary design attributes. The FDA, moreover, is focusing on the cleanability of processing equipment during strengthened inspections of food processing facilities.

Author details

Pooja Nikhanj[1*], Mohini Prabha Singh[2], Simran Saini[2], Gurliin Kaur[2] and Juhi Kumari[2]

1 Punjab Horticultural Postharvest Technology Centre, Punjab Agricultural University Campus, Ludhiana, India

2 Department of Microbiology, Punjab Agricultural University, Ludhiana, India

*Address all correspondence to: poojanikhanj@pau.edu

IntechOpen

© 2022 The Author(s). Licensee IntechOpen. This chapter is distributed under the terms of the Creative Commons Attribution License (http://creativecommons.org/licenses/by/3.0), which permits unrestricted use, distribution, and reproduction in any medium, provided the original work is properly cited.

References

[1] Abadias M, Usall J, Oliveira M, Alegre I, Viñas I. Efficacy of neutral electrolyzed water (NEW) for reducing microbial contamination on minimally-processed vegetables. International Journal of Food Microbiology. 2008;**123**:151-158

[2] Abbott JA, Harker FR. The commercial storage of fruits, vegetables, and florist and nursery stocks. In: Gross KC, Wang CY, Saltveit ME, editors. Texture. Washington, DC: Agricultural Research Service; 2004. Accepted (In public review), Available from: http://usna.usda.gov/hb66-021texture.pdf

[3] Aguiló-Aguayo I, Charles F, Renard CMGC, Page D, Carlin F. Pulsed light effects on surface decontamination, physical qualities and nutritional composition of tomato fruit. Postharvest Biology and Technology. 2014;**86**:29-36

[4] Anonymous (2019). National Horticulture Board. 2017. Available from: http://www.nhb.gov.in.

[5] Artes F, Allende A. Minimal Fresh Processing of Vegetables, Fruits and Juices. Spain: Technical University of Cartagena; 2005. pp. 677-716

[6] Bachmann J, Earles R. Post-Harvest Handling of Fruits and Vegetables. US: ATTRA.; 2000. pp. 24-38

[7] Beaulieu JC, Gorny JL. Fresh-cut fruits. In: Gross KC, Wang CY, Saltveit M, editors. The Commercial Storage of Fruits, Vegetables, and Florist and Nursery Crops. California, US: USDA; 2002. Available from: http://www.ba.ars.usda.gov/hb66/index:html

[8] Beaulieu JC. Volatile changes in cantaloupe during growth, maturation, and in storage conditions and in stored fresh-cuts prepared from fruit harvested at various maturities. Journal of the American Society for Horticultural Science. 2006;**131**(1):127-139

[9] Bhargava K, Conti DS, da Rocha SRP, Zhang Y. Application of an oregano oil nanoemulsion to the control of foodborne bacteria on fresh lettuce. Food Microbiology. 2015;**47**:69-73

[10] Botondi R, Moscetti R, Massantini R. A comparative study on the effectiveness of ozonated water and peracetic acid in the storability of packaged fresh-cut melon. Journal of Food Science and Technology. 2016;**53**:2352-2360

[11] Bourne MC. Physical properties and structure of horticultural crops. In: Peleg M, Bagley EB, editors. Physical Properties of Foods. Westport: AVI; 1983. pp. 207-228

[12] Charles F, Vidal V, Olive F, Filgueiras H, Sallanon H. Pulsed light treatment as new method to maintain physical and nutritional quality of fresh-cut mangoes. Innovative Food Science & Emerging Technologies. 2013;**18**:190-195

[13] Chen C, Hu W, He Y, Jiang A, Zhang R. Effect of citric acid combined with UV-C on the quality of fresh-cut apples. Postharvest Biology and Technology. 2011;**111**:126-131

[14] Choi IL, Yoo T, Kang HM. UV-C treatments enhance antioxidant activity, retain quality and microbial safety of fresh-cut paprika in MA storage. Horticulture, Environment and Biotechnology. 2015;**56**:324-329

[15] Cook R. Trends in the marketing of fresh produce and fresh-cut

products. 2009. Available from: http://www.agecon.ucdavis.edu/people/faculty/facultydocs/Cook/Articles/freshcut2009Cook090922.pdf

[16] Ding T, Rahman ME, Oh DH. Inhibitory effects of low concentration electrolyzed water and other sanitizers against foodborne pathogens on oyster mushroom. Food Control. 2011;**22**:318-322

[17] Fang Z, Bhandari B. Encapsulation of polyphenols—A review. Trends in Food Science and Technology. 2015;**21**(10):510-523

[18] Fernández A, Noriega E, Thompson A. Inactivation of *Salmonella entericaserovar* typhimurium on fresh produce by cold atmospheric gas plasma technology. Food Microbiology. 2013;**33**:24-29

[19] Fernández A, Picouet P, Lloret E. Cellulose-silver nanoparticle hybrid materials to control spoilage-related microflora in absorbent pads located in trays of fresh-cut melon. International Journal of Food Microbiology. 2010;**142**:222-228

[20] Garg N, Churey JJ, Splittstoesser DF. Microflora of fresh-cut vegetables stored at refrigerated and abuse temperatures. Journal of Food Science. 1993;**30**(5):385-386

[21] Gayán E, Condón S, Álvarez I. Biological aspects in food preservation by ultraviolet light: A review. Food and Bioprocess Technology. 2014;7:1-20

[22] Gil MI, Selma MV, Lopez-Galvez F, Allende A. Fresh-cut product sanitation and wash water disinfection: Problems and solutions. International Journal of Food Microbiology. 2009;**134**:37-45

[23] Gil M, Gómez-López V, Hung YC, Allende A. Potential of electrolyzed water as an alternative disinfectant agent in the fresh-cut industry. Food and Bioprocess Technology. 2015;**8**:1336-1348

[24] Gómez P, Salvatori D, García-Loredo A, Alzamora S. Pulsed light treatment of cut apple: Dose effect on color, structure, and microbiological stability. Food and Bioprocess Technology. 2012;**5**:2311-2322

[25] González-Aguilar G, Ayala-Zavala JF, Olivas GI, de la Rosa LA, Álvarez-ParrillaE. Preserving quality of fresh-cut products using safe technologies. Journal für Verbraucherschutz und Lebensmittelsicherheit. 2010;**5**:65-72

[26] Gutiérrez DR, Chaves AR, Rodríguez SDC. Use of UV-C and gaseous ozone as sanitizing agents for keeping the quality of fresh-cut rocket (*Eruca sativa* mill). Journal of Food Processing and Preservation. 2016;**41**(3):203-208

[27] Horvitz S, Cantalejo MJ. Application of ozone for the postharvest. Treatment of fruits and vegetables. Critical Reviews in Food Science and Nutrition. 2014;**54**:312-339

[28] Hurst WC. Sanitation of lightly processed fruits and vegetables. Journal of HortScience. 1995;**30**(1):22-24

[29] Jacxsens L, Devlieghere F, Van der Steen C, Debevere J. Effect of high oxygen modified atmosphere packaging on microbial growth and sensorial qualities of fresh-cut produce. International Journal of Food Microbiology. 2001;**71**:197-210

[30] James JB, Ngarmsak T. Processing of Fresh-Cut Tropical Fruits and Vegetables: A Technical Guide. Bangkok: Food and Agriculture Organization of the United Nations Regional Office for Asia and the Pacific. FAO; 2010. pp. 1-84

[31] Kader AA, Mitcham EJ. Optimum procedure for ripening mangoes. In: Kader AA, editor. Management of Fruit Ripening, Post-Harvest Horticultural Series No. 9. California, US: University of California; 1999. pp. 20-21

[32] Kader A. Quality parameters of fresh-cut fruit and vegetable products. In: Lamikanra O editor. Fresh-Cut Fruits and Vegetables. Science, Technology and Market. Boca Raton, FL, CRC Press. LLA; 2002

[33] Kader A, Cantwell M, Mitcham B. Methods for determining quality of fresh commodities. Perishables Handling Newsletter. 2007;**8**:1-5

[34] Kalia A, Parshad VR. Novel trends to revolutionize preservation and packaging of fruits/fruit products: Microbiological and nanotechnological perspectives. Critical Reviews in Food Science and Nutrition. 2013;**55**:159-182

[35] Karaca H, Velioglu YS. Effects of ozone treatments on microbial quality and some chemical properties of lettuce, spinach, and parsley. Postharvest Biology and Technology. 2014;**88**:46-53

[36] Kasim MU, Kasim R, Erkal S. UV-C treatments on fresh-cut green onions enhanced antioxidant activity, maintained green color and controlled 'telescoping'. Journal of Food, Agriculture and Environment. 2008;**6**:63-67

[37] Kim DH, Kim HB, Chung HS, Moon KD. Browning control of fresh-cut lettuce by phytoncide treatment. Food Chemistry. 2014;**159**:188-192

[38] Kong Q, Wu A, Qi W, Qi R, Carter JM, Rasooly R, et al. Effects of electron-beam irradiation on blueberries inoculated with *Escherichia coli* and their nutritional quality and shelf life. Postharvest Biology and Technology. 2014;**95**:28-35

[39] Lacombe A, Niemira BA, Gurtler JB, Fan X, Sites J, Boyd G, et al. Atmospheric cold plasma inactivation of aerobic microorganisms on blueberries and effects on quality attributes. Food Microbiology. 2015;**46**:479-484

[40] Lante A, Tinello F, Nicoletto M. UV-A light treatment for controlling enzymatic browning of fresh-cut fruits. Innovative Food Science & Emerging Technologies. 2016;**34**:141-147

[41] Li X, Jiang Y, Li W, Tang Y, Yun J. Effects of ascorbic acid and high oxygen modified atmosphere packaging during storage of fresh-cut eggplants. Food Science and Technology International. 2014;**20**:99-108

[42] López-Gálvez F, Ragaert P, Haque MA, Eriksson M, Van Labeke MC, Devlieghere F. High oxygen atmospheres can induce russet spotting development in minimally processed iceberg lettuce. Postharvest Biology and Technology. 2015;**100**:168-175

[43] MamiY PG, Ziaie F, Ghasemnezhad M, Salmanpour V. Improvement of shelf life and postharvest quality of white button mushroom by electron beam irradiation. Journal of Food Processing and Preservation. 2014;**38**:1673-1681

[44] Manzocco L, Da Pieve S, Bertolini A, Bartolomeoli I, Maifreni M, Vianello A, et al. Surface decontamination of fresh-cut apple by UV-C light exposure: Effects on structure, colour and sensory properties. Postharvest Biology and Technology. 2011;**61**:165-171

[45] Mastromatteo M, Conte A, Lucera A, Saccotelli MA, Buonocore GG, Zambrini AV, et al. Packaging solutions to prolong the shelf life of Fiordilatte

cheese: Bio-based nanocomposite coating and modified atmosphere packaging. LWT—Food Science and Technology. 2015;**60**:230-237

[46] Park EJ, Alexander E, Taylor GA, Costa R, Kang DH. The decontaminative effects of acidic electrolyzed water for *Escherichia coli* O_{157}:H_7, *Salmonella typhimurium*, and *Listeria monocytogenes* on green onions and tomatoes with differing organic demands. Food Microbiology. 2009;**26**:386-390

[47] Meng X, Zhang M, Adhikari B. Extending shelf-life of fresh-cut green peppers using pressurized argon treatment. Postharvest Biology and Technology. 2012;**71**:13-20

[48] Misra NN, Keener KM, Bourke P, Mosnier JP, CullenP J. In-package atmospheric pressure cold plasma treatment of cherry tomatoes. The Journal of Bioscience and Bioengineering. 2014;**118**:177-182

[49] Monnin A, Lee J, Pascall MA. Efficacy of neutral electrolyzed water for sanitization of cutting boards used in the preparation of foods. Journal of Food Engineering. 2012;**110**:541-546

[50] Niemira BA. Cold plasma decontamination of foods. Annual Review of Food Science and Technology. 2012;**3**:125-142

[51] Nishida C, Uauy R, Kumanyika S, Shetty P. The joint WHO/FAO expert consultation on diet, nutrition and the prevention of chronic diseases: Process, product and policy implications. Public Health Nutrition. 2003;**7**(1A):245-250. DOI: 10.1079/PHN2003592

[52] Oms-Oliu G, Soliva-Fortuny R, Martin-Belloso O. Using polysaccharide-based edible coatings to enhance quality and antioxidant properties of fresh-cut melon. LWT—Food Science and Technology. 2008;**10**:1862-1870

[53] QadrilOS YB, Srivastava AK, YildizF. Fresh-cut fruits and vegetables: Critical factors influencing microbiology and novel approaches to prevent microbial risks—A review. Cogent Food & Agriculture. 2015;**1**:1-4

[54] Ramazzina I, Tappi S, Rocculi P, Sacchetti G, Berardinelli A, Marseglia A, et al. Effect of cold plasma treatment on the functional properties of fresh-cut apples. Journal of Agricultural and Food Chemistry. 2016;**64**:8010-8018

[55] Ramos-Villarroel AY, Aron-Maftei N, Martín-Belloso O, Soliva-Fortuny R. Influence of spectral distribution on bacterial inactivation and quality changes of fresh-cut watermelon treated with intense light pulses. Postharvest Biology and Technology. 2012;**69**:32-39

[56] Reid DS, Fennema OR. Water and ice. In: Fennema's Food Chemistry. 4th ed. Florida, US: CRC Press; 2008. pp. 17-82

[57] Reyes LF, Villarreal JE, Cisneros-Zevallos L. The increase in antioxidant capacity after wounding depends on the type of fruit or vegetable tissue. Journal of Food Chemistry. 2007;**101**(3):1254-1262

[58] Roberts PB. Food irradiation is safe: Half a century of 1562 studies. Radiation Physics and Chemistry. 2014;**105**:78-82

[59] Rocculi P, Romani S, Rosa MD. Evaluation of physico-chemical parameters of minimally processed apples packed in non-conventional modified atmosphere. Food Research International. 2004;**37**:329-335

[60] Salvia-Trujillo L, Rojas-Graü MA, Soliva-Fortuny R, Martín-Belloso O. Use of antimicrobial nanoemulsions as

edible coatings: Impact on safety and quality attributes of fresh-cut fuji apples. Postharvest Biology and Technology. 2015;**105**:8-16

[61] Sams CE. Preharvest factors affecting postharvest texture. Postharvest Biology and Technology. 1999;**15**:249-254

[62] Sanguanpuag K, Kanlayanarat S, Tanprasert K. Trends of fresh-cut produce in Thai retail markets for identification of packaging for shredded green papaya. Journal of Acta Horticulturae. 2007;**46**:481-483

[63] Sapers GM. Washing and sanitizing raw materials for minimally processed fruits and vegetables. In: Novak JS, Sapers GM, Juneja VK, editors. Microbial Safety of Minimally Processed Foods. Boca Raton, FL: CRC Press; 2003. pp. 221-253

[64] Sun SH, Kim SJ, Kwak SJ, Yoon KS. Efficacy of sodium hypochlorite and acidified sodium chlorite in preventing browing and microbial on fresh-cut produce. Preservation Nutrition Food Science. 2012;**17**:210-216

[65] Silveira AC, Araneda C, Hinojosa A, Escalona VH. Effect of non-conventional modified atmosphere packaging on fresh cut watercress (Nasturtium officinale R. Br.) quality. Postharvest Biology and Technology. 2014;**92**:114-120

[66] Singh S, Alam MS. Preservation of fresh cut fruits and vegetables: Current status and emerging technologies. Stewart Postharvt Review. 2012;**8**:1-10

[67] Tauxe RV. Emerging foodborne diseases: An evolving public health challenge. Dairy, Food and Environmental Sanitation. 1997;**17**(12):788-795

[68] Tzortzakis N, Singleton I, Barnes J. Deployment of low-level ozone-enrichment for the preservation of chilled fresh produce. Postharvest Biology and Technology. 1997;**43**:261-270

[69] Varoquaux P, Mazollie J. Overview of the European fresh-cut produce industry. In: Lamikanra O, editor. Fresh-Cut Fruits and Vegetables. Science, Technology and Market, Boca Raton, FL. CRC Press; 2002 LLA

[70] Watada AE, Qi L. Quality of fresh-cut produce. Postharvest Biology and Technology. 1999;**15**(3):201-205

[71] Wright KP, Kader AA. Effect of slicing and controlled-atmosphere storage on the ascorbate content and quality of strawberries and persimmons. Postharvest Biology and Technology. 1997;**10**:39-48

[72] Wu Z-S, Zhang M, Adhikari B. Application of high pressure argon treatment to maintain quality of fresh-cut pineapples during cold storage. Journal of Food Engineering. 2012;**110**:395-404

[73] Wu Z-S, Zhang M, Adhikari B. Effects of high pressure argon and xenon mixed treatment on wound healing and resistance against the growth of *Escherichia coli* or *Saccharomyces cerevisiae* in fresh-cut apples and pineapples. Food Control. 2013;**30**:265-271

[74] Yeoh WK, Ali A, Forney CF. Effects of ozone on major antioxidants and microbial populations of fresh-cut papaya. Postharvest Biology and Technology. 2014;**89**:56-58

[75] Zagory D. Effects of post-processing handling and packaging on microbal populations. Postharvest Bology and Technology. 1999;**15**(3):313-321

[76] Zhang B-Y, Samapundo S, Pothakos V, Sürengil G, Devlieghere F. Effect of high oxygen and high carbon dioxide atmosphere packaging on the microbial

spoilage and shelf-life of fresh-cut honeydew melon. International Journal of Food Microbiology. 2013;**166**:378-390

[77] Zhang M, Zhan ZG, Wang SJ, Tang JM. Extending the shelf-life of asparagus spears with a compressed mix of argon and xenon gases. LWT—Food Science and Technology. 2008;**41**:686-691

[78] Ziuzina D, Patil S, Cullen PJ, Keener KM, Bourke P. Atmospheric cold plasma inactivation of *Escherichia coli*, *Salmonella entericaserovar typhimurium* and *Listeria monocytogenes* inoculated on fresh produce. Food Microbiology. 2014;**42**:109-116

[79] Sun SH, Kim SJ, Kwak SJ, Yoon KS. Efficacy of sodium hypochlorite and acidified sodium chlorite in preventing browning and microbial growth on fresh-cut produce. Preservation Nutrition Food Science. 2012;**17**(3):210-216

[80] Plesoianu AM, Nour V, Tutulescu F, Elena IM. Quality of fresh-cut apples as affected by dip wash treatments with organic acids and acidic electrolyzed water. Food Science and Technology. 2021;**42**:1-8. DOI: 10.1590/fst.62620

[81] Count A. A new example of nanotechnology applied to minimally processed fruit. The Case of Fresh-Cut Melon. 2015;**6**(4):1-4. DOI: 10.4172/2157-7110.1000439

[82] Li X, Li W, Jiang Y, Ding Y, Yun J, Tang Y, et al. Effect of nano-ZnO-coated active packaging on quality of fresh-cut 'Fuji' apple. 2011;**46**(9):154-162. DOI: 10.1111j.1365-2621.2011.02706.x

[83] Sarron E, Widehem PG, Ausseunac T. Ozone treatment for preserving fresh vegetables quality: A critical review. Food. 2021;**13**(3):605-609. DOI: 10.3390/foods10030605

[84] Siddiq R, Auras R, Siddiq M, Dolan KD. Effect of modified atmosphere packaging (MAP) and NatureSeal® treatment on the physico-chemical, microbiological, and sensory quality of fresh-cut d'Anjou pears. Food Packaging and Shelf Life. 2020;**23**:100454

[85] Parmo DB, Pedro M, Oria R. Effects of electron-beam irradiation on the shelf life, microbial populations and sensory characteristics of summer truffles (tuber aestivum) packaged under modified atmospheres. Food Microbiology. 2011;**28**(1):141-148

[86] Moreira MR, Tomadoni B, Belloso OM, Fortuny RS. Preservation of fresh-cut apple quality attributes by pulsed light in combination with gellan gum-based prebiotic edible coatings. LWT—Food Science and Technology. 2015;**64**:1130-1137

[87] Misra NN, Keener KM, Bourke P, Mosnier JP, Cullen PJ. In-package atmospheric pressure cold plasma treatment of cherry tomatoes. Journal of Bioscience and Bioengineering. 2014;**118**(2):177-182

[88] Tappi S, Berardinelli A, Ragni L, Dalla Rosa M, Guarnieri A, Rocculi P. Atmospheric gas plasma treatment of fresh-cut apples. Innovative Food Science and Emerging Technologies. 2014;**21**:114-122

[89] Misra NN, Moiseev T, Patil S, et al. Cold plasma in modified atmospheres for post-harvest treatment of strawberries. Food and Bioprocess Technology. 2014;**7**(10):3045-3054

[90] Cullen M, Sheu FH, Tsai MJ, Chu YH. The effects of dielectric barrier discharge plasma gas and plasma-activated water on texture, color, and bacterial characteristics of shiitake mushroom. Journal of Food Processing and Preservation. 2019 Article ID e14316

Chapter 4

Using Nanotechnology for Enhancing the Shelf Life of Fruits

Ravi Kondle, Kushagra Sharma, Gurpreet Singh and Amit Kotiyal

Abstract

Edible coatings are thin layers of edible materials formed directly on fruits, usually by immersing the fruits in a coating material solution, and they are one of the most intriguing food developments in recent years. Colorants, flavours, nutrients, and anti-browning and antimicrobial agents can all be carried by edible coatings, extending shelf life and reducing pathogen growth on food surfaces. To manage moisture transfer, gas exchange, or oxidative processes, edible coatings can be applied using various procedures such as dipping, spraying, or coating. Because these systems have a larger surface area, nanoparticles may help to improve the barrier characteristics and functionality of fruit preservation coatings. Antimicrobial nanoparticles (NPs) are employed as matrixes in edible coatings and films (ECF), which are then applied to fruits to extend shelf life and improve storage quality. Nano Chitosan is one of the most prevalent polysaccharides, protein, and lipid-based edible coatings. These are characterised by poor gas and water barrier qualities, and they are frequently used as moisture loss sacrifice agents. Therefore, the purpose of this book chapter is to study the effect of nano edible coatings such as chitosan/tripolyphosphate (TPP), chitosan-methyl cellulose/silica (SiO_2), gelatin-fiber/titanium dioxide (TiO_2), gelatin-chitosan/ (Ag/ZnO), Gelatin/kafirin to quality attributes and prolonging the shelf life of fruits.

Keywords: edible coatings, nanoparticles, chitosan, titanium dioxide, fruits, nano-silver, silicon dioxide, nanogold, nano liposome, colloidosomes

1. Introduction

Consumers these days are demanding high-quality fruits and vegetables that are rich in health-promoting compounds and also prefer fresh foods because of their high-value nutrition. To achieve more beneficial health, moreover, these foods should keep up the good physicochemical and sensory quality, as well as be safe for consumption and should be free from contaminants and pathogenic microorganisms [1]. This growing demand has now challenged the food and horticulture industries to develop relevant preservation practices. This has prompted a sense of urgency for scientists and food processing industries to evaluate different approaches to enhance the freshness, quality, shelf-life, and food safety, through the use of natural, edible, and biodegradable polymers [2]. These edible coatings and films reduce the loss of quality attributes by forming a semipermeable safety barrier around fruits. Edible films and coatings can consist of 3 types of biological materials: polysaccharides, lipids, and proteins [3]. Many biopolymers such as alginate, carrageenan, chitosan, pectin, starch, and xanthan gum have been widely used to form edible films and food coatings. Their film-forming properties allow the synthesis of membranes (thickness > 30 μm) and coatings (<30 μm), which are successfully used to preserve foodstuffs.

Edible coatings can also be used as a carrying matrix of antimicrobial agents which can increase its functioning by substituting the microbial spoilage and increasing the shelf life of the product [4]. The edible coating idea helps to increase the use of raw and perishable vegetables and fruits and control the horticulture loss of crops. Edible coating base material is made of polysaccharides, proteins, and lipids. In view of Zaragoza et al. [1] edible coating controls the respiration rate, it also controls the extension of microbes during the preservation of fruits and vegetables. Chitosan is to be used in organic-based coating for stopping food spoil and defile. It has a very impressive capacity for heavy metal adsorption. It is used as edible coatings to extend fruit shelf life by reducing transpiration and respiration rates. TiO_2 nanoparticles have excellent photocatalytic activity, which is very effective for the removal of organic pollutants. Therefore, the integration of TiO_2 and chitosan can complement each other with their own advantages, and the chemical grafting of antioxidant molecules (such as chitosan) directly on the surface of TiO_2 nanoparticles has the best effect on the treatment of wastewater pollutants.

Mostly the traditional packaging is derived from petroleum plastics such as polypropylene, polyethylene, and polystyrene, which later after the product utilization becomes a major worry due to environmental damage they cause regarding their difficult degradation. Nevertheless, serious environmental problems are created due to the non-biodegradability of these materials, thus enlarging the attentiveness of researchers in biodegradable packaging production utilizing natural polymers extracted from renewable sources for application in food packaging. In this sense, many researchers have shown interest in coatings and edible films which represent an environmentally friendly alternative for food packaging. Edible packaging is known as a future alternative to protecting food quality and improving shelf life by slowing down microbial spoilage and providing moisture and gas barrier properties. In 2016, the edible packaging market was valued at $697 million and by the year 2023, it is expected to hit $1097 million increasing at a compound annual growth rate (CGAR) of 6.81% from 2017 to 2023 on an international level.

Nanotechnology works well for food packaging, and edible coatings made of both inorganic and organic nanoparticles are also possible [5]. Inorganic nanoparticles including silver oxide Np, titanium oxide, zinc oxide, silver and are mostly employed in food packaging as a detector, to prevent foodborne illness, and to lengthen shelf life. In 2006, FDA said that nanomaterials are particles with dimensions less than a micrometre scale that exhibit special properties [6]. In recent years, using edible coatings (ECs) to extend the shelf life of fresh foods has shown to be a successful and environmentally beneficial alternative. An edible coating is a thin layer of food that is applied directly on a food surface and has filmogenic qualities. The inclusion of compounds with antibacterial activity inside the polymeric matrix is one of the most fascinating aspects of coating design. The edible coating's non-toxic anti-fungal ingredients may prevent fungal deterioration, which is the principal cause of postharvest losses of fruit and vegetable goods [7].

2. Edible coatings

These are thin layers of edible material applied to a product surface in addition to or instead of natural protective wax coatings and act as a barrier to moisture, oxygen, and solute movement for food [8]. Fruits and vegetables continue to respire even after harvest and use up all the oxygen within the product, which is not replaced as quickly as by edible coating and produces carbon dioxide, which accumulates within the product because it cannot escape as easily through the coating. Therefore, fruits and vegetables stay firm, fresh, and nutritious for longer, and their shelf life is almost doubled. The amount of coating affects the extent to which the internal atmosphere (O_2 and CO_2) is modified and the degree of minimization in weight loss. Since ancient times, edible coatings and films have been used in the food business to protect food goods. This is not a novel method of preservation. The first edible fruit coating was wax. In the twelfth and thirteenth centuries, the Chinese coated lemons and oranges with war [9].

2.1 Traits of edible coating

The edible coating properties mostly depend on molecular structure than chemical constitution and size.

i. The edible coating should be water-resistant so that it remains intact and adequately covers the product when applied

ii. It should not deplete O_2 or build up excess carbon dioxide. At least 1–3% oxygen is needed around the object to avoid the transition from aerobic to anaerobic respiration

iii. It should reduce water vapor permeability

iv. It should improve appearance, maintain structural integrity, improve mechanical handling properties, carry phytochemicals (antioxidants, vitamins, etc.) and contain volatile flavor compounds.

v. It should melt above 40°C without decomposing

vi. It should be easily emulsifiable, not tacky or sticky and should have effective drying performance

vii. It should never interfere with the quality of fresh fruits or vegetables and should not impart undesirable texture

viii. It should have low viscosity and be economical

ix. It should be translucent but not glassy and should be able to withstand slight pressure

2.2 Benefits of edible coatings

1. Improves external appearance by giving extra shine to fruit surface

2. Reduces weight loss and keeps the fruit firm so it can maintain its fresh appearance

3. Reduces respiration rate and ethylene production, thereby delaying aging

4. Prevents fruits from chilling injury and storage disorders

5. Act as a barrier to free gas exchange

6. Provides a carrier for post-harvest chemical treatments

7. Adds nutrients such as aroma compounds, antioxidants, pigments, ions that stop browning reactions, and vitamins

8. In some countries, taxes on packaging material transport can be saved by using edible coatings and films.

3. Nano edible coatings

Presently edible coatings are being developed using organic nanomaterials which are effective in maintaining post-harvest quality and controlling crop loss. The most explored nanoparticles in fruits are zinc oxide, silver, and chitosan, considering their high antimicrobial activity and stability (**Figure 1**). However, other nanoparticles such as Fe, TiO_2, cerium oxide, and Cu have been used in various sectors of the food industry.

A very effective strategy is to elaborate edible coatings consisting of nanoparticles mixed with organic and inorganic materials producing nanocomposites. The organic nano coatings are chitosan, Alginate, nano liposome, colloidosomes, casein micelles,

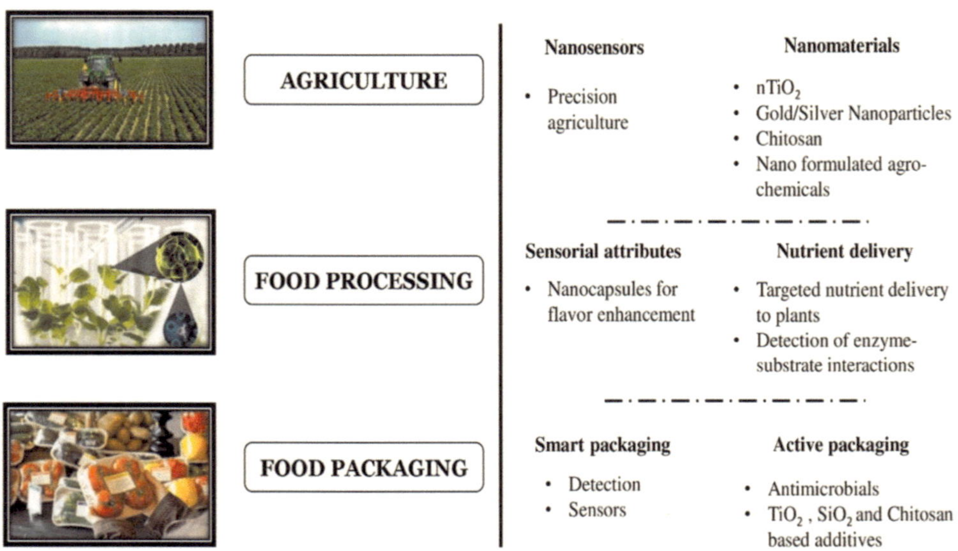

Figure 1.
Applications for nanotechnology in agriculture, food processing, and packaging.

and nano cochleate, however, inorganic coating materials are nano silver, gold, silicon dioxide, TiO_2, nano Fe, ZnO, and carbon nanotube.

3.1 Chitosan and chitin-based coatings and films

Insects, crustaceans, and fungi use the naturally occurring mucopolysaccharide chitin, which is composed of 2-acetamido-2-deoxy-β-D-glucose molecules connected by -bonds, as a structural component (1-4). Chitosan, a concentrated alkali-induced N-deacetylated derivative of chitin, is produced in this environment. Chitosan and chitin are alike to cellulose in their excessive insolubility degree and low chemical reactivity. Solubility of chitosan depends on N-acetylation degree and molecular weight, also it can be dissolved in acid solutions where pH is less than 6.3 even at a concentration above 2% (w/v). Rheological properties, solubility, and appearance, among other properties of chitosan properties, are also dependent on the N-acetylation degree.

Similar to cellulose, chitosan and chitin has a high insolubility degree and minimal chemical reactivity. Chitosan may dissolve in acid solutions with a pH lower than 6.3 at concentrations of more than 2% (w/v), and its solubility is proportional to its N-acetylation degree and molecular weight. Chitosan's N-acetylation level affects not only its rheological qualities but also its solubility and appearance. The potential of chitosan-based films and coatings to act as natural preservatives is widely known.

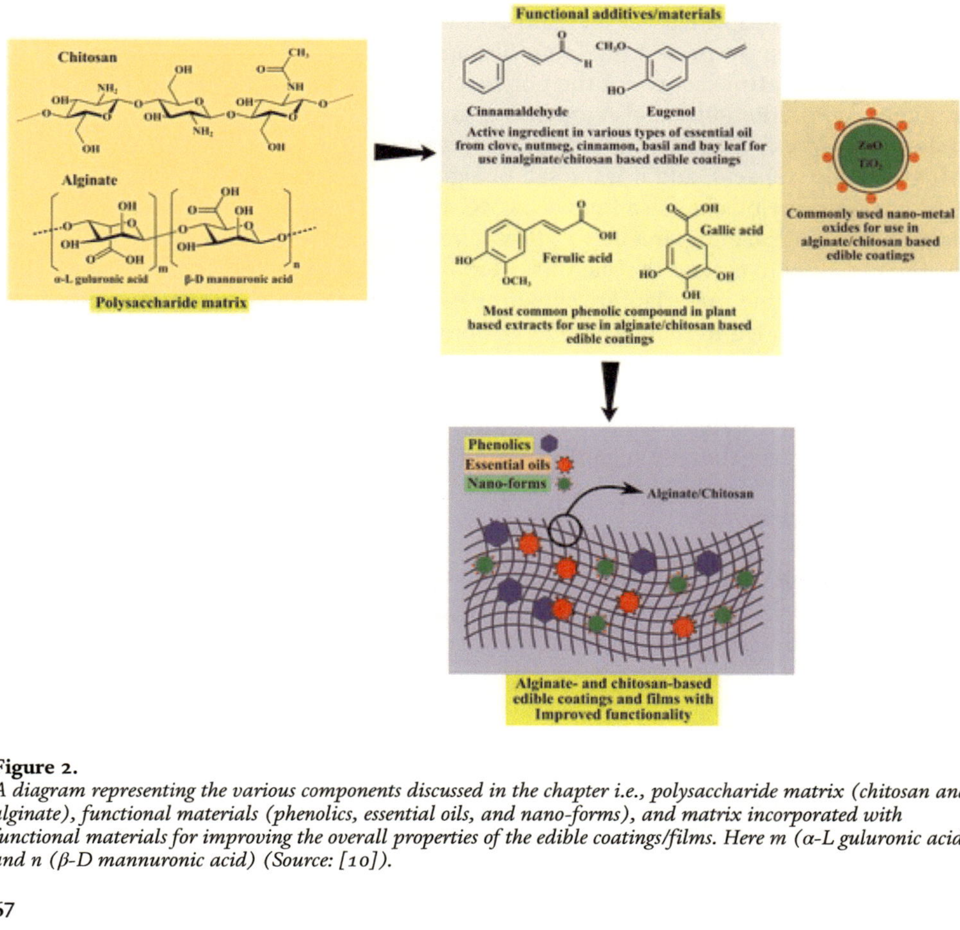

Figure 2.
A diagram representing the various components discussed in the chapter i.e., polysaccharide matrix (chitosan and alginate), functional materials (phenolics, essential oils, and nano-forms), and matrix incorporated with functional materials for improving the overall properties of the edible coatings/films. Here m (α-L guluronic acid) and n (β-D mannuronic acid) (Source: [10]).

Coatings and films made of chitosan are permeable to gases (O_2 and CO_2), have excellent mechanical properties, and also have high permeability to water vapor, which limits their use in humid environments, since controlling moisture transfer is a desirable property. For this main reason, several plans have been made to improve the functional properties of chitosan coatings and films. As can be seen, its functional coating qualities may be improved by adjusting factors such as the degree of solvent, pH, deacetylation, and the addition of surfactants, proteins, lipids, or polysaccharides. Essential oils boost water vapour permeability and give antibacterial and antioxidant benefits. A number of reagents, such as ferulic acid, genipin, glutaraldehyde, formaldehyde, cinnamaldehyde, and sodium trimetaphosphate, are added to the formulation to slow down the dissolving or swelling and enhance the characteristics of chitosan-based coatings. Their main use is in food processing as a functional food and helps in Encapsulation antimicrobial agent and also works as plant growth-promoting agent.

For the effective use of chitosan coating, the chitosan was to be combined with other substances. As seen, the single chitosan coating was oftentimes combined with physical methods that are short heating, short gas fumigation, and modified atmosphere packaging (**Figure 2**) [11].

3.2 Alginate-based films and coatings

Gels or insoluble polymers may be formed from sodium alginate, a well-known polysaccharide, due to its strong reactivity to polyvalent metal cations. Since it may create a semipermeable barrier on fruits and vegetables, it has been widely utilized as an edible covering for preserving foods like apples and peaches [10]. Marine brown algae are a rich natural source of the polysaccharide alginate (Phaeophyceae, majorly Laminaria). Pseudomonas and Azotobacter are two bacterial families that contribute to its development. Alginate is a linear copolymer of (1–4) β-D-mannuronic (M) and α-L-guluronic (G) acid, and it is the salt of alginic acid. These acid residues are located in M or G-residue blocks (also known as MG-blocks) or in MG-residue blocks. M: G residue distribution and percentage differ across algal species. Because nano coating has the potential to extend the shelf life of many food goods, it is seeing rapid growth.

Silver-containing materials are shown to exhibit bactericidal or anti-microbial properties, which led to their strong development in the last few years [12]. Alginates are infamous for their good film-forming properties and performance [13]. Alginates are globally used in edible coatings because of their good availability and regulatory status. The United States Food and Drug Administration (FDA) classifies food-grade sodium alginate as generally regarded as safe to use (**Figure 3**) [14].

3.3 Nano liposome

Its main application is in the Food processing industry and its main use is in specific delivery of nutraceuticals and active and passive delivery of genes, protein & peptides, and also the delivery of pesticides and fertilizers.

3.4 Colloidosomes

Colloidosomes also called Pickering emulsion capsules, have gained a lot of attention for the Encapsulation of hydrophilic and hydrophobic activities [15]. These are microcapsules whose shell comprises tightly packed colloidal particles. Their physical properties like permeability, mechanical strength, and biocompatibility can also be

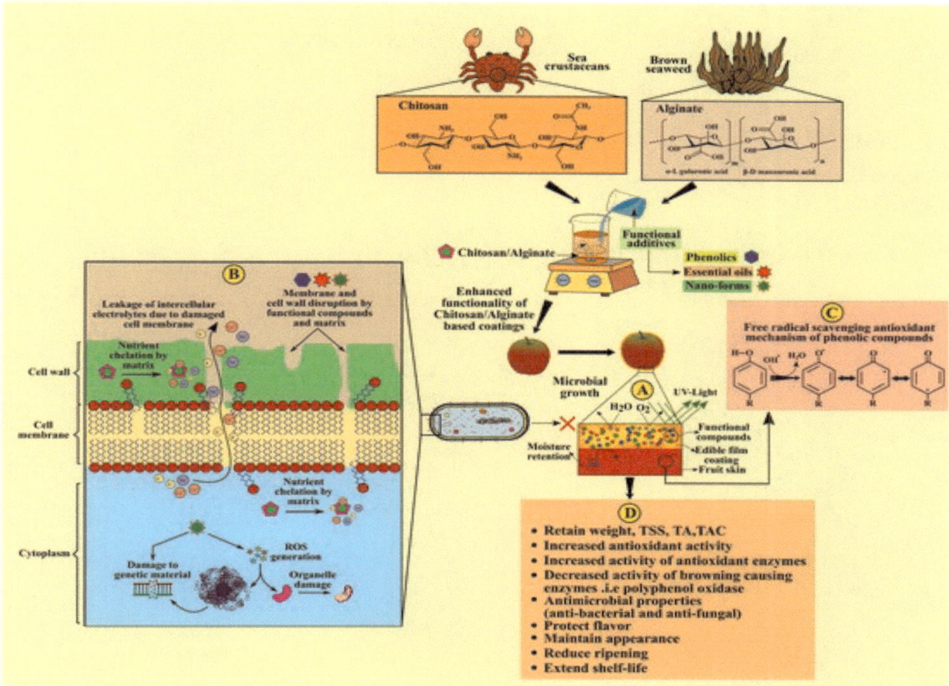

Figure 3.
The role and mechanism of action of functional materials in enhancing the shelf-life of fruits. (A) The matrix of alginate/chitosan along with functional additives acts as a water and O_2 barrier for inhibiting respiration and eventually reactive oxygen species generation. The specialized coating also inhibits the spoilage of fruits by microbes and UV light. (B) Phenolics, essential oils, and nanoparticles destabilize microbial membranes, which can perforate cells and block protein synthesis, causing electrolyte leakage and ultimately cell death. Nano-metallic forms generate (ROS) like hydroxyl radicals, and SOD and result in organelle damage. They further restrict the synthesis of DNA, RNA, and lipids required for the survival of the microbes. (C) These functional materials mainly phenolics act as antioxidants and prevent fruits from being damaged by reactive oxygen species. (D) The synergistic effect of alginate/chitosan-based coatings with functional additives maintains the appearance, flavour, and extends the shelf-life of fruits and vegetables. Here ROS (reactive oxygen species), total soluble solids, titratable acidity, and total ascorbic acid content.

controlled by the proper choice of colloids and preparation conditions for their assembly [16]. Colloidosomes are also used in the Food processing unit and their main application is increasing the nutrient content of food.

3.5 Casein micelles

Casein micelles show hydrophobic and hydrophilic properties that make them ideal for encapsulation of food bioactive [17]. Its main application is as a nutritional supply that helps in the delivery of sensitive products.

3.6 Alginate and chitosan

Alginate and chitosan are used to coat the nanoparticles and make negative and positive charges available on the particles [18]. Their main application is as a target delivery supply that supplies B-carotene, lycopene, vitamins A, D, E, and omega-3-fatty acids.

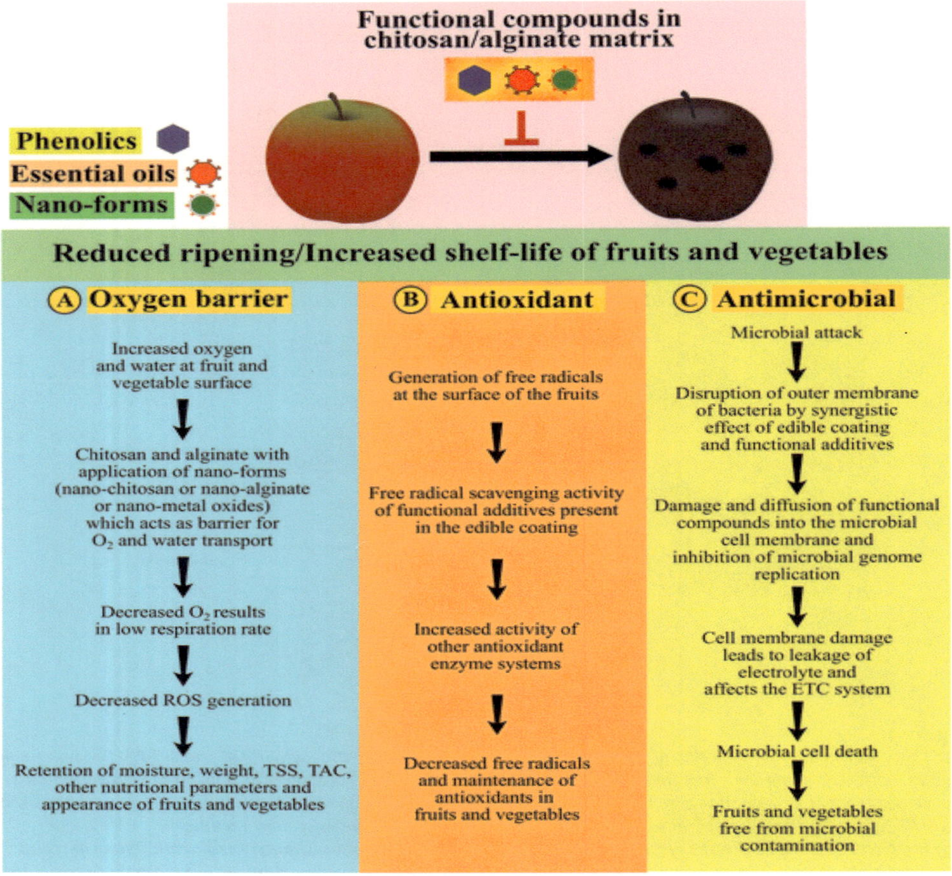

Figure 4.
*Overview of the mechanism of the synergistic effect of alginate/chitosan matrix and functional additives (phenolics, essential oils, nano-forms). (A) Acting as an O_2 barrier. (B) Acting as an antioxidant. (C) Acting as anti-microbial. The overall action of (A), (B), and (C) results in a prolonged shelf-life of fruits and vegetables (source: (**Table 1**) [10]).*

3.7 Nano cochleate

Nano cochleates' main application is in Nutritional supply where Nutrients are efficiently delivered without affecting colour and taste (**Figure 4**) [6].

4. Edible films and coatings for fruit preservation

Fruits are acknowledged as the main sources of vitamins, minerals, antioxidants, and fibre in the diet of consumers. At a similar time, their short shelf life is well known, due to their high percentage of moisture content (75–95%), which is one of the major causes of their fast degradation. Many types of nanocomposite films and edible coatings are used nowadays very much on fresh fruits and vegetables for increasing their shelf life using the same methods as modified atmosphere methods, which have already shown good results for preserving fruits and vegetable quality [23]. Total soluble solids content (°Brix) is a very important maturity index for fruit and vegetables. Edible coatings are

S. no	Nano material	Application and properties	Examples	References
1	Nano silver	It has the property of strong disinfection & storage. It is also rich in 22 essential vitamins and minerals and is also used for sterilization & quality control. Sometimes it acts as an antibacterial agent	Used as nano-silver salad bowl, storage box, daily vitamin boost, nano colloidal silver, nanosilver sol, and also in food packaging	[6, 19]
2	Silicon dioxide	Its main application is as biosensor which prevents UV light and is used as pesticides & herbicides and as edible coating. It can detect food colorant hygroscopic	Acts as a drying agent also used in food packaging of fruits and vegetables	[6]
3	Titanium dioxide	Titania exists as an oxide of titanium and is recognized as NPs of metal oxides comprising unique optical, thermal, electric, and magnetic properties	It works as a biosensor which is used as a whitener in dairy for products like milk and cheese	[20]
4	Nanoclusters iron np 30 nm	Its main application is for the development of functional food	In nano cuticles slim shake vanilla and fortified fruit juice	[6, 19]
5	Zinc oxide	It is used as a Food preservative and as an edible coating having its main properties as an antimicrobial agent.	Improve fruit properties of strawberry, banana, and others	[6]
6	Nano gold	AuNPs are being used in nano packaging industries for their properties like possessing therapeutic and antibacterial characteristics as well as their inert and nontoxic nature	It acts as a pathogen and glucose detector	[21]
7	Carbon nanotube	Carbon nanotubes have recommendable electronic conductivity, vigorous stability, and environmental friendliness	Application is in the food packaging industry e.g., wine and honey-making industry	[22]

Table 1.
Inorganic nano materials.

also very operative in lowering TSS, or we can say in lowering the ripening rates of fruit and vegetable products [24]. Edible coatings and films (ECF) are employed as matrixes for including antimicrobial nanoparticles (NPs), and then they are used on fruits and vegetables to increase their shelf life and enhance quality in storage. The preservation ability of the quality of fruits and vegetables indicates that many ECFs with NPs could be used as the ideal materials for food application. Looking at the introduction to these characteristics, an attempt is made to look out for future trends in this field [25].

5. Future trends of ECF with NPs

In the last few years, nanosized particles as anti-microbial agents incorporated into edible coatings are the subject of many studies. However, additional research including the interactions between coating materials and nanosized particles is very much important. Also, further work about the effect of these nanosized particle's addition on the properties of coating materials should be understood. These investigation results may provide insights regarding upcoming improvements to their physical and antimicrobial properties for practical applications. In order to determine the requirements for its production and application, it is crucial that future research concentrate on improving the uniformity of composite coating qualities and keeping an eye out for its impacts on the storage quality of fruits and vegetables. It's crucial to do further research on economical methods of creating and using ECF [25].

6. Application methods of edible coatings

The application technique will be selected based on the nature of the food to be coated, the surface qualities, the rheological properties of the solution, and the primary purpose of the coating, all of which influence how well the coating will preserve fresh fruits and vegetables. Coatings' ability to stick to food surfaces is crucial to their serving their purpose. Interfacial contact between food surface and coating may be evaluated by measuring wettability. This factor is essential to consider when monitoring the effectiveness of the coating solution on the food's surface. Coating fresh fruits and vegetables with edible substances are often done by dipping, spraying, or hand-coating. Fluidized bed processing and foaming are two more methods; however, they are seldom employed outside of research settings [26]. Edible coatings depend on numerous parameters such as type of coating used, amount, viscosity, and also surface tension. The coating method also affects the efficiency and quality of the coating (**Table 2**) [34].

S. no	Fruit	Edible coatings used	Findings	References
1	Fresh cut apple	Sodium alginate	Nanoemulsion-based edible coatings presented higher *E. coli* inactivation and slower psychrophilic bacteria growth compared to conventional emulsions at the same concentration	[27]
2	Grape berry	Chitosan	The use of the nanoemulsion effectively reduced the initial growth of S. Typhimurium, total aerobic mesophiles, yeasts, and molds, and showed retention of antioxidant capacity	[27]
3	Fresh cut apple	Nopal mucilage	The coatings formed with the nano-emulsion had a significant inhibitory effect on PME and PPO activity, in contrast to conventional emulsions	[27]
4	Guava (*Psidium guajava* L.)	SLN (solid lipid nanoparticles) Candeubawa S wax (carnauba wax and candelilla wax) Poloxamer 407	The potential use of SLNs in edible coatings could be applied easily to minimize the senescence of several products	[27]
5	Guava (*Psidium guajava* L.)	Poloxamer 407	The application of candeuba wax (SLN) helps to conserve the natural maturation process but at a slower rate	[27]

S. no	Fruit	Edible coatings used	Findings	References
6	Apples	Nano-SiO_2	The preparation of edible coating by ultrasonic processing and incorporation into an SPI matrix results in a decreased respiration rate, maintenance of firmness, and extension of shelf life	[28]
7	Fresh-cut papaya, pear	Montmorillonite (MMT)	Adding 15 g/L of montmorillonite at 80°C and essential oil of oregano decreased weight loss and maintained the quality of papaya; moreover, the edible coating helped slow microbial growth	[29]
8	Strawberries	Montmorillonite (MMT)	This edible coating contained 70% WPI, 0.5% potassium sorbate, 3.75% calcium caseinate and 0.375% MMT. It was effective in limiting mold growth for at least 12 days and maintained the quality of the fresh coated strawberries	[27]
9	Ready-to-eat pomegranate	Nano-ZnO_2	Edible coatings with 0.2% ZnO_2 reduced yeast and mould development at 6 and 12 days of storage, although bacterial load increased. CMC and nano-ZnO_2 helped preserve pomegranate bioactive	[30]
10	Citrus fruit	Silver nanoparticles (AgNPs)	AgNPs caused cell deformation, cytoplasmic leakage, and cell death of *P. italicum*. AgNPs also showed significant activity on *E. coli* and *S. aureus* with beneficial effects for citrus fruit preservation	[28]
11	Kinnow (*Citrus reticulata*)	Silver nanoparticles (AgNPs)	Silver nanoparticles were added to a coating emulsion base together with either CMC or guar gum at a ratio of 1:1	[31]
12	Strawberries	Limonene	limonene liposomes showed significantly lower fungal growth as compared to the control on the 14th day of storage	[32]
13	Cantaloupe	Chitosan/nano-silica/ nisin	Their combination was found to be perfect which increased shelf life by maintaining color, Vit-C, and Peroxidase Activity for up to storage time of days	[33]
14	Blueberries	Chitosan/nano-TiO_2	They could maintain the nutrient composition while preserving quality at zero degrees	[25]
15	Blackberry	Chitosan	Showed best antifungal effect over racemosus	[23]
16	Fresh Fruit (Redberry) *Arbutus unedo*	Sodium alginate	AL 1% + Eug 0.20% showed the best results in terms of preservation	[23]
17	Fresh cut pineapples	Sodium alginate	The edible coating containing 0.5% and 1% citral nano-emulsion improved physicochemical attributes and reduced microbial growth	[23]

Table 2.
Use of different nano edible coatings of fruits and its findings.

7. Nanoparticle synthesis techniques

There are two general methods of nanoparticle synthesis:
1. Top-down method and 2. Bottom-up method is shown in (**Figure** 5).

7.1 Top-down method

In this method, larger compounds are broken down into nano-scaled materials by using mechanical and chemical forces. Mechanical milling, lithography, electrospinning, etching, sputtering, and laser ablation are the most common top-down approaches to nanoparticle synthesis.

7.1.1 Mechanical milling

Mechanical milling is a method of placing elements in a high-power mill with or without a medium (wet and dry milling) to reduce the particle size of the element. The rolling ball transfers its kinetic energy to the milled elements, resulting in the size reduction of the elements to nanoscale dimensions [29, 35–41]. This energy transfer depends on various factors such as the type of mill, packing of balls, milling speed, type of milling (wet or dry), duration, and milling temperature. This method is more reliable than traditional methods it can be used for both wet and dry materials and large-scale synthesis of nanomaterials due to their inexpensiveness.

7.1.2 Nanolithography

It is a nanofabrication technique for developing nanopatterns with a size range between 1 and 100 nm. It can be divided into two types:

i. Mask lithography includes soft lithography, nanoimprint lithography, and photolithography—In these masks, molds or templates are used in

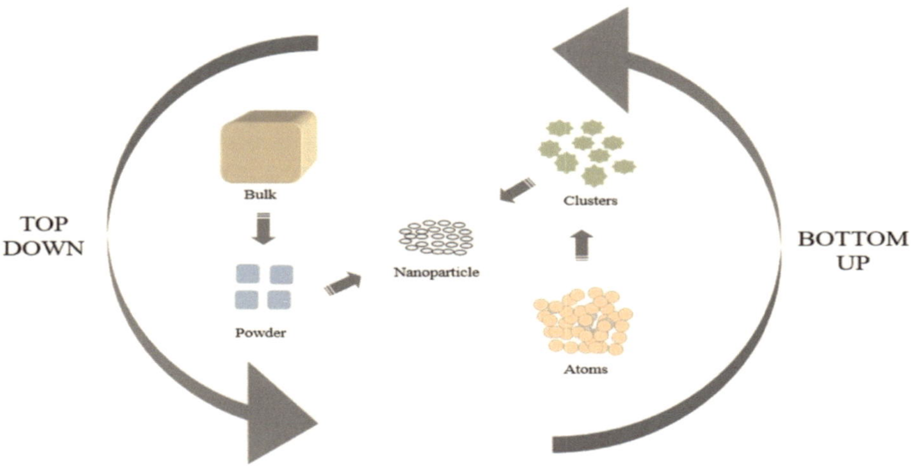

Figure 5.
Schematic diagram of top-down and bottom up method.

nanopattern fabrication, while in maskless lithography nanopatterning is performed without the involvement of masks.

ii. Mask-less lithography including electron beam lithography, focused ion beam lithography, and scanning probe lithography.

7.1.3 Sputtering

The method relies on using a high-energy plasma or gas to generate nanoparticles that travel and strike the surface to form a layer.

This method is based on the use of high-energy plasma or gas to produce nanomaterials that travels and strikes the surface to form the layer. Sputtering is carried out in different ways: DC sputtering, reactive sputtering, RF sputtering, and magnetron sputtering. In this technique, the target surface is bombarded with highly energetic gas ions, resulting in the ejection of surface molecules or small clusters. Sputtering is advantageous because the composition of the deposited nanofilm is the same as the target source.

7.1.4 Laser ablation

This method is a complex process in which a laser beam is used to remove microscopic material from a target source. This is a method used to produce highly refined nanoparticles whose properties such as size and distribution depend on laser focusing parameters, laser pulse parameters, and the medium used. Recently, pulsed laser ablation in liquid is an emerging technique used in the synthesis of monodispersed colloidal nanoparticles without the use of complex chemicals. Laser ablation in liquid is advantageous in reducing the thermal effect on the pattern source, reducing preparation time, and being environmentally friendly.

7.1.5 Electrospinning

This technique is used to develop fibers of metals, ceramics, composites, and polymers of a few microns to the nanoscale range by aligning the fibers, thereby reducing Gibb's free energy. Coaxial electrospinning is used to develop ultrathin fibers up to a length of a few centimeters.

7.1.6 Etching

This method is mainly used in nanotechnology to chemically remove material from a sample surface. The two main types of etching are wet etching (liquid chemicals or etchants are used to remove the layer from the sample surface) and dry etching (etchant gases or plasmas are used to remove the layer from the sample surface). Metal nanoparticles generated after etching the metal surface can be converted into usable material.

7.2 Bottom-up method

The bottom-up method is to synthesize the nanomaterial from atomic or molecular species via various processes. Chemical vapor deposition, sol-gel, solvothermal and hydrothermal methods, and reverse micelle methods are various methods used for nanoparticle preparation.

7.2.1 Chemical vapor deposition (CVD)

It is a widely used bottom-up method to deposit nanomaterials and thin film on a pre-selected substrate. This is a widely used bottom-up method to deposit nanomaterials and thin film on the preselected substrate. In this technique, the chemical reaction takes place between precursor, gas, or vapor and the preselected substrate at high temperatures. This reaction causes the deposition of desired product on the selected surface. This technique provides nanocrystals with high purity, quality, and minimum defects on the substrate but the disadvantage of this technique is its high production cost and the toxicity of gaseous by-products.

7.2.2 Sol-gel method

This method is mainly based on the precursor hydrolysis and polycondensation reactions of the hydrolyzed products resulting in the formation of the polymeric network. The method derives its name from the process by which a liquid precursor in the preparation of nanoparticles is first transformed into a sol and then into a final structural network known as a gel. The sol-gel method is widely used in the preparation of metal oxides such as ZnO, TiO_2, SnO_2, and WO_3 nanoparticles due to their effective control over the shape and size of the nanoparticles.

7.2.3 Solvothermal and hydrothermal method

In this method nanoparticles are obtained by heterogeneous reaction in a solvent in a closed vessel at high temperature and pressure near its critical point. This method is carried out in an aqueous medium, whereas the solvothermal method is carried out in a non-aqueous medium. Solvothermal and hydrothermal methods are very helpful in engineering nanomaterials such as nanosheets, nanorods, nanospheres and nanowires.

7.2.4 Reverse micelle method

In oil-in-water microemulsion, the hydrophilic head region of surfactant molecules orients outwards and hydrophobic tails towards the core trapping the oil droplets while in water-in-oil microemulsion the surfactant inverts its orientation and results in the formation of reverse micelles water droplets. The size of the nanoscale water droplets trapped in the core of reverse micelles, known as the "water-pool", can be changed by changing the ratio of water to surfactants. The type of surfactant in reverse micelles helps in the variation of nanoparticle properties depending on their size and morphology.

8. Commercially available nanomaterials and companies

S. no	Trade	Company	Type of material	Type of product	Application	Form
1.	Aegis HFX Resin and OXCE Resin	Honeywell International Inc., USA	Nylon 6-nanoclay composite	Beer and flavored alcoholic beverage bottles, PET	O_2 scavenging	Barrier nylon resins

S. no	Trade	Company	Type of material	Type of product	Application	Form
2.	OMAC® Imperm®	Mitsubishi Gas Chemical Inc., Japan	Cerium oxide	Retort product and hot fill of meat and fish products	Oxygen scavenging	Film
3.	Oxy Guard®	Clariant Ltd., Swaziland	Iron oxidation	Fried snacks	Oxygen scavenging	Sachets & Film
4.	ATCOR DE 10S/100 OS/ 200 OS	Emco Packaging Systems, UK		Cooked meat	Oxygen scavenging	Labels
5.	Cryovac® OS Systems	Cryovac Div., Sealed Air Corporation, USA	Polymer oxidation	Strawberries, eggplant	Oxygen scavenging	Tray, Films
6	Ageless®E	Mitsubishi Gas Chemical Inc., Japan	Sodium carbonate/ sodium glycinate	Ham, ready-to-eat meat product	CO_2 Scavenger	Sachets and label
7	UltraZap R Xtenda Pak pads	Paper Pak Industries, Canada	Allyl isothiocyanate (AIT) or scavenging molecular O_2 (Listeria populations)	Meat, poultry, fish, dairy, confectioneries, and baked goods	CO_2 emitter and antimicrobial pad	Tray pads
8	Microspheres	Bernard Technologies, Inc., USA	Chlorine dioxide	Fruits	Microbial Contamination	
9	RipeSense™ Sensor	Ripesense limited, New Zealand	Changing color based on aromatic compounds	Seafood, Oysters	Freshness Indicators	Stickers
10	TimeStrip®	TimeStrip UK Ltd, UK	TTI based on enzyme, lipase, and pH	Dried fruits, cheeses, coffee	Freshness (based on color)	Stickers
11	N-coat	Multifilm Packing	Nanoclay	Fruits Vegetables	Gas barrier	Film
12	Biomaster	Corporation, USA Addmaster Limited, USA	Nanosilver	Fruits Vegetables	Antimicrobial	Bag, Spray
13	Ethysorb ®	Stay Fresh Ltd	PE-Nanoclay composite			
14	Tip Top bread	George Weston Foods	Nanosized self-assembled liquid structure		Ethylene scavenger with nano capsule with tuna fish oil	Bag
15	Carnation Instant food	Carnation Breakfast Essential, Switzerland	Titanium Dioxide (Nanoencapsulation)	Powdered milk-based products	Anticaking	Spray

9. Conclusions

Fruits are a major part of the human diet, supplying essential minerals and vitamins for human health. Acceptability of fruits by consumers depends on quality parameters such as color, texture, absence of decay, and especially the nutritional and health benefit they provide. Edible coatings, driven by their low cost and non-toxic nature, are one of the most well-studied natural polymers and their application has proven promising for fruit preservation. The application of nanoparticles to extend the shelf life of fruits appears most promising in the field of harvest storage. Current materials widely used for coating fruits are zinc oxide, silver, and chitosan nanoparticles because they show good results in preserving post-harvest quality. Another promising area of research is the combination of nanoparticle-enriched edible coatings with the use of current technologies such as low-temperature storage and controlled atmosphere storage. The specifics of the food, the substance to be added, and the intended extension of shelf life all influence the best sub-micron technology to use. As we can see, the ingredients should be non-toxic and be obtained from natural sources, such that the functional nanosystem allows the controlled release of active substances with low solubility. Moreover, the research looked out here, it has become clear that much more work is needed. In particular, we must also understand the behaviour of these materials after consumption, in order to make safe nanosystems that can be used freely in commercial products. Studies on this subject are limited and more information is needed to develop new coating applications with better functionality and higher sensory performance.

Author details

Ravi Kondle*, Kushagra Sharma, Gurpreet Singh and Amit Kotiyal
Lovely Professional University, Phagwara, Punjab, India

*Address all correspondence to: kondleravi27@gmail.com

IntechOpen

© 2022 The Author(s). Licensee IntechOpen. This chapter is distributed under the terms of the Creative Commons Attribution License (http://creativecommons.org/licenses/by/3.0), which permits unrestricted use, distribution, and reproduction in any medium, provided the original work is properly cited.

References

[1] Zaragoza MLZ, Guerrero DQ, Reza RMG, Villegas MAC, Gomez GL, Morlan ZU. Effects of UV-C and edible nano-coating as a combined strategy to preserve fresh-cut cucumber. Polymers. 2021;**13**:3705

[2] Tahir HE, Xiaobo Z, Mahunu GK, Arslan M, Abdalhai M, Zhihua L. Recent developments in gum edible coating applications for fruits and vegetables preservation: A review. Carbohydrate Polymers. 2019;**224**(15):115141. DOI: 10.1016/j.carbpol.2019.115141

[3] Lima AM, Miguel A, et al. New edible coatings composed of galactomannans and collagen blends to improve the postharvest quality of fruits—Influence on fruits gas transfer rate. Journal of Food Engineering. 2010;**97**:101-109

[4] Prakasha A, Baskaranb R, Vadive V. Citral nanoemulsion incorporated edible coating to extend the shelf life of fresh cut pineapple. LWT—Food Science and Technology. 2019;**118**:108851

[5] Saravanakumar K, Sathiyaseelan A, Mariadoss AVA, Xiaowen H, Wang MH. Physical and bioactivities of biopolymeric films incorporated with cellulose, sodium alginate and copper oxide nanoparticles for food packaging application. International Journal of Biological Macromolecules. 2020;**153**: 207-214. DOI: 10.1016/j.ijbiomac. 2020.02.250

[6] Dholariya PK, Borkar S, Borah A. Prospect of nanotechnology in food and edible packaging: A review. The Pharma Innovation Journal. 2021;**10**(5):197-203

[7] Moreno MI, Sánchez JAR, Santoyo MC. An extensive review of natural polymers used as coatings for postharvest shelf-life extension: Trends and challenges. Polymers. 2021;**13**:3271

[8] McHugh TH, Senesi E. Apple wraps: A novel method to improve the quality and extend the shelf life of fresh-cut apples. Journal of Food Science. 2000;**65**: 480-485

[9] Madhu B, Patel S, Rao PVK. Use of edible coatings to increase the shelf life of jaggery: A review. International Journal of Current Microbiology and Applied Sciences. 2018;7(06):2319-7706

[10] Nair MS, Tomar M, Punia S, Kochd WK, Kumar M. Enhancing the functionality of chitosan- and alginate-based active edible coatings/films for the preservation of fruits and vegetables: A review. International Journal of Biological Macromolecules. 2020;**81**:1-17

[11] Jianglian D, Shaoying Z. Application of chitosan based coating in fruit and vegetable preservation: A review. Journal of Food Processing & Technology. 2013;**4**:5

[12] Villafane F. Edible coatings for carrots. Food Reviews International. 2017;**33**(1):84-103

[13] Jaramillo CM, Pimiento CQ, Hoyos CG, Gallego RZ, Córdoba AL. Alginate edible coatings for application on wild andean blueberries (Vaccinium meridionale Swartz): Effect of the addition of nanofibrils isolated from cocoa by-products. Polymers. 2020; **12**:824

[14] Pirozzi A, Grosso VD, Ferrari G, Donsi F. Edible coatings containing oregano essential oil nanoemulsion for improving postharvest quality and shelf life of tomatoes. Food. 2020;**9**:1605

[15] Brossault D, McCoy TM, Routh AF. Preparation of multicore colloidosomes:

Nanoparticle-assembled capsules with adjustable size, internal structure, and functionalities for oil encapsulation. ACS Appled Polymer, Composite, and Edible Mater. 2021;**13**(43):51495-51503

[16] Lee D, Weitz DA. Double emulsion templated nanoparticle colloidosomes with selective permeability. Advanced Materials. 2008;**20**(18):3498-3503. DOI: 10.1002/adma.200800918

[17] Sadiq U, Gill H, Chandrapala J. Casein micelles as an emerging delivery system for bioactive food components. Food. 2021;**10**:1965

[18] Davoudi Z, Peroutka-Bigus N, Bellaire B, Jergens A, Wannemuehler M, Wang Q. Gut organoid as a new platform to study alginate and chitosan mediated PLGA nanoparticles for drug delivery. Marine Drugs. 2021;**19**:282

[19] Nile SH, Baskar V, Selvaraj D, Nile A, Xiao J, Kai G. Nanotechnologies in Food Science: Applications, Recent Trends, and Future Perspectives. Singapore: Springer; 2020

[20] Irshad MA, Nawaz R, Rehman MZ, Adrees M, Rizwan M, Ali S, et al. Synthesis, characterization and advanced sustainable applications of titanium dioxide nanoparticles: A review. Ecotoxicology and Environmental Safety. 2021; **212**:111978

[21] Paidari S, Ibrahim SA. Potential application of gold nanoparticles in food packaging: A mini-review. Gold Bulletin. 2021;**54**:31-36

[22] Cao B, Hu M, Cheng Y, Jing P, Liu B, Zhou B, et al. Tailoring the d-band center of N-doped carbon nanotube arrays with Co4N nanoparticles and single-atom Co for a superior hydrogen evolution reaction NPG. Asia Materials. 2021;**13**:1

[23] Mitelut AC, Popa EE, Dr Aghici MC, Popescu PA, Popa VI, Bujor OC, et al. Latest developments in edible coatings on minimally processed fruits and vegetables: A review. Food. 2021;**10**:2821

[24] Davidovic S, Miljkovic M, Gordic M, Barjas GC, Nesic A, Brankovic SD. Dextran-based edible coatings to prolong the shelf life of blueberries. Polymers. 2021;**13**:4252

[25] Xing Y, Li W, Wang Q, Li X, Xu Q, Guo X, et al. Antimicrobial nanoparticles incorporated in edible coatings and films for the preservation of fruits and vegetables. Molecules. 2019;**24**:1695

[26] Zhang R, Hoffmann T, Stosas E. Novel technique for coating of fine particles using fluidized bed and aerosol atomizer. PRO. 2020;**8**(12):1525

[27] Zaragoza MLZ, Reza RG, Muñoz NM, Linares VM, Couoh TFB, Elvira SM, et al. Nanosystems in edible coatings: A novel strategy for food preservation. International Journal of Molecular Sciences. 2018;**19**:705

[28] Liu R, Liu D, Liu Y, Song Y, Wu T, Zhang M. Using soy protein SiOx nanocomposite film coating to extend the shelf life of apple fruit. International Journal of Food Science and Technology. 2017;**52**:2018-2030

[29] Cortez-Vega WR, Pizato S, de Souza JTA, Prentice C. Using edible coatings from Whitemouth croaker (*Micropogonias furnieri*) protein isolate and organo-clay nanocomposite to improve the conservation properties of fresh-cut "Formosa" papaya. Innovative Food Science and Emerging Technologies. 2014;**22**:197-202

[30] Saba K, Amini R. Nano-ZnO/ carboxymethyl cellulose-based active coating impact on ready-to-use

pomegranate during cold storage. Food Chemistry. 2017;**232**:721-726

[31] Shah RW, Aisar M, Jahangir M, Abbasi KS, Khan SU, Ali N, et al. Influence of CMC- and guar gum-based silver nanoparticle coatings combined with low temperature on major aroma volatile components and the sensory quality of kinnow (*Citrus reticulata*). International Journal of Food Science and Technology. 2016;**512**:345-352

[32] Dhital R, Joshi P, Becerra N, Umagiliyage A, Chai T, Kohli P, et al. Integrity of edible nano-coatings and its effects on quality of strawberries subjected to simulated in-transit vibrations. Food Science and Technology. 2017;**2017**:257-264

[33] Sami R, Almatrafi M, Elhakem A, Alharbi M, Benajiba N, Helal M. Effect of nano silicon dioxide coating films on the quality characteristics of fresh-cut cantaloupe. Membranes. 2021;**11**:140

[34] Salha BA, Aharon G. Extending the shelf life of strawberries by the sonochemical coating of their surface with nanoparticles of an edible anti-bacterial compound. Applied Nanoscience. 2021;**2**:14-24

[35] Chaudhary P, Fatima F, Kumar A. Relevance of nanomaterials in food packaging and its advanced future prospects. Journal of Inorganic and Organometallic Polymers and Materials. 2000;**30**(12):5180-5192

[36] Bratovcic A. Physical, chemical, mechanical, and antimicrobial properties of bio-nanocomposite films and edible coatings. International Journal for Research in Applied Sciences and Biotechnology. 2021;**8**(5):2349-8889

[37] Filho JGO, Miranda M, Ferreira MD, Plotto A. Nanoemulsions as edible coatings: A potential strategy for fresh fruits and vegetable preservation. Food. 2021;**10**:2438

[38] Pinzon MI, Sanchez LT, Garcia OR, Gutierrez R, Luna JC, Villa CC. Increasing shelf life of strawberries (Fragaria spp) by using a banana starch-chitosan-Aloe vera gel composite edible coating. International Journal of Food Science and Technology. 2020;**55**:92-98

[39] Qiao G, Xiao Z, Ding W, Rok A. Effect of chitosan/nano-titanium dioxide/thymol and tween films on ready-to-eat cantaloupe fruit quality. Coatings. 2019;**9**:828

[40] Sneha Naira M, Maharishi T, Punia S. Enhancing the functionality of chitosan- and alginate-based active edible coatings/films for the preservation of fruits and vegetables: A review. International Journal of Biological Macromolecules. 2020;**2020**: 1-17

[41] Zapata MADL, Castro LP, Rodríguez MLR, Pérez OBA, Herrera RR. Candelilla wax nano emulsion coating with phytomolecules of tarbus improves the shelf life and quality of apples. Scientia Horticulturae. 2019;**231**(32): 43-48

Chapter 5

Perspective Chapter: Technological Strategies to Increase Insect Consumption – Transformation of Commodities Meal and Oil into Food/Functional Ingredients

Valeria Villanueva, Yanelis Ruiz, Fabrizzio Valdés, Marcela Sepúlveda and Carolina Valenzuela

Abstract

Insects have been proposed as an alternative source of nutrients to conventional foods, mainly protein sources because they have excellent nutritional quality and are sustainable. However, there are multiple barriers to mass consumption of insects, primarily the rejection and neophobia they provoke in individuals from Western cultures. Several studies have indicated that the acceptance of insects as food ingredients could be improved "if insects did not look like insects." Therefore, the focus of current research is to transform commodity-type ingredients such as insect flour and oil through various technologies applied in the food industry such as protein concentration, encapsulation, hydrolysis, fermentation, deodorization, to develop food ingredients with better sensory and technological properties are better accepted by people as a part of their diet. Interestingly, some food ingredients obtained from insects also have functional properties that could increase interest in consumption. These aspects will be reviewed in this chapter for further consideration of insects as food ingredients of the future.

Keywords: insect, functional ingredients, sensorial properties, food, feed

1. Introduction

The size of the world population and its accelerated growth are the greatest threat to humanity in terms of sustainability. The world population is expected to increase to 9.8 billion people by 2050 [1], requiring a 70–100% increase in food production to feed the world. Population growth could soon outstrip food production [2, 3]. Among the foods produced to feed humans and animals, those of animal origin are recognized as the least sustainable. For example, the production of 1 kg of beef protein has a carbon footprint between 45 and 640 kg CO_2 equivalents and a land use of 37–2100 m^2 [4, 5]. Enteric-derived methane from ruminant livestock accounts for 17–37% of the methane emitted to the atmosphere from human activities [6–8]. Ingredients of animal origin are the most complex to replace in animal and human diets in terms of nutritional needs because: i) They have high crude protein content (20–23% for meat and fish and 40–70% for animal meals); ii) have highly digestible amino acids (close to 85–90% for meals and even higher for meat) [9–13]; iii) have a high content of essential amino acids [14, 15], iv) have highly bioavailable organic minerals, such as heme iron and zinc [14, 16], and v) have a high concentration of vitamins. Vitamin B12 is only found in foods of animal origin [17]. Several of these characteristics are not present in plant sources [18–21]. In addition, projections indicate that the price of animal-derived meat and meals will increase steadily [22].

For these reasons, there is an urgent search for new sustainable and moderate-cost protein ingredients with nutritional properties similar to those from an animal origin. Among the available alternatives are protein ingredients obtained from non-conventional raw vegetable materials such as chickpeas, lentils, beans, peas, broad beans, and others [23–25]. However, they do not always meet the demanding amino acid requirements (in terms of digestibility and essential amino acid supply) of animals and humans [26, 27]. Other alternatives are the development of protein ingredients from microalgae, algae, yeasts, fungi, microorganisms, and the re-processing of animal or marine waste [28–33]. The drawback of these alternatives is their low productive volume, which is extremely variable, and their high cost. Technological strategies have also been applied to protein ingredients, such as fermentation [34] and hydrolysis [35, 36], which increase protein digestibility, but do not modify the amino acid profile [26, 37].

The Food and Agriculture Organization of the United Nations (FAO) has proposed insects as food ingredients of the future to feed humans and animals [38]. Their use is based on the fact that insects have similar nutritional characteristics to ingredients of animal origin, in terms of protein contribution, amino acid profile, amino acid digestibility, and the presence of minerals and vitamins [39–44]. The most widely used insects worldwide for the development of food ingredients for humans and animals are black soldier fly larvae (BSFL, *Hermetia illucens*), mealworm larvae (ML, *Tenebrio molitor*), and adult house crickets (*Acheta domesticus*) [45–48], because they are produced industrially in mini-farms. From these, basic ingredients, commodities such as whole meal, defatted meal, and insect oil are obtained using simple technologies commonly used in the food industry [48–50]. However, there is great potential for obtaining other food ingredients from insects, with better sensory, and technological and even functional properties that have only been scarcely studied and have few existing industrial applications. The objective of this chapter is to analyze the potential of transforming insect flour and oil commodities into food/functional ingredients with improved sensory, technological and functional properties for massification and use as food ingredients for the future.

2. Development

2.1 Sensory barriers to the massification of insect consumption

There are several insect-based foods on the market such as cereal bars, drinks, pastas, candies, snacks, hamburgers, for human consumption [51, 52] and made with different concentrations of insect flour and oil [53–55]. The main problem lies with the acceptance of this type of food by people who are not familiar with entomophagy (insect consumption), such as people from Western cultures, who often feel disgust, perceive insects as unpleasant, and reject their consumption [56]. In several survey-type studies, it was found that insect consumption could be better accepted if "insects did not look like insects" [57–59].

People who would be willing to consume insects describe some unpleasant sensory characteristics, such as: i) unpleasant odors [60, 61], identified as smelling of fungi, algae, fishy, and earthy [62–64], ii) unpleasant tastes of fish, fungi, bitterness [65, 66], iii) dark brown color of flour causing rejection [67], iv) grainy and rough texture of flour and whole insects [62, 63], and soft and oily texture in larvae [68] (**Figure 1A and B**), and v) unpalatable appearance of whole insects (**Figure 1C**) and meals, such as BSFL meal (**Figure 1E**) [69]. Additionally, processing to convert whole larvae/insects into meal (**Figure 1D**) can worsen these sensory perceptions, as Maillard reactions occur during the thermal process [70]. These reactions alter the color, odor, and flavor of insect flours in a negative way [71] and reduce the availability of some nutrients such as vitamin B12, potassium, phosphorus, sodium, some amino acids such as lysine [66]. There is also generation of unpleasant volatiles such as aldehydes, ketones, alcohols, esters, hydrocarbons, sulfur compounds and phenols, which generate unpleasant aromas [65, 72]. Thermal processing also generates darkening; for example, fly and mealworm larvae are cream-colored with yellow and orange shades (**Figure 1A**) and the meal obtained from these is dark brown (**Figure 1E**), due to the generation of brown and black coloring pigments such as melanoidins [66, 73]. The chitinous exoskeleton of insects is resistant to crushing [74]; therefore, the flour obtained after the milling process has granular texture, due to the large particle size (1.0–1.4 mm) of insect flours [53, 75], compared with flours of plant origin for example, wheat flour, which has a small particle size of about 100–150 μm [76].

The color of insect oil varies in yellow shades, and their melting point is variable depending on the profile and fatty acid content. For example, oil from BSFL contains lauric acid as the main fatty acid (21–29% of the total fatty acids, depending on the larval diet) [77]. Lauric acid is a saturated fatty acid, which gives the oil a high melting temperature (≈ 43°C). The oil is solid at room temperature, limiting its use as a food ingredient and making the incorporation into feed and/or diet formulations complex [78]. During oil processing, negative sensory changes also occur, mainly in oils with higher polyunsaturated fatty acid content, which tend to oxidation, producing odors and flavors described as "rancid and unpleasant" [60, 61]. Crude oil contains various components such as gums, free fatty acids, aromatic residues, and pigments, which negatively affect flavor, nutritional value, appearance, and stability [79].

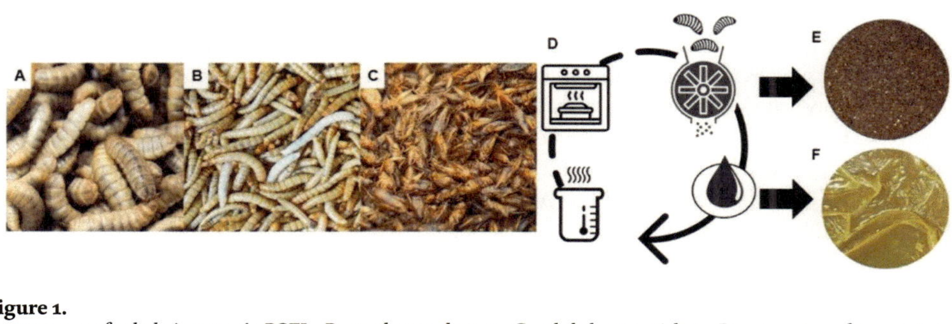

Figure 1.
Appearance of whole insects, A: BSFL, B: mealworm larvae, C: adult house crickets, D: processing of insects into food ingredients, such as flour (E) and oil (F).

The addition of whole or processed insects to a food negatively affects its sensory quality, even if added in small amounts (<5% for flour), because they contribute to characteristic flavors and aromas, considered unpalatable to people, affect the appearance and texture, and darken the product [54, 80–82]. Therefore, the addition of insect-based ingredients to foods remains a major challenge.

2.2 Common food ingredients from insects: meal and oil

The main insect-based ingredients produced in the world have been whole meal and defatted meal and oil, which are obtained by relatively simple processing and are widely used in the food industry [50]. The following processes are used to obtain flour: blanching, drying, grinding, and addition of additives [83, 84]. Blanching is the process where whole insects (larvae and/or adults) are placed in boiling water, and then removed and immersed in ice water to stop the thermal process. Blanching is used as a pretreatment to reduce the microbial load of bacteria and fungi and inactivate the degradative enzymes responsible for spoilage, but does not affect bacterial spores [85–88]. Blanching time can be from seconds to 16 minutes, with a 5-minute average, and this process can be repeated several times for differing periods of time. The ratio of insects/water used has been 1/10–1/12. The time of immersion in ice water is from 30 seconds to 5 minutes. Between the blanching and cooling processes in water, the insects can be drained and crushed. Sterilizing solutions of 5% NaCl can also be used in this process [50]. The second process the insects receive is drying to reduce total water content and water activity, decreasing degradation reactions, including enzymatic reactions and those produced by microorganisms [89, 90]. The drying methods used include air convention dryer, solar drying, oven drying, smoke drying, frying pan, freeze drying, microwave-assisted drying, fluidized bed drying, oven drying with air circulation, and ultrasound-assisted aqueous extraction. Of all these methods, the most widely used for the industrial production of insect meal is oven drying in conventional hot air drying, using temperature ranges between 40 and 80°C for 8 to 48 hours until the sample reaches constant weight [50]. The last process is milling, which mechanically reduces the whole insect to the consistency of powder or flour [91]. For grinding, the use of a roller mill [75], blade mill [92, 93], colloid mill [94], or mechanical disruptor [60] has been described with times varying between 2 and 10 minutes, depending on the method chosen [60, 93].

To obtain defatted meal, it is necessary to extract the oil. Oil extraction is commonly performed with organic solvents, such as hexane, ethanol, isopropanol, methanol, petroleum ether, acetone, diethyl ether, and their mixtures [94–102]. Solvent extraction techniques involve partitioning between two immiscible liquids, continuous extractions, or batch extraction of solids. This process consists of three stages: pretreatment, desolventization, and solvent refining [103]. Although extraction with organic solvents has been the most widely used for oil extraction, other methods recognized as "green" for being more innocuous have also been studied, such as extraction with supercritical CO_2 [104–106]. The latter is a promising process, with a good percentage of defatting; however, it is more expensive. High hydrostatic pressure extraction has also been investigated [107, 108]. Insect oil has been used in the formulation of human food [109, 110], salmonid diets [111, 112], complete feeds, and pet snacks [48].

The two most important nutritional components in insect meal are protein and fat. The ranges of crude protein and crude fat content of whole and defatted meal of the insects most commonly used in human food development and most consumed by animals are presented in **Table 1** [14, 94, 99, 113–120]. Of the three insects analyzed, crickets have the highest protein content in the complete meal, followed by mealworm and BSFL. The insects with the least amount of fat content in the complete meal are mealworm followed by BSFL and then crickets. Complete meal has been widely used to formulate diets for productive animals, mainly in aquaculture [121–123], pet diets and snacks [48], and human food [44, 124]. Defatted meal has a significant increase in total protein content (20–23%) and a reduction in fat content [99]. Defatted meal has been used to develop new ingredients that concentrate insect protein (hydrolysates, isolates, protein concentrates) for humans [44], specialized pet foods, such as hypoallergenic foods [48], and bioactive extracts with potential nutraceutical use [125, 126].

The protein and fat content is variable for each insect, so **Table 1** presents ranges. The primary factors influencing fat content are intrinsic variability of each insect species, developmental stage (larvae, pupae or adults), the diets used to feed the insects during the rearing and fattening period, and environmental conditions [127, 128].

2.3 Transformation of insect meal and oil into new food/functional ingredients

For mass consumption of insects to become the food of the future, it is necessary to transform insects into food ingredients of greater acceptability for the human population, using various technologies used by the food industry [129]. For animals, this is not necessary as insect meals and oil have high acceptability by aquaculture species [130], productive animals (pigs, hens, and chickens) [131–133], and domestic pets such as dogs and exotic animals [48]. **Table 2** and **Figure 2** present the new food and/or functional ingredients based on insect oil (**Figure 2A**) and meal (**Figure 1B**) commodities developed for humans. The main ingredients used as a base have been whole meal, defatted meal, and oil [172, 173]. More insect meal-based ingredients have been developed than insect oil. The developed oil-based ingredients are refined and deodorized oils, with better sensory properties (better odor and lighter yellow color). Emulsion technology has been applied to change the physical appearance of some insect oils, primarily BSFL, which as indicated above is solid at room temperature (**Figure 2A**), making it difficult to use in

Ingredients	BSFL	Mealworm	Cricket
	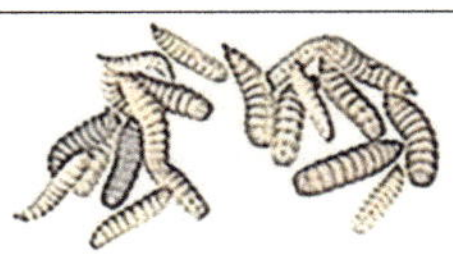		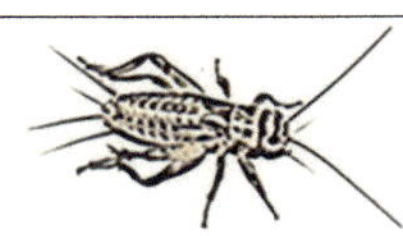
Whole meal			
Protein (%)	40–43 [14]	48–57 [14]	58–69 [14]
Fat (%)	17–34 [14]	32–40 [14]	11–23 [14]
Defatted meal			
Protein (%)	46–60 [113–116]	62–71 [94, 117–119]	79–81 [99, 107]
Fat (%)	5–11 [113–116]	1–14 [94, 117–119]	1–5 [99, 107]

BSFL: Black soldier fly larvae (Hermetia Illucens), mealworm (T. molitor), cricket (A. domesticus).

Table 1.
Protein and fat ranges in dry basis of whole and defatted meal, of common insects used in human and animal feed.

Insect	Technology	New ingredient	Properties
From insect oil			
BSFL[79, 134]	Oil purification	Refined oil	Reduction of oil viscosity, turbidity and density Oil with high oxidative stability and better quality
ML[110]	Oil deodorization	Deodorized oil	Improvement of the organoleptic properties of the oil, such as appearance, color, and odor
ML[135]	Oleogelation with waxes	Oleogel as solid fat replacer in cookies	The replacement with carnauba wax/insect oleogel showed a desirable cookie quality in terms of spreadability and texture properties
BSFL[136]	Homogenization	Fat emulsions	Emulsions showed twofold lower consistency compared to the lecithin solutions of the same concentration
BSFL[137, 138]	Pre-homogenization and ultrasonication	Nanoemulsions	Nanoemulsions with high value-added for several applications in food industry Applications as drug delivery vehicles
BSFL, BM[139]	Solvent extraction	Oils with antimicrobial activity	Antimicrobial activity against *B. subtilis* and *S. aureus*
BSFL[140–145]	Solvent extraction	Lauric acid	Reduction of *E. coli*, *Streptococcus* spp., *Yersinia enterocolitica* and *Enterobacteriaceae* Increased number of *Lactobacillus* and *Bifidobacteriu.* Higher concentrations of total volatile fatty acids. Reduction of cytokine (IL-10 and 6) Positive effect on the gut microbiota composition and intestinal morphology
From insect meal			
AD, BSFL, ML, SG, AM[92, 95, 101, 146]	Alkaline solubilization and isoelectric precipitation	Protein concentrates	Lighter colored powders (beige-brown) were obtained High protein content (62–85%) Better aroma and appearance
ML, AL[147, 148]	Alkaline extraction, acid precipitation and salting out procedures	Protein isolates	Higher protein content (65–87%). Similar total amino acid content Better technological properties
AD, ML[64, 149, 150]	Biological and enzymatic hydrolysis and high hydrostatic pressures	Protein hydrolysates	Enhanced flavor (by hydrolysis and Maillard reaction) Lower chitin content obtaining antimicrobial substances Good sensory properties

Insect	Technology	New ingredient	Properties
MB, SB[151]	Acid/alkaline/water extraction	Gelatin	Very good gelling ability
ML, ADI, PB[152]	Homogenization of larvae with pork fat and freezing	Emulsions as meat replacement	ML was the most suitable candidate for use as a meat replacement due to its physicochemical and rheological properties
ML[153]	Extrusion	Snack	Extrusion improved the digestibility of ML proteins and starch Snacks with 10% YLM meal showed good expansion properties and pore structure, obtaining acceptable textural qualities; at 20%, the snacks showed poor expansion properties due to the higher fat content
BSFL[154]	Extrusion	Pellets	Extrudates of pure insect meal had the lowest water absorption index and the highest water solubility index. Insect meal in the corn blends negatively affected the pasting properties of the extrudates
HFL[155]	Spray drying	Micro-powder	Better appearance, similar to meals of vegetable origin Reduced emission of volatile compounds. Smaller particle size (9 μm) than other insect meals (355–1400 μm) [53, 75] Low protein content (5.1% db)
HFL[156]	Ionic gelation	Beads	Black "caviar" looking beads, darkening the color of HFL meal. Low protein content (27%) compared with HFL meal (54%) High antioxidant capacity of 1235–6903 μmol TE/100 g
AD, ML[157–159]	Yeast/lactic acid fermentation	Fermented powder	Flavor was improved The intensity of indole, pyrazines, 1-octen-3-ol and 3-octanol, which have unpleasant odors, was reduced. Pleasant volatiles, such as ethyl acetate, isopentyl acetate and 2-butanone, were increased Lactic fermentation resulted in successful acidification and increased shelf-life and safety by the control of Enterobacteria and bacterial spores

Insect	Technology	New ingredient	Properties
AD, ML, TE, OC, PB[125, 160–163]	Ultrasound-assisted extraction Pressurized liquid extraction	Functional extract	All extracts exhibited antioxidant activity and showed lipase inhibitory activity Potent hemolytic activity and anticoagulation activities
BSFL[164–166]	Solvent extraction, RNA isolation, cDNA cloning, solid-phase extraction and reverse-phase chromatography	Antimicrobial peptides (AMP)	The highest levels of AMP were induced by larvae diets supplemented with protein or sunflower oil. AMPs demonstrated activities against a spectrum of bacteria AMP exhibited antibacterial activity against both *Escherichia coli* and methicillin-resistant *Staphylococcus aureus*

BSFL: black soldier fly larvae, HFL: house fly larvae, ML: mealworm larvae, ADL: Allomyrina dichotoma larvae, AD: Acheta domesticus, BM: Bombyx mori; SG: Schistocerca gregaria; AM: Apis mellifera; AL: Anastrepha ludens; ADI: Allomyrina dichotoma; PB: Protaetia brevitarsis; TE: Teleogryllus emma; OC: Oxya chinensis; MB: Melon bug; SB: Sorghum bug.

Table 2.
New food and/or functional ingredients developed from insect meal and oil.

the formulation of diets, since it is complex to homogenize with the other ingredients and tends to form aggregates when combined with ingredients in powder form. After the emulsion process, liquid formulations are obtained (**Figure 2Ai**), with a milky appearance (**Figure 2Ai-iv**), which could be converted to powder by spray drying, to facilitate their use as a food ingredient and increase shelf life. These emulsions have been proposed as value-added ingredients for the food industry and as potential nutrient and drug vehicles for the pharmaceutical industry (in the case of nanoemulsions). Some nanoemulsions retain a milky appearance, while others tend to be transparent (**Figure 2Av**). BSFL fat has a similar fatty acid composition as coconut and palm oil, making it one of the most promising alternative fat sources for the food industry, where these lipid sources are used in a large number of processed foods [136].

The antimicrobial capacity of BSFL oil has been studied and demonstrated in *in vitro* studies against Gram-positive and Gram-negative bacteria [139, 174]. The antimicrobial capacity of BSFL is due to its high concentration of lauric acid [14]. The mechanisms of lauric acid antimicrobial processes are still being studied, but three have been described: 1) destruction of the cell membrane of gram-positive bacteria and lipid-coated viruses by physicochemical processes, 2) interference with cellular processes, such as signal transduction and transcription, and 3) destabilization of cell membranes [175], through inhibition of the enzyme MurA [176]. Very few *in vivo* investigations have been performed in animals to study the antimicrobial property of BSFL. The inclusion of BSFL oil in the diet does not affect the microbiota, improves intestinal morphology, and increases beneficial microorganism populations [140–142].

BSFL oil has the ability to regulate blood cholesterol levels due to its lauric acid content. In an *in silico* study animals fed lauric acid had increased cholesterol metabolism due to reduced HMG-CoA enzyme activity [177]. BSFL oil may also affect markers and coagulation factors, inhibiting platelet aggregation, prolonging the activated partial thromboplastin time. In *ex vivo* and *in vivo* studies, the compounds extracted

Figure 2.
Appearance of traditional insect ingredients, BSFL oil (A) and BSFL meal (B) and new insect-based ingredients, such as emulsions (Ai-Aiv [167, 168]), nanoemulsions (Av, [137]), protein extracts (Bi, [169]), protein concentrates (Bii, [170]), alginate-insect meal beads (Biii, [156]), micro-powders (Biv, [155]), and aqueous extracts (Bv, [171]).

from three insects, *Protaetia brevitarsis* seulensis, *Tenebrio mollitor* and *Oxya chinensis* sinuosa, succeeded in reducing platelet aggregation and the rate and size of arterial and pulmonary thrombus formation in mice [160–162].

Studies on new ingredients based on whole and defatted meals have focused on concentrating protein by developing protein concentrates, protein isolates, protein hydrolysates, and protein fermentates, using methods such as alkaline extraction and isoelectric precipitation (**Table 2**). The development of protein concentrates and protein isolates is focused on because i) protein is one of the most expensive nutrients in human and animal diets; and projections indicate that the price of protein ingredients of animal and plant origin will increase steadily [22]. ii) Proteins of animal origin are complex to replace in human and animal diets and are not very sustainable [178, 179]. Insect proteins represent a sustainable replacement alternative to animal proteins [180, 181]. iii) The protein content of insect meals, especially defatted meal, is very high (**Table 1**) and similar to meals of vegetable origin (40–55%), meat/bone meal (40–50%), and offal (40–60%) [182]. The protein content of defatted cricket, mealworm, and BSFL meal is similar to meals of marine origin such as fish meal (60–75%) [182]. iv) The protein quality of insects, in terms of essential amino acid content (good source of lysine, methionine, threonine, leucine, alanine, valine) and amino acid digestibility (80–93%), is excellent. v) Insect proteins tend to be high in glutamic acid, which is the main amino acid in BSFL and mealworm meal [48], and is related to umami taste, highly preferred by animals and humans [183, 184]. vi) Insect proteins have technological properties suitable for the processing of certain foods such as meat substitutes [185], jerky meat analog [186], extruded cereals [153], rusks [187], and pastas [188]. vii) Products that concentrate insect proteins as concentrates and isolates have better sensory properties than insect meals, such as lighter colors [65], better taste [64], better volatile profile [149], and better emulsifying and foaming properties [98, 102]. Insect protein concentrates and isolates have been used in human food and are commercially available. Some examples are Becrit® and Trillions®, under the concept of protein shakes; Isaac nutrition®, protein powder; AdalbaPro IPC®, protein concentrate; and AdalbaPro FTIP®, protein concentrate powder with fiber texture. **Figure 2** shows that the main change in appearance of protein concentrates (**Figure 2Bi-ii**), made from insect meals (**Figure 2B**), is a lighter coloration of the brown shades of the meals.

Hydrolysates and fermentates have been developed for the purpose of reducing some antinutritional factors of insect meals such as chitin [149], improving organoleptic properties, increasing shelf life [189, 190], providing antioxidant properties [149, 191, 192], increased nutrient digestibility, production of antimicrobial substances, and health-promoting molecules [149]. Enzymatic hydrolysis using a variety of enzymes, such as alkalase, papain, peptidase, protease; and alcalase, papain, peptidase, protease; and biological hydrolysis using yeasts (*Yarrowia lipolytica* and *Debaryomyces hansenii*) have been studied to obtain peptides with bioactive properties and to improve protein digestibility, mainly. Fermentation of insect meal with lactic acid bacteria or yeast (*Saccharomyces cerevisiae*) has been used to improve sensory properties, mainly by modifying the volatile profiles, decreasing indole, pyrazines, 1-octen-3-ol, and 3-octanol, and increasing propanol, ethanol, acetone and 2-butanone, reducing fecal, toasted, earthy, mushroom, and bitter taste [157, 158].

Other technologies applied to develop new ingredients have been extrusion to produce pellets and snacks based mainly on insect meal mixed with cereals. The main result has been the increase in the digestibility of some nutrients such as protein and starch. In these studies, it was indicated that the insect meal content used alters some important properties of the extruded products (**Table 2**). This technology has been widely used for the development of complete foods and snacks for dogs and cats [48]. Some examples are Eat Small Mindfulness®, Brit Care Immunity®, CircularPet®, Yora®, Insecta®, buggybigs®.

Encapsulation is a technology widely used in the food industry, especially spray drying [193], because it improves the sensory characteristics of the compounds to be encapsulated [194]. In addition, controlled release formulations can be developed to add the encapsulated compounds in complex foods such as yogurt, beverages, dairy, and others [195–197]. Although, this technology is widely used to improve the sensory properties of ingredients, it has not been thoroughly studied for encapsulating insect meal and oil. Among the existing works, spray drying has been used to develop insect meal micro-powders (**Figure 2Biv**), which presented better appearance and color (similar to wheat flour), better texture (with smaller particle size), and better aroma than unencapsulated house fly larvae meal (**Table 2**). However, the protein content of the micro-powders was low at about 5.1 g per 100 g of powder, whereas the meal that gave rise to the micro-powders contained 54 g of protein per 100 g of flour. Even so, micro-powders are considered a "source of protein" according to the Codex Alimentarius [198]. The challenge for this technology is to concentrate on the nutrients, especially protein, from insect meals, being able to use previously described technologies such as protein isolates and concentrates. In another study, house fly meal was encapsulated by ionic gelation, obtaining alginate-insect meal beads with an appearance similar to black "caviar" (**Figure 2Biii**), with better aroma than the unencapsulated meal (**Table 2**). The application of this type of product in human food is complex, due to the rejection that its appearance could cause, but it is possible to incorporate it as food for exotic pets (such as water turtles, fish, ferrets, hedgehogs), which consume live and dehydrated insect larvae [48]. The pet industry has a high level of innovation in food products and consumption of innovative foods and snacks is on the rise [199, 200].

In **Table 2**, the development of oil nanoemulsions is described. The technique is considered an encapsulation process, since the oil is protected and separated from the water by a dynamic surfactant layer formed by emulsifying agents. This technique has been widely used in the food industry for the following reasons: 1) to improve the stability of some lipophilic active compounds such as vitamin D3 [201], carotenoids [202], and α-tocopherol [203], 2) to improve the absorption, bioavailability, and bioactivity of lipophilic bioactive compounds with low absorption such as curcumin [204] and astaxanthin [205], and 3) to provide the ability to release encapsulated actives in a controlled manner [206, 207].

The functional properties of insect-based food ingredients, such as antioxidant capacity, antimicrobial activity, inhibition of platelet aggregation, enzymatic inhibition, and antidiabetic potential, have been less studied than their nutritional properties as food ingredients [208, 209]. The literature shows that ingredients obtained from insects such as aqueous extracts [125], meals [191, 210, 211], and proteins and peptides [126, 212, 213], exhibit high antioxidant capacity, so they could have potential use in health disorders associated with oxidative stress [214]. The antioxidant capacity is due to the presence of phenolic compounds, proteins, peptides, chitin, fatty acids, and others [215, 216].

A great diversity of bioactive compounds has been isolated from insects, such as free fatty acids, amino acids, organic acids, carbohydrates, hydrocarbons, sterols, and others [125]. The methodologies for their extraction have been ultrasound-assisted extraction (UAE) and pressurized liquid extraction (PLE), using ethanol or a mixture of ethanol and water [125] (**Table 2**). The appearance of these extracts is presented in **Figure 2Bv**, observing different colorations that depend on several factors, such as concentration of the extracts and extraction technique. Extracts have anti-inflammatory, antimicrobial, antiangiogenic, antiproliferative, and antioxidant properties. The ability to inhibit the activity of certain enzymes has also been studied. In *in vitro* studies, extracts of *A. domesticus* and *T. molitor* were able to inhibit pancreatic lipase [125]. These extracts could have an application in the treatment and prevention of obesity [217]. Proteins from the insects such as *B. mori*, *T. molitor*, *Alphitobius diaperinus*, and *Gryllus bimaculatus* [218–221] were able to inhibit angiotensin-converting enzyme (ACE), dipeptidyl peptidase-4 (DPP-IV) [220], and α-glucosidase activity [218, 221], with antidiabetic potential. In animal studies, supplementation with ethanolic extract of *B. mori* improved glycemic status in obese mice with type 2 diabetes, reducing glycemia and restoring pancreatic functionality [222, 223]. Soluble extracts obtained from 12 insect species showed antioxidant activity, the highest in extracts from grasshoppers, silkworms, and crickets [216]. Antioxidant activity was also found in aqueous extract of *Vespa affinis* L. [214]. Antibacterial substances such as N-beta-alanyl-5-S-glutathionyl-5-S-glutathionyl-3,4-dihydroxyphenylalanine from *Sarcophaga peregrina* and p-hydroxycinnamaldehyde from *Acantholyda parki* larvae were isolated from extracts of immunized insects [224, 225].

Other insect-based functional compounds extensively studied in the recent years are antimicrobial peptides (AMPs), which are extracted and purified by different technologies, such as reverse phase high-performance liquid chromatography (RP-HPLC), DNA extraction, RNA extraction, fast performance liquid chromatography (FPLC), and gel filtration chromatography [226] (**Table 2**). AMPs are peptides with low molecular weight, high thermal stability, and a broad antimicrobial spectrum [227, 228]. A large number of AMPs derived from *Acalolepta luxuriosa*, *A. mellifera*, *B. mori*, *Galleria mellonella*, *Heterometrus spinifer*, *Holotrichia diomphalia*, *Hyalophora cecropia*, *Oxysternon conspicillatum*, *Pandinus imperator*, and *Sarcophaga peregrine* have been investigated and are effective against a wide range of Gram-negative and Gram-positive bacteria [229, 230]. Their mechanism of action depends on the type of AMP and the target pathogen. AMPs can interact with the microbial membrane surface, alter permeability and induce cell lysis, enter the cell, and damage bacterial components such as DNA and RNA, and promote the bacteriostatic effects [226, 228]. The use of AMPs as an alternative to antimicrobials in human and animal health could help reduce antimicrobial resistance [228].

The challenge for the future of insects as food for humans and animals is to increase research on technologies that can be used to transform common insect-based ingredients such as meal and oil into ingredients with higher added value, functional properties, and optimal sensory properties for greater acceptance and consumption of insects by humans, so these ingredients can be included in a greater number of foods.

Acknowledgements

The authors acknowledge the support of FONDEF IDea I + D ID22I10030 and the valuable help of Susan Cleveland, who proofread the chapter.

Conflict of interest

The authors declare no conflict of interest.

Author details

Valeria Villanueva, Yanelis Ruiz, Fabrizzio Valdés, Marcela Sepúlveda and Carolina Valenzuela*
Facultad de Ciencias Veterinarias y Pecuarias, Universidad de Chile, Santiago, Chile

*Address all correspondence to: cvalenzuelav@u.uchile.cl

IntechOpen

© 2022 The Author(s). Licensee IntechOpen. This chapter is distributed under the terms of the Creative Commons Attribution License (http://creativecommons.org/licenses/by/3.0), which permits unrestricted use, distribution, and reproduction in any medium, provided the original work is properly cited.

References

[1] Food and Agriculture Organization of the United Nations (FAO). FAOSTAT Food and agriculture data [Internet]. 2022. Available from: https://www.fao.org/faostat/es/#home. [Accessed: August 4, 2022]

[2] Bruinsma J. The resource outlook to 2050: by how much do land, water and crop yields need to increase by 2050? [Internet]. 2009. Available from: https://www.fao.org/publications/card/es/c/f70d2098-c9ad-5b05-8972-cf987e74d681/. [Accessed: August 4, 2022]

[3] Myers SS, Smith MR, Guth S, Golden CD, Vaitla B, Mueller ND, et al. Climate change and global food systems: Potential impacts on food security and undernutrition. Annual Review of Public Health. 2017;**38**(1):259-277. DOI: 10.1146/annurev-publhealth-031816-044356

[4] Nijdam D, Rood T, Westhoek H. The price of protein: Review of land use and carbon footprints from life cycle assessments of animal food products and their substitutes. Food Policy. 2012;**37**(6):760-770. DOI: 10.1016/j.foodpol.2012.08.002

[5] Röös E, Ekelund L, Tjärnemo H. Communicating the environmental impact of meat production: Challenges in the development of a Swedish meat guide. Journal of Cleaner Production. 2014;**73**(6):154-164. DOI: 10.1016/j.jclepro.2013.10.037

[6] Steinfeld H, Wassenaar T. The role of livestock production in carbon and nitrogen cycles. Annual Review of Environment and Resources. 2007;**32**(1):271-294. DOI: 10.1146/annurev.energy.32.041806.143508

[7] Lassey KR. Livestock methane emission and its perspective in the global methane cycle. Australian Journal of Experimental Agriculture. 2008;**48**(2):114-118. DOI: 10.1071/EA07220

[8] Cai Y, Tang R, Tian L, Chang SX. Environmental impacts of livestock excreta under increasing livestock production and management considerations: Implications for developing countries. Current Opinion in Environmental Science & Health. 2021;**24**(3):100300. DOI: 10.1016/j.coesh.2021.100300

[9] Murray SM, Patil AR, Fahey GC, Merchen NR, Hughes DM. Raw and rendered animal by-products as ingredients in dog diets. Journal of Animal Science. 1997;**75**(9):2497-2505. DOI: 10.2527/1997.7592497x

[10] Johnson ML, Parsons CM, Fahey GC, Merchen NR, Aldrich CG. Effects of species raw material source, ash content, and processing temperature on amino acid digestibility of animal by-product meals by cecectomized roosters and ileally cannulated dogs. Journal of Animal Science. 1998;**76**(4):1112-1122. DOI: 10.2527/1998.7641112x

[11] Wang X, Parsons CM. Effect of raw material source, processing systems, and processing temperatures on amino acid digestibility of meat and bone meals. Poultry Science. 1998;**77**(6):834-841. DOI: 10.1093/ps/77.6.834

[12] Shirley RB, Parsons CM. Effect of ash content on protein quality of meat and bone meal. Poultry Science. 2001;**80**(5):626-632. DOI: 10.1093/ps/80.5.626

[13] Dozier WA, Dale NM, Dove CR. Nutrient composition of feed-grade and pet-food-grade poultry by-product meal. Journal of Applied Poultry Research. 2003;**12**(4):526-530. DOI: 10.1093/japr/12.4.526

[14] Makkar HPS, Tran G, Heuzé V, Ankers P. State-of-the-art on use of insects as animal feed. Animal Feed Science and Technology. 2014;**197**(11):1-33. DOI: 10.1016/j.anifeedsci.2014.07.008

[15] Bosch G, Zhang S, Oonincx DGAB, Hendriks WH. Protein quality of insects as potential ingredients for dog and cat foods. Journal of Nutritional Science. 2014;**3**(3):e29. DOI: 10.1017/jns.2014.23

[16] Ordoñez-Araque R, Egas-Montenegro E. Edible insects: A food alternative for the sustainable development of the planet. International Journal of Gastronomy and Food Science. 2021;**23**(6):100304. DOI: 10.1016/j.ijgfs.2021.100304

[17] Schmidt A, Call L-M, Macheiner L, Mayer HK. Determination of vitamin B12 in four edible insect species by immunoaffinity and ultra-high performance liquid chromatography. Food Chemistry. 2019;**281**(3):124-129. DOI: 10.1016/j.foodchem.2018.12.039

[18] Bednar GE, Murray SM, Patil AR, Flickinger EA, Merchen NR, Fahey GC. Selected animal and plant protein sources affect nutrient digestibility and fecal characteristics of ileally cannulated dogs. Arch für Tierernaehrung. 2000;**53**(2):127-140. DOI: 10.1080/17450390009381942

[19] Donadelli RA, Jones CK, Beyer RS. The amino acid composition and protein quality of various egg, poultry meal by-products, and vegetable proteins used in the production of dog and cat diets. Poultry Science. 2019;**98**(3):1371-1378. DOI: 10.3382/ps/pey462

[20] Félix AP, Menezes Souza CM, Bastos TS, Kaelle GCB, Oliveira SG, Maiorka A. Digestibility of raw soybeans in extruded diets for dogs determined by different methods. Italian Journal of Animal Science. 2020;**19**(1):95-102. DOI: 10.1080/1828051X.2019.1698324

[21] Acuff HL, Dainton AN, Dhakal J, Kiprotich S, Aldrich G. Sustainability and pet food: Is there a role for veterinarians? The Veterinary Clinics of North America. Small Animal Practice. 2021;**51**(3):563-581. DOI: 10.1016/j.cvsm.2021.01.010

[22] OECD/FAO. OECD-FAO Agricultural outlook (Edition 2021), OECD Agriculture Statistics (database) [Internet]. 2021. DOI: 10.1787/4bde2d83-en. [Accessed: March 26, 2022]

[23] Øverland M, Sørensen M, Storebakken T, Penn M, Krogdahl Å, Skrede A. Pea protein concentrate substituting fish meal or soybean meal in diets for Atlantic salmon (Salmo salar)—Effect on growth performance, nutrient digestibility, carcass composition, gut health, and physical feed quality. Aquaculture. 2009;**288**(3-4):305-311. DOI: 10.1016/j.aquaculture.2008.12.012

[24] Freeman LM, Stern JA, Fries R, Adin DB, Rush JE. Diet-associated dilated cardiomyopathy in dogs: What do we know? Journal of the American Veterinary Medical Association. 2018;**253**(11):1390-1394. DOI: 10.2460/javma.253.11.1390

[25] ReillyLM,vonSchaumburgPC,HokeJM, DavenportGM,UtterbackPL,ParsonsCM, et al. Macronutrient composition, true metabolizable energy and amino acid digestibility, and indispensable amino acid scoring of pulse ingredients for

use in canine and feline diets. Journal of Animal Science. 2020;**98**(6):skaa149. DOI: 10.1093/jas/skaa149

[26] Dodd SAS, Adolphe JL, Verbrugghe A. Plant-based diets for dogs. Journal of the American Veterinary Medical Association. 2018;**253**(11):1425-1432. DOI: 10.2460/javma.253.11.1425

[27] Dodd SAS, Shoveller AK, Fascetti AJ, Yu ZZ, Ma DWL, Verbrugghe A. A comparison of key essential nutrients in commercial plant-based pet foods sold in Canada to American and European canine and feline dietary recommendations. Animals. 2021;**11**(8):2348. DOI: 10.3390/ani11082348

[28] Brain-Isasi S, Carú C, Lienqueo ME. Valorization of the green seaweed *Ulva rigida* for production of fungal biomass protein using a hypercellulolytic terrestrial fungus. Algal Research. 2021;**59**(3):102457. DOI: 10.1016/j.algal.2021.102457

[29] Dale N, Valenzuela C. Nutritional properties of dried salmon silage for broiler feeding. Animal Science Journal. 2016;**87**(6):791-795. DOI: 10.1111/asj.12480

[30] Luigia DS. New protein sources: Novel foods. In: Ferranti P, Berry EM, Anderson JR, editors. Encyclopedia of Food Security and Sustainability. Amsterdam: Elsevier; 2019. pp. 276-279. DOI: 10.1016/B978-0-08-100596-5.22129-X

[31] Fasolin LH, Pereira RN, Pinheiro AC, Martins JT, Andrade CCP, Ramos OL, et al. Emergent food proteins – Towards sustainability, health and innovation. Food Research International. 2019;**125**(5):108586. DOI: 10.1016/j.foodres.2019.108586

[32] Food and Agriculture Organization of the United Nations (FAO). The state of world fisheries and aquaculture sustainability in action [Internet]. 2020. Available from: https://www.fao.org/3/ca9229en/ca9229en.pdf. [Accessed: November 16, 2021]

[33] Mosna D, Bottani E, Vignali G, Montanari R. Environmental benefits of pet food obtained as a result of the valorisation of meat fraction derived from packaged food waste. Waste Management. 2021;**125**(3):132-144. DOI: 10.1016/j.wasman.2021.02.035

[34] Thirunathan P, Manickavasagan A. Processing methods for reducing alpha-galactosides in pulses. Critical Reviews in Food Science and Nutrition. 2019;**59**(20):3334-3348. DOI: 10.1080/10408398.2018.1490886

[35] Tavano OL. Protein hydrolysis using proteases: An important tool for food biotechnology. Journal of Molecular Catalysis B: Enzymatic. 2013;**90**(1):1-11. DOI: 10.1016/j.molcatb.2013.01.011

[36] Marciniak A, Suwal S, Naderi N, Pouliot Y, Doyen A. Enhancing enzymatic hydrolysis of food proteins and production of bioactive peptides using high hydrostatic pressure technology. Trends in Food Science and Technology. 2018;**80**(11):187-198. DOI: 10.1016/j.tifs.2018.08.013

[37] Kanakubo K, Fascetti AJ, Larsen JA. Assessment of protein and amino acid concentrations and labeling adequacy of commercial vegetarian diets formulated for dogs and cats. Journal of the American Veterinary Medical Association. 2015;**247**(4):385-392. DOI: 10.2460/javma.247.4.385

[38] Food and Agriculture Organization of the United Nations (FAO). Edible Insects. Future Prospects for Food and

Feed Security. Rome, Italy: Wageningen UR; 2013. p. 201

[39] Rumpold BA, Schlüter OK. Nutritional composition and safety aspects of edible insects. Molecular Nutrition & Food Research. 2013;**57**(5):802-823. DOI: 10.1002/mnfr.201200735

[40] Belluco S, Losasso C, Maggioletti M, Alonzi CC, Paoletti MG, Ricci A. Edible insects in a food safety and nutritional perspective: A critical review. Comprehensive Reviews in Food Science and Food Safety. 2013;**12**(3):296-313. DOI: 10.1111/1541-4337.12014

[41] Zielińska E, Baraniak B, Karaś M, Rybczyńska K, Jakubczyk A. Selected species of edible insects as a source of nutrient composition. Food Research International. 2015;**77**:460-466. DOI: 10.1016/j.foodres.2015.09.008

[42] Williams JP, Williams JR, Kirabo A, Chester D, Peterson M. Chapter 3 - Nutrient content and health benefits of insects. In: Dossey AT, Morales-Ramos JA, Rojas G, editors. Insects as Sustainable Food Ingredients. USA: Academic Press; 2016. pp. 61-84. DOI: 10.1016/B978-0-12-802856-8/00003-X

[43] Mishyna M, Glumac M. So different, yet so alike Pancrustacea: Health benefits of insects and shrimps. Journal of Functional Foods. 2021;**76**:104316. DOI: 10.1016/j.jff.2020.104316

[44] Borges MM, da Costa DV, Trombete FM, Câmara AKFI. Edible insects as a sustainable alternative to food products: An insight into quality aspects of reformulated bakery and meat products. Current Opinion in Food Science. 2022;**46**(3):100864. DOI: 10.1016/j.cofs.2022.100864

[45] van Huis A. Environmental sustainability of insects as human food. In: Reference Module in Food Science. Ámsterdam: Elsevier; 2019. pp. 1-5. DOI: 10.1016/B978-0-08-100596-5.22589-4

[46] van Huis A. Insects as food and feed, a new emerging agricultural sector: A review. Journal of Insects as Food and Feed. 2020;**6**(1):27-44. DOI: 10.3920/JIFF2019.0017

[47] van Huis A. Nutrition and health of edible insects. Current Opinion in Clinical Nutrition and Metabolic Care. 2020;**23**(3):228-231. DOI: 10.1097/MCO.0000000000000641

[48] Valdés F, Villanueva V, Durán E, Campos F, Avendaño C, Sánchez M, et al. Insects as feed for companion and exotic pets: A current trend. Animals. 2022;**12**(11):1450. DOI: 10.3390/ani12111450

[49] Tzompa-Sosa DA, Yi L, van Valenberg HJF, Lakemond CMM. Four insect oils as food ingredient: Physical and chemical characterisation of insect oils obtained by an aqueous oil extraction. Journal of Insects as Food and Feed. 2019;**5**(4):279-292. DOI: 10.3920/JIFF2018.0020

[50] Melgar-Lalanne G, Hernández-Álvarez A, Salinas-Castro A. Edible insects processing: Traditional and innovative technologies. Comprehensive Reviews in Food Science and Food Safety. 2019;**18**(4):1166-1191. DOI: 10.1111/1541-4337.12463

[51] Caparros Megido R, Gierts C, Blecker C, Brostaux Y, Haubruge É, Alabi T, et al. Consumer acceptance of insect-based alternative meat products in Western countries. Food Quality and Preference. 2016;**52**(3):237-243. DOI: 10.1016/j.foodqual.2016.05.004

[52] Motoki K, Ishikawa S, Spence C, Velasco C. Contextual acceptance of insect-based foods. Food Quality and Preference. 2020;**85**(3):103982. DOI: 10.1016/j.foodqual.2020.103982

[53] de Oliveira LM, da Silva Lucas AJ, Cadaval CL, Mellado MS. Bread enriched with flour from cinereous cockroach (*Nauphoeta cinerea*). Innovative Food Science and Emerging Technologies. 2017;**44**(6):30-35. DOI: 10.1016/j.ifset.2017.08.015

[54] Osimani A, Milanović V, Cardinali F, Roncolini A, Garofalo C, Clementi F, et al. Bread enriched with cricket powder (*Acheta domesticus*): A technological, microbiological and nutritional evaluation. Innovative Food Science and Emerging Technologies. 2018;**48**(6):150-163. DOI: 10.1016/j.ifset.2018.06.007

[55] Cicatiello C, Vitali A, Lacetera N. How does it taste? Appreciation of insect-based snacks and its determinants. International Journal of Gastronomy and Food Science. 2020;**21**(5):100211. DOI: 10.1016/j.ijgfs.2020.100211

[56] Jensen NH, Lieberoth A. We will eat disgusting foods together – Evidence of the normative basis of Western entomophagy-disgust from an insect tasting. Food Quality and Preference. 2019;**72**(4):109-115. DOI: 10.1016/j.foodqual.2018.08.012

[57] Orsi L, Voege LL, Stranieri S. Eating edible insects as sustainable food? Exploring the determinants of consumer acceptance in Germany. Food Research International. 2019;**125**:108573. DOI: 10.1016/j.foodres.2019.108573

[58] Tuccillo F, Marino MG, Torri L. Italian consumers' attitudes towards entomophagy: Influence of human factors and properties of insects and insect-based food. Food Research International. 2020;**137**(20):109619. DOI: 10.1016/j.foodres.2020.109619

[59] Higa JE, Ruby MB, Rozin P. Americans' acceptance of black soldier fly larvae as food for themselves, their dogs, and farmed animals. Food Quality and Preference. 2021;**90**(4):104119. DOI: 10.1016/j.foodqual.2020.104119

[60] Larouche J, Deschamps M-H, Saucier L, Lebeuf Y, Doyen A, Vandenberg GW. Effects of killing methods on lipid oxidation, colour and microbial load of black soldier fly (*Hermetia illucens*) larvae. Animals. 2019;**9**(4):182. DOI: 10.3390/ani9040182

[61] Singh Y, Cullere M, Kovitvadhi A, Chundang P, Dalle ZA. Effect of different killing methods on physicochemical traits, nutritional characteristics, in vitro human digestibility and oxidative stability during storage of the house cricket (*Acheta domesticus* L.). Innovative Food Science and Emerging Technologies. 2020;**65**(5):102444. DOI: 10.1016/j.ifset.2020.102444

[62] Dion-Poulin A, Turcotte M, Lee-Blouin S, Perreault V, Provencher V, Doyen A, et al. Acceptability of insect ingredients by innovative student chefs: An exploratory study. International Journal of Gastronomy and Food Science. 2021;**24**(6):100362. DOI: 10.1016/j.ijgfs.2021.100362

[63] Sogari G, Menozzi D, Mora C. Sensory-liking expectations and perceptions of processed and unprocessed insect products. International Journal of Food System Dynamics. 2018;**9**(4):314-320. DOI: 10.18461/ijfsd.v9i4.942

[64] Grossmann KK, Merz M, Appel D, De Araujo MM, Fischer L. New insights into the flavoring potential of cricket (*Acheta domesticus*) and mealworm

(*Tenebrio molitor*) protein hydrolysates and their Maillard products. Food Chemistry. 2021;**364**(3):130336. DOI: 10.1016/j.foodchem.2021.130336

[65] Lee JH, Cha J-Y, Kim T-K, Choi Y-S, Jang HW. Effects of a defatting process on the thermal stabilities and volatile compound profiles of proteins isolated from *Protaetia brevitarsis* larvae. LWT. 2021;**151**(1):112095. DOI: 10.1016/j.lwt.2021.112095

[66] Ssepuuya G, Nakimbugwe D, De Winne A, Smets R, Claes J, Van Der Borght M. Effect of heat processing on the nutrient composition, colour, and volatile odour compounds of the long-horned grasshopper *Ruspolia differens* serville. Food Research International. 2020;**129**(4):108831. DOI: 10.1016/j.foodres.2019.108831

[67] Kim H-W, Setyabrata D, Lee YJ, Jones OG, Kim YHB. Pre-treated mealworm larvae and silkworm pupae as a novel protein ingredient in emulsion sausages. Innovative Food Science and Emerging Technologies. 2016;**38**(5):116-123. DOI: 10.1016/j.ifset.2016.09.023

[68] Wang Y-S, Shelomi M. Review of black soldier fly (*Hermetia illucens*) as animal feed and human food. Food. 2017;**6**(10):91. DOI: 10.3390/foods6100091

[69] Mishyna M, Chen J, Benjamin O. Sensory attributes of edible insects and insect-based foods – Future outlooks for enhancing consumer appeal. Trends in Food Science and Technology. 2020;**95**(4):141-148. DOI: 10.1016/j.tifs.2019.11.016

[70] David-Birman T, Raften G, Lesmes U. Effects of thermal treatments on the colloidal properties, antioxidant capacity and *in-vitro* proteolytic degradation of cricket flour. Food Hydrocolloids. 2018;**79**(6):48-54. DOI: 10.1016/j.foodhyd.2017.11.044

[71] Mishyna M, Haber M, Benjamin O, Martinez JJI, Chen J. Drying methods differentially alter volatile profiles of edible locusts and silkworms. Journal of Insects as Food and Feed. 2020;**6**(4):405-415. DOI: 10.3920/JIFF2019.0046

[72] Khatun H, Claes J, Smets R, De Winne A, Akhtaruzzaman M, Van Der Borght M. Characterization of freeze-dried, oven-dried and blanched house crickets (*Acheta domesticus*) and Jamaican field crickets (*Gryllus assimilis*) by means of their physicochemical properties and volatile compounds. European Food Research and Technology. 2021;**247**(5):1291-1305. DOI: 10.1007/s00217-021-03709-x

[73] Kouřimská L, Adámková A. Nutritional and sensory quality of edible insects. NFS Journal. 2016;**4**(5):22-26. DOI: 10.1016/j.nfs.2016.07.001

[74] Bassett FS, Dunn ML, Pike OA, Jefferies LK. Physical, nutritional, and sensory properties of spray-dried and oven-roasted cricket (*Acheta domesticus*) powders. Journal of Insects as Food and Feed. 2021;**7**(6):987-1000. DOI: 10.3920/JIFF2020.0107

[75] Purschke B, Brüggen H, Scheibelberger R, Jäger H. Effect of pre-treatment and drying method on physico-chemical properties and dry fractionation behaviour of mealworm larvae (*Tenebrio molitor* L.). European Food Research and Technology. 2018;**244**(2):269-280. DOI: 10.1007/s00217-017-2953-8

[76] Liu T, Hou GG, Lee B, Marquart L, Dubat A. Effects of particle size on the quality attributes of reconstituted whole-wheat flour and tortillas made from it. Journal of Cereal Science.

2016;**71**(4):145-152. DOI: 10.1016/j.jcs.2016.08.013

[77] Matsue M, Mori Y, Nagase S, Sugiyama Y, Hirano R, Ogai K, et al. Measuring the antimicrobial activity of lauric acid against various bacteria in human gut microbiota using a new method. Cell Transplantation. 2019;**28**(12):1528-1541. DOI: 10.1177/0963689719881366

[78] Zeng C, Liu Y, Ding Z, Xia H, Guo S. Physicochemical properties and antibacterial activity of hydrophobic deep eutectic solvent-in-water nanoemulsion. Journal of Molecular Liquids. 2021;**338**(11):116950. DOI: 10.1016/j.molliq.2021.116950

[79] Mai HC, Dao ND, Lam TD, Nguyen BV, Nguyen DC, Bach LG. Purification process, physicochemical properties, and fatty acid composition of black soldier fly (*Hermetia illucens* Linnaeus) larvae oil. Journal of the American Oil Chemists' Society. 2019;**96**(11):1303-1311. DOI: 10.1002/aocs.12263

[80] Biró B, Fodor R, Szedljak I, Pásztor-Huszár K, Gere A. Buckwheat-pasta enriched with silkworm powder: Technological analysis and sensory evaluation. LWT. 2019;**116**(3):108542. DOI: 10.1016/j.lwt.2019.108542

[81] Choi Y-S, Kim T-K, Choi H-D, Park J-D, Sung J-M, Jeon K-H, et al. Optimization of replacing pork meat with yellow worm (*Tenebrio molitor* L.) for frankfurters. Korean Journal for Food Science of Animal Resources. 2017;**37**(5):617-625. DOI: 10.5851/kosfa.2017.37.5.617

[82] Haber M, Mishyna M, Martinez JJI, Benjamin O. The influence of grasshopper (*Schistocerca gregaria*) powder enrichment on bread nutritional and sensorial properties. LWT. 2019;**115**(3):108395. DOI: 10.1016/j.lwt.2019.108395

[83] Mishyna M, Keppler JK, Chen J. Techno-functional properties of edible insect proteins and effects of processing. Current Opinion in Colloid & Interface Science. 2021;**56**:101508. DOI: 10.1016/j.cocis.2021.101508

[84] Meshulam-Pascoviche D, David-Birman T, Refael G, Lesmes U. Big opportunities for tiny bugs: Processing effects on the techno-functionality and digestibility of edible insects. Trends in Food Science and Technology. 2022;**122**(4):265-274. DOI: 10.1016/j.tifs.2022.02.012

[85] Vandeweyer D, Lenaerts S, Callens A, Van Campenhout L. Effect of blanching followed by refrigerated storage or industrial microwave drying on the microbial load of yellow mealworm larvae (*Tenebrio molitor*). Food Control. 2017;**71**:311-314. DOI: 10.1016/j.foodcont.2016.07.011

[86] Wynants E, Crauwels S, Verreth C, Gianotten N, Lievens B, Claes J, et al. Microbial dynamics during production of lesser mealworms (*Alphitobius diaperinus*) for human consumption at industrial scale. Food Microbiology. 2018;**70**:181-191. DOI: 10.1016/j.fm.2017.09.012

[87] Xiao H-W, Bai J-W, Sun D-W, Gao Z-J. The application of superheated steam impingement blanching (SSIB) in agricultural products processing – A review. Journal of Food Engineering. 2014;**132**:39-47. DOI: 10.1016/j.jfoodeng.2014.01.032

[88] Severini C, Baiano A, De Pilli T, Carbone BF, Derossi A. Combined treatments of blanching and dehydration: Study on potato cubes. Journal of Food

Engineering. 2005;**68**(3):289-296. DOI: 10.1016/j.jfoodeng.2004.05.045

[89] Grabowski NT, Klein G. Microbiology of processed edible insect products – Results of a preliminary survey. International Journal of Food Microbiology. 2017;**243**:103-107. DOI: 10.1016/j.ijfoodmicro.2016.11.005

[90] Grabowski NT, Klein G. Microbiology of cooked and dried edible Mediterranean field crickets (*Gryllus bimaculatus*) and superworms (*Zophobas atratus*) submitted to four different heating treatments. Food Science and Technology International. 2017;**23**(1):17-23. DOI: 10.1177/1082013216652994

[91] Liceaga AM. Processing insects for use in the food and feed industry. Current Opinion in Insect Science. 2021;**48**:32-36. DOI: 10.1016/j.cois.2021.08.002

[92] Caligiani A, Marseglia A, Leni G, Baldassarre S, Maistrello L, Dossena A, et al. Composition of black soldier fly prepupae and systematic approaches for extraction and fractionation of proteins, lipids and chitin. Food Research International. 2018;**105**:812-820. DOI: 10.1016/j.foodres.2017.12.012

[93] Scholliers J, Steen L, Glorieux S, Van de Walle D, Dewettinck K, Fraeye I. The effect of temperature on structure formation in three insect batters. Food Research International. 2019;**122**:411-418. DOI: 10.1016/j.foodres.2019.04.033

[94] Son Y-J, Lee J-C, Hwang I-K, Nho CW, Kim S-H. Physicochemical properties of mealworm (*Tenebrio molitor*) powders manufactured by different industrial processes. LWT. 2019;**116**:108514. DOI: 10.1016/j.lwt.2019.108514

[95] Azagoh C, Ducept F, Garcia R, Rakotozafy L, Cuvelier M-E, Keller S, et al. Extraction and physicochemical characterization of *Tenebrio molitor* proteins. Food Research International. 2016;**88**:24-31. DOI: 10.1016/j.foodres.2016.06.010

[96] Bußler S, Rumpold BA, Jander E, Rawel HM, Schlüter OK. Recovery and techno-functionality of flours and proteins from two edible insect species: Meal worm (*Tenebrio molitor*) and black soldier fly (*Hermetia illucens*) larvae. Heliyon. 2016;**2**(12):e00218. DOI: 10.1016/j.heliyon.2016.e00218

[97] Choi BD, Wong NAK, Auh JH. Defatting and sonication enhances protein extraction from edible insects. Korean Journal for Food Science of Animal Resources. 2017;**37**(6):955-961. DOI: 10.5851/kosfa.2017.37.6.955

[98] Mishyna M, Martinez J-JI, Chen J, Benjamin O. Extraction, characterization and functional properties of soluble proteins from edible grasshopper (*Schistocerca gregaria*) and honey bee (*Apis mellifera*). Food Research International. 2019;**116**:697-706. DOI: 10.1016/j.foodres.2018.08.098

[99] Ribeiro JC, Lima RC, Maia MRG, Almeida AA, Fonseca AJM, Cabrita ARJ, et al. Impact of defatting freeze-dried edible crickets (*Acheta domesticus* and *Gryllodes sigillatus*) on the nutritive value, overall liking and sensory profile of cereal bars. LWT. 2019;**113**(5):108335. DOI: 10.1016/j.lwt.2019.108335

[100] Zhao X, Vázquez-Gutiérrez JL, Johansson DP, Landberg R, Langton M. Yellow mealworm protein for food purposes - extraction and functional properties. PLoS One. 2016;**11**(2):e0147791. DOI: 10.1371/journal.pone.0147791

[101] Amarender RV, Bhargava K, Dossey AT, Gamagedara S. Lipid and

protein extraction from edible insects – Crickets (Gryllidae). LWT. 2020;**125**:109222. DOI: 10.1016/j.lwt.2020.109222

[102] Kim T-K, Yong HI, Kim Y-B, Jung S, Kim H-W, Choi Y-S. Effects of organic solvent on functional properties of defatted proteins extracted from *Protaetia brevitarsis* larvae. Food Chemistry. 2021;**336**:127679. DOI: 10.1016/j.foodchem.2020.127679

[103] Birch EJ. FATS | extraction by solvent based methods. In: Wilson ID, editor. Encyclopedia of Separation Science. London: Academic Press; 2000. pp. 2794-2801. DOI: 10.1016/B0-12-226770-2/06661-8

[104] Pan W-J, Liao A-M, Zhang J-G, Dong Z, Wei Z-J. Supercritical carbon dioxide extraction of the oak silkworm (*Antheraea pernyi*) pupal oil: Process optimization and composition determination. International Journal of Molecular Sciences. 2012;**13**(2):2354-2367. DOI: 10.3390/ijms13022354

[105] Kim SW, Jung TS, Ha YJ, Gal SW, Noh CW, Kim IS, et al. Removal of fat from crushed black soldier fly larvae by carbon dioxide supercritical extraction. Journal of Animal and Feed Sciences. 2019;**28**(1):83-88. DOI: 10.22358/jafs/105132/2019

[106] Purschke B, Stegmann T, Schreiner M, Jäger H. Pilot-scale supercritical CO2 extraction of edible insect oil from *Tenebrio molitor* L. larvae – Influence of extraction conditions on kinetics, defatting performance and compositional properties. European Journal of Lipid Science and Technology. 2017;**119**(2):1600134. DOI: 10.1002/ejlt.201600134

[107] Bolat B, Ugur AE, Oztop MH, Alpas H. Effects of high hydrostatic pressure assisted degreasing on the technological properties of insect powders obtained from *Acheta domesticus* & *Tenebrio molitor*. Journal of Food Engineering. 2021;**292**:110359. DOI: 10.1016/j.jfoodeng.2020.110359

[108] Laroche M, Perreault V, Marciniak A, Gravel A, Chamberland J, Doyen A. Comparison of conventional and sustainable lipid extraction methods for the production of oil and protein isolate from edible insect meal. Food. 2019;**8**(11):572. DOI: 10.3390/foods8110572

[109] Delicato C, Schouteten JJ, Dewettinck K, Gellynck X, Tzompa-Sosa DA. Consumers' perception of bakery products with insect fat as partial butter replacement. Food Quality and Preference. 2020;**79**:103755. DOI: 10.1016/j.foodqual.2019.103755

[110] Tzompa-Sosa DA, Dewettinck K, Gellynck X, Schouteten JJ. Replacing vegetable oil by insect oil in food products: Effect of deodorization on the sensory evaluation. Food Research International. 2021;**141**:110140. DOI: 10.1016/j.foodres.2021.110140

[111] Belghit I, Waagbø R, Lock E-J, Liland NS. Insect-based diets high in lauric acid reduce liver lipids in freshwater Atlantic salmon. Aquaculture Nutrition. 2019;**25**(2):343-357. DOI: 10.1111/anu.12860

[112] Kumar V, Fawole FJ, Romano N, Hossain MS, Labh SN, Overturf K, et al. Insect (black soldier fly, *Hermetia illucens*) meal supplementation prevents the soybean meal-induced intestinal enteritis in rainbow trout and health benefits of using insect oil. Fish & Shellfish Immunology. 2021;**109**:116-124. DOI: 10.1016/j.fsi.2020.12.008

[113] Kamarudin MS, Rosle S, Md Yasin IS. Performance of defatted black

soldier fly pre-pupae meal as fishmeal replacement in the diet of lemon fin barb hybrid fingerlings. Aquaculture Reports. 2021;**21**:100775. DOI: 10.1016/j.aqrep.2021.100775

[114] Li M, Li M, Wang G, Liu C, Shang R, Chen Y, et al. Defatted black soldier fly (*Hermetia illucens*) larvae meal can partially replace fish meal in diets for adult Chinese soft-shelled turtles. Aquaculture. 2021;**541**:736758. DOI: 10.1016/j.aquaculture.2021.736758

[115] Stejskal V, Tran HQ, Prokesová M, Zare M, Gebauer T, Policar T, et al. Defatted black soldier fly (*Hermetia illucens*) in pikeperch (*Sander lucioperca*) diets: Effects on growth performance, nutrient digestibility, fillet quality, economic and environmental sustainability. Animal Nutrition. 2022;**12**:7-19. DOI: 10.1016/j.aninu.2022.06.022

[116] Kishawy ATY, Mohammed HA, Zaglool AW, Attia MS, Hassan FAM, Roushdy EM, et al. Partial defatted black solider larvae meal as a promising strategy to replace fish meal protein in diet for Nile tilapia (*Oreochromis niloticus*): Performance, expression of protein and fat transporters, and cytokines related genes and economic efficiency. Aquaculture. 2022;**555**:738195. DOI: 10.1016/j.aquaculture.2022.738195

[117] Basto A, Matos E, Valente LMP. Nutritional value of different insect larvae meals as protein sources for European sea bass (*Dicentrarchus labrax*) juveniles. Aquaculture. 2020;**521**:735085. DOI: 10.1016/j.aquaculture.2020.735085

[118] Tran HQ, Van Doan H, Stejskal V. Does dietary *Tenebrio molitor* affect swimming capacity, energy use, and physiological responses of European perch *Perca fluviatilis*? Aquaculture. 2021;**539**:736610. DOI: 10.1016/j.aquaculture.2021.736610

[119] Tran HQ, Prokešová M, Zare M, Matoušek J, Ferrocino I, Gasco L, et al. Production performance, nutrient digestibility, serum biochemistry, fillet composition, intestinal microbiota and environmental impacts of European perch (*Perca fluviatilis*) fed defatted mealworm (*Tenebrio molitor*). Aquaculture. 2022;**547**:737499. DOI: 10.1016/j.aquaculture.2021.737499

[120] Tang C, Yang D, Liao H, Sun H, Liu C, Wei L, et al. Edible insects as a food source: A review. Food Production, Processing and Nutrition. 2019;**1**(1):8. DOI: 10.1186/s43014-019-0008-1

[121] Ferrer Llagostera P, Kallas Z, Reig L, Amores de Gea D. The use of insect meal as a sustainable feeding alternative in aquaculture: Current situation, Spanish consumers' perceptions and willingness to pay. Journal of Cleaner Production. 2019;**229**:10-21. DOI: 10.1016/j.jclepro.2019.05.012

[122] Hidalgo MC, Morales AE, Pula HJ, Tomás-Almenar C, Sánchez-Muros MJ, Melenchón F, et al. Oxidative metabolism of gut and innate immune status in skin and blood of tench (Tinca tinca) fed with different insect meals (*Hermetia illucens* and *Tenebrio molitor*). Aquaculture. 2022;**558**(1):738384. DOI: 10.1016/j.aquaculture.2022.738384

[123] Mastoraki M, Katsika L, Enes P, Guerreiro I, Kotzamanis YP, Gasco L, et al. Insect meals in feeds for juvenile gilthead seabream (Sparus aurata): Effects on growth, blood chemistry, hepatic metabolic enzymes, body composition and nutrient utilization. Aquaculture. 2022;**561**(1):738674. DOI: 10.1016/j.aquaculture.2022.738674

[124] González CM, Garzón R, Rosell CM. Insects as ingredients for bakery goods.

A comparison study of *H. illucens*, *A. domestica* and *T. molitor* flours. Innovative Food Science and Emerging Technologies. 2019;**51**:205-210. DOI: 10.1016/j.ifset.2018.03.021

[125] Navarro del Hierro J, Gutiérrez-Docio A, Otero P, Reglero G, Martin D. Characterization, antioxidant activity, and inhibitory effect on pancreatic lipase of extracts from the edible insects *Acheta domesticus* and *Tenebrio molitor*. Food Chemistry. 2020;**309**:125742. DOI: 10.1016/j.foodchem.2019.125742

[126] Montiel-Aguilar LJ, Torres-Castillo JA, Rodríguez-Servin R, López-Flores AB, Aguirre-Arzola VE, Méndez-Zamora G, et al. Nutraceutical effects of bioactive peptides obtained from *Pterophylla beltrani* (Bolivar & Bolivar) protein isolates. Journal of Asia-Pacific Entomology. 2020;**23**(3):756-761. DOI: 10.1016/j.aspen.2020.06.006

[127] Weru J, Chege P, Kinyuru J. Nutritional potential of edible insects: A systematic review of published data. International Journal of Tropical Insect Science. 2021;**41**(3):2015-2037. DOI: 10.1007/s42690-021-00464-0

[128] Hernández-Álvarez A-J, Mondor M, Piña-Domínguez I-A, Sánchez-Velázquez O-A, Melgar LG. Drying technologies for edible insects and their derived ingredients. Drying Technology. 2021;**39**(13):1991-2009. DOI: 10.1080/07373937.2021.1915796

[129] Menozzi D, Sogari G, Veneziani M, Simoni E, Mora C. Eating novel foods: An application of the theory of planned behaviour to predict the consumption of an insect-based product. Food Quality and Preference. 2017;**59**:27-34. DOI: 10.1016/j.foodqual.2017.02.001

[130] Rawski M, Mazurkiewicz J, Kierończyk B, Józefiak D. Black soldier fly full-fat larvae meal as an alternative to fish meal and fish oil in Siberian sturgeon nutrition: The effects on physical properties of the feed, animal growth performance, and feed acceptance and utilization. Animals. 2020;**10**(11):2119. DOI: 10.3390/ani10112119

[131] de Souza VJ, Andronicos NM, Kolakshyapati M, Hilliar M, Sibanda TZ, Andrew NR, et al. Black soldier fly larvae in broiler diets improve broiler performance and modulate the immune system. Animal Nutrition. 2021;7(3):695-706. DOI: 10.1016/j.aninu.2020.08.014

[132] Håkenåsen IM, Grepperud GH, Hansen JØ, Øverland M, Ånestad RM, Mydland LT. Full-fat insect meal in pelleted diets for weaned piglets: Effects on growth performance, nutrient digestibility, gastrointestinal function, and microbiota. Animal Feed Science and Technology. 2021;**281**:115086. DOI: 10.1016/j.anifeedsci.2021.115086

[133] Zhao Q, Xie T, Hong X, Zhou Y, Fan L, Liu Y, et al. Modification of functional properties of perilla protein isolate by high-intensity ultrasonic treatment and the stability of o/w emulsion. Food Chemistry. 2022;**368**:130848. DOI: 10.1016/j.foodchem.2021.130848

[134] Matthäus B, Piofczyk T, Katz H, Pudel F. Renewable resources from insects: Exploitation, properties, and refining of fat obtained by cold-pressing from *Hermetia illucens* (black soldier fly) larvae. European Journal of Lipid Science and Technology. 2019;**121**(7):1800376. DOI: 10.1002/ejlt.201800376

[135] Kim D, Oh I. The characteristic of insect oil for a potential component of oleogel and its application as a solid fat replacer in cookies. Gels. 2022;**8**(6):355. DOI: 10.3390/gels8060355

[136] Ruban AA, Novikova (Zakharova) MB, Kostin AA. Effective viscosity of lecithin solutions and fat emulsions of black soldier fly larvae with different lecithin content. Food Systems. 2021;**4**(3):220-225. DOI: 10/21323/2618-9771-2021-4-3-220-225

[137] Chou T-H, Nugroho DS, Cheng Y-S, Chang J-Y. Development and characterization of nano-emulsions based on oil extracted from black soldier fly larvae. Applied Biochemistry and Biotechnology. 2020;**191**(1):331-345. DOI: 10.1007/s12010-019-03210-y

[138] Chou T-H, Nugroho DS, Chang J-Y, Cheng Y-S, Liang C-H, Deng M-J. Encapsulation and characterization of nanoemulsions based on an anti-oxidative polymeric amphiphile for topical apigenin delivery. Polymers (Basel). 2021;**13**(7):1016. DOI: 10.3390/polym13071016

[139] Saviane A, Tassoni L, Naviglio D, Lupi D, Savoldelli S, Bianchi G, et al. Mechanical processing of *Hermetia illucens* larvae and *Bombyx mori* pupae produces oils with antimicrobial activity. Animals. 2021;**11**(3):783. DOI: 10.3390/ani11030783

[140] Spranghers T, Michiels J, Vrancx J, Ovyn A, Eeckhout M, De Clercq P, et al. Gut antimicrobial effects and nutritional value of black soldier fly (*Hermetia illucens* L.) prepupae for weaned piglets. Animal Feed Science and Technology. 2018;**235**:33-42. DOI: 10.1016/j.anifeedsci.2017.08.012

[141] Yu M, Li Z, Chen W, Rong T, Wang G, Ma X. Hermetia illucens larvae as a potential dietary protein source altered the microbiota and modulated mucosal immune status in the colon of finishing pigs. Journal of Animal Science and Biotechnology. 2019;**10**(1):50. DOI: 10.1186/s40104-019-0358-1

[142] Dabbou S, Ferrocino I, Gasco L, Schiavone A, Trocino A, Xiccato G, et al. Antimicrobial effects of black soldier fly and yellow mealworm fats and their impact on gut microbiota of growing rabbits. Animals. 2020;**10**(8):1292. DOI: 10.3390/ani10081292

[143] Yu M, Li Z, Chen W, Wang G, Rong T, Liu Z, et al. Hermetia illucens larvae as a fishmeal replacement alters intestinal specific bacterial populations and immune homeostasis in weanling piglets. Journal of Animal Science. 2020;**98**(3):1-13. DOI: 10.1093/jas/skz395/5810268

[144] Józefiak A, Nogales-Mérida S, Rawski M, Kierończyk B, Mazurkiewicz J. Effects of insect diets on the gastrointestinal tract health and growth performance of Siberian sturgeon (Acipenser baerii Brandt, 1869). BMC Veterinary Research. 2019;**15**(1):348. DOI: 10.1186/s12917-019-2070-y

[145] Sypniewski J, Kierończyk B, Benzertiha A, Mikołajczak Z, Pruszyńska-Oszmałek E, Kołodziejski P, et al. Replacement of soybean oil by Hermetia illucens fat in Turkey nutrition: Effect on performance, digestibility, microbial community, immune and physiological status and final product quality. British Poultry Science. 2020;**61**(3):294-302. DOI: 10.1080/00071668.2020.1716302

[146] Zielińska E, Karaś M, Baraniak B. Comparison of functional properties of edible insects and protein preparations thereof. LWT. 2018;**91**:168-174. DOI: 10.1016/j.lwt.2018.01.058

[147] Jiang Y, Zhu Y, Zheng Y, Liu Z, Zhong Y, Deng Y, et al. Effects of salting-in/out-assisted extractions on structural, physicochemical and functional properties of *Tenebrio molitor*

larvae protein isolates. Food Chemistry. 2021;**338**:128158. DOI: 10.1016/j.foodchem.2020.128158

[148] Valle FR, Mena MH, Bourges H. An investigation into insect protein. Journal of Food Processing & Preservation. 1982;**6**(2):99-110. DOI: 10.1111/j.1745-4549.1982.tb00645.x

[149] Patrignani F, Parrotta L, Del Duca S, Vannini L, Camprini L, Dalla Rosa M, et al. Potential of *Yarrowia lipolytica* and *Debaryomyces hansenii* strains to produce high quality food ingredients based on cricket powder. LWT. 2020;**119**:108866. DOI: 10.1016/j.lwt.2019.108866

[150] Dion-Poulin A, Laroche M, Doyen A, Turgeon SL. Functionality of cricket and mealworm hydrolysates generated after pretreatment of meals with high hydrostatic pressures. Molecules. 2020;**25**(22):5366. DOI: 10.3390/molecules25225366

[151] Mariod AA, Fadul H. Extraction and characterization of gelatin from two edible Sudanese insects and its applications in ice cream making. Food Science and Technology International. 2015;**21**(5):380-391. DOI: 10.1177/1082013214541137

[152] Kim T-K, Yong HI, Jung S, Sung J-M, Jang HW, Choi Y-S. Physicochemical and textural properties of emulsions prepared from the larvae of the edible insects *Tenebrio molitor*, *Allomyrina dichotoma*, and *Protaetia brevitarsis seulensis*. Journal of Animal Science and Technology. 2021;**63**(2):417-425. DOI: 10.5187/jast.2021.e25

[153] Azzollini D, Derossi A, Fogliano V, Lakemond CMM, Severini C. Effects of formulation and process conditions on microstructure, texture and digestibility of extruded insect-riched snacks. Innovative Food Science and Emerging Technologies. 2018;**45**:344-353. DOI: 10.1016/j.ifset.2017.11.017

[154] Alam MR, Scampicchio M, Angeli S, Ferrentino G. Effect of hot melt extrusion on physical and functional properties of insect based extruded products. Journal of Food Engineering. 2019;**259**:44-51. DOI: 10.1016/j.jfoodeng.2019.04.021

[155] Sánchez M, Gómez C, Avendaño C, Harmsen I, Ortiz D, Ceballos R, et al. House fly (*Musca domestica*) larvae meal as an ingredient with high nutritional value: Microencapsulation and improvement of organoleptic characteristics. Food Research International. 2021;**145**(5):110423. DOI: 10.1016/j.foodres.2021.110423

[156] Sánchez M, Villamizar-Sarmiento MG, Harmsen I, Valdés F, Villanueva V, Ceballos R, et al. Encapsulation of house fly larvae (*Musca domestica*) meal by ionic gelation as a strategy to develop a novel nutritive food ingredient with improved aroma and appearance. LWT. 2022;**163**:113597. DOI: 10.1016/j.lwt.2022.113597

[157] Kim J, Lee HE, Kim Y, Yang J, Lee S-J, Jung YH. Development of a post-processing method to reduce the unique off-flavor of *Allomyrina dichotoma*: Yeast fermentation. LWT. 2021;**150**(5):111940. DOI: 10.1016/j.lwt.2021.111940

[158] Lee HE, Kim J, Kim Y, Bang WY, Yang J, Lee S-J, et al. Identification and improvement of volatile profiles of *Allomyrina dichotoma* larvae by fermentation with lactic acid bacteria. Food Bioscience. 2021;**43**:101257. DOI: 10.1016/j.fbio.2021.101257

[159] Klunder HC, Wolkers-Rooijackers J, Korpela JM, Nout MJR. Microbiological aspects of processing and storage

of edible insects. Food Control. 2012;**26**(2):628-631. DOI: 10.1016/j.foodcont.2012.02.013

[160] Lee W, Kim M-A, Park I, Hwang JS, Na M, Bae J-S. Novel direct factor Xa inhibitory compounds from *Tenebrio molitor* with anti-platelet aggregation activity. Food and Chemical Toxicology. 2017;**109**(6):19-27. DOI: 10.1016/j.fct.2017.08.026

[161] Lee W, Lee H, Kim M-A, Choi J, Kim K-M, Hwang JS, et al. Evaluation of novel factor Xa inhibitors from *Oxya chinensis sinuosa* with anti-platelet aggregation activity. Scientific Reports. 2017;**7**(1):7934. DOI: 10.1038/s41598-017-08330-1

[162] Lee J, Lee W, Kim M-A, Hwang JS, Na M, Bae J-S. Inhibition of platelet aggregation and thrombosis by indole alkaloids isolated from the edible insect *Protaetia brevitarsis seulensis* (Kolbe). Journal of Cellular and Molecular Medicine. 2017;**21**(6):1217-1227. DOI: 10.1111/jcmm.13055

[163] Pyo S-J, Kang D-G, Jung C, Sohn H-Y. Anti-thrombotic, anti-oxidant and haemolysis activities of six edible insect species. Food. 2020;**9**(4):401. DOI: 10.3390/foods9040401

[164] Vogel H, Müller A, Heckel DG, Gutzeit H, Vilcinskas A. Nutritional immunology: Diversification and diet-dependent expression of antimicrobial peptides in the black soldier fly *Hermetia illucens*. Developmental and Comparative Immunology. 2018;**78**:141-148. DOI: 10.1016/j.dci.2017.09.008

[165] Shin HS, Park S-I. Novel attacin from *Hermetia illucens*: cDNA cloning, characterization, and antibacterial properties. Preparative Biochemistry & Biotechnology. 2019;**49**(3):279-285. DOI: 10.1080/10826068.2018.1541807

[166] Park S-I, Kim J-W, Yoe SM. Purification and characterization of a novel antibacterial peptide from black soldier fly (*Hermetia illucens*) larvae. Developmental and Comparative Immunology. 2015;**52**(1):98-106. DOI: 10.1016/j.dci.2015.04.018

[167] Baigts-Allende D, Doost AS, Ramírez-Rodrigues M, Dewettinck K, Van der Meeren P, de Meulenaer B, et al. Insect protein concentrates from Mexican edible insects: Structural and functional characterization. LWT. 2021;**152**:112267. DOI: 10.1016/j.lwt.2021.112267

[168] Tello A, Aganovic K, Parniakov O, Carter A, Heinz V, Smetana S. Product development and environmental impact of an insect-based milk alternative. Future Foods. 2021;**4**:100080. DOI: 10.1016/j.fufo.2021.100080

[169] Zhang L, Xiao Q, Wang Y, Hu J, Xiong H, Zhao Q. Effects of sequential enzymatic hydrolysis and transglutaminase crosslinking on functional, rheological, and structural properties of whey protein isolate. LWT. 2022;**153**:112415. DOI: 10.1016/j.lwt.2021.112415

[170] Ndiritu AK, Kinyuru JN, Kenji GM, Gichuhi PN. Extraction technique influences the physico-chemical characteristics and functional properties of edible crickets (*Acheta domesticus*) protein concentrate. Journal of Food Measurement and Characterization. 2017;**11**(4):2013-2021. DOI: 10.1007/s11694-017-9584-4

[171] Janssen RH, Canelli G, Sanders MG, Bakx EJ, Lakemond CMM, Fogliano V, et al. Iron-polyphenol complexes cause blackening upon grinding *Hermetia illucens* (black soldier fly) larvae. Scientific Reports. 2019;**9**(1):2967. DOI: 10.1038/s41598-019-38923-x

[172] Akpossan R, Digbeu Y, Koffi M, Kouadio J, Dué E, Kouamé P. Protein fractions and functional properties of dried *Imbrasia oyemensis* larvae full-fat and defatted flours. International Journal of Biochemistry Research & Review. 2015;**5**(2):116-126. DOI: 10.9734/IJBCRR/2015/12178

[173] dos Santos OV, Dias PCS, Soares SD, da Conceição LRV, Teixeira-Costa BE. Artisanal oil obtained from insects' larvae (*Speciomerus ruficornis*): Fatty acids composition, physicochemical, nutritional and antioxidant properties for application in food. European Food Research and Technology. 2021;**247**(7):1803-1813. DOI: 10.1007/s00217-021-03752-8

[174] Borrelli L, Varriale L, Dipineto L, Pace A, Menna LF, Fioretti A. Insect derived lauric acid as promising alternative strategy to antibiotics in the antimicrobial resistance scenario. Frontiers in Microbiology. 2021;**12**(3):620798. DOI: 10.3389/fmicb.2021.620798

[175] Dayrit FM. The properties of lauric acid and their significance in coconut oil. Journal of the American Oil Chemists' Society. 2015;**92**(1):1-15. DOI: 10.1007/s11746-014-2562-7

[176] Herdiyati Y, Astrid Y, Shadrina AAN, Wiani I, Satari MH, Kurnia D. Potential fatty acid as antibacterial agent against oral bacteria of *Streptococcus mutans* and *Streptococcus sanguinis* from basil (*Ocimum americanum*): *In vitro* and *in silico* studies. Current Drug Discovery Technologies. 2021;**18**(4):532-541. DOI: 10.2174/1570163817666200712171652

[177] Lekshmi Sheela D, Nazeem PA, Narayanankutty A, Manalil JJ, Raghavamenon AC. *In silico* and wet lab studies reveal the cholesterol lowering efficacy of lauric acid, a medium chain fat of coconut oil. Plant Foods for Human Nutrition. 2016;**71**(4):410-415. DOI: 10.1007/s11130-016-0577-y

[178] Alexander P, Berri A, Moran D, Reay D, Rounsevell MDA. The global environmental paw print of pet food. Global Environmental Change. 2020;**65**:102153. DOI: 10.1016/j.gloenvcha.2020.102153

[179] Okin GS. Environmental impacts of food consumption by dogs and cats. PLoS One. 2017;**12**(8):e0181301. DOI: 10.1371/journal.pone.0181301

[180] Swanson KS, Carter RA, Yount TP, Aretz J, Buff PR. Nutritional sustainability of pet foods. Advances in Nutrition. 2013;**4**(2):141-150. DOI: 10.3945/an.112.003335

[181] Henchion M, Hayes M, Mullen A, Fenelon M, Tiwari B. Future protein supply and demand: Strategies and factors influencing a sustainable equilibrium. Food. 2017;**6**(7):53. DOI: 10.3390/foods6070053

[182] Batal A, Dale N. Feedstuffs Ingredient Analysis Table: 2016 Edition. [Internet]. 2016. Available from: https://feedstuffs.farmcentric.com/mdfm/Feeess50/author/427/2015/11/Feedstuffs_RIBG_Ingredient_Analysis_Table_2016.pdf. [Accessed: March 15, 2021]

[183] Luna D, Carrasco C, Álvarez D, González C, Egaña JI, Figueroa J. Exploring anhedonia in kennelled dogs: Could coping styles affect hedonic preferences for sweet and umami flavours? Animals. 2020;**10**(11):2087. DOI: 10.3390/ani10112087

[184] Wang W, Zhou X, Liu Y. Characterization and evaluation of umami taste: A review. TrAC Trends in

Analytical Chemistry. 2020;**127**:115876. DOI: 10.1016/j.trac.2020.115876

[185] Smetana S, Ashtari Larki N, Pernutz C, Franke K, Bindrich U, Toepfl S, et al. Structure design of insect-based meat analogs with high-moisture extrusion. Journal of Food Engineering. 2018;**229**:83-85. DOI: 10.1016/j.jfoodeng.2017.06.035

[186] Kim T-K, Yong HI, Cha JY, Park S-Y, Jung S, Choi Y-S. Drying-induced restructured jerky analog developed using a combination of edible insect protein and textured vegetable protein. Food Chemistry. 2022;**373**:131519. DOI: 10.1016/j.foodchem.2021.131519

[187] Roncolini A, Milanović V, Aquilanti L, Cardinali F, Garofalo C, Sabbatini R, et al. Lesser mealworm (*Alphitobius diaperinus*) powder as a novel baking ingredient for manufacturing high-protein, mineral-dense snacks. Food Research International. 2020;**131**:109031. DOI: 10.1016/j.foodres.2020.109031

[188] Pasini G, Cullere M, Vegro M, Simonato B, Dalle ZA. Potentiality of protein fractions from the house cricket (*Acheta domesticus*) and yellow mealworm (*Tenebrio molitor*) for pasta formulation. LWT. 2022;**164**:113638. DOI: 10.1016/j.lwt.2022.113638

[189] Borremans A, Lenaerts S, Crauwels S, Lievens B, Van Campenhout L. Marination and fermentation of yellow mealworm larvae (*Tenebrio molitor*). Food Control. 2018;**92**:47-52. DOI: 10.1016/j.foodcont.2018.04.036

[190] Kewuyemi YO, Kesa H, Chinma CE, Adebo OA. Fermented edible insects for promoting food security in Africa. Insects. 2020;**11**(5):283. DOI: 10.3390/insects11050283

[191] Mintah BK, He R, Dabbour M, Xiang J, Agyekum AA, Ma H. Techno-functional attribute and antioxidative capacity of edible insect protein preparations and hydrolysates thereof: Effect of multiple mode sonochemical action. Ultrasonics Sonochemistry. 2019;**58**(4):104676. DOI: 10.1016/j.ultsonch.2019.104676

[192] Rossi S, Parrotta L, Del Duca S, Rosa MD, Patrignani F, Schluter O, et al. Effect of Yarrowia lipolytica RO25 cricket-based hydrolysates on sourdough quality parameters. LWT. 2021;**148**:111760. DOI: 10.1016/j.lwt.2021.111760

[193] Ray S, Raychaudhuri U, Chakraborty R. An overview of encapsulation of active compounds used in food products by drying technology. Food Bioscience. 2016;**13**:76-83. DOI: 10.1016/j.fbio.2015.12.009

[194] Siccama JW, Pegiou E, Zhang L, Mumm R, Hall RD, Boom RM, et al. Maltodextrin improves physical properties and volatile compound retention of spray-dried asparagus concentrate. LWT. 2021;**142**:111058. DOI: 10.1016/j.lwt.2021.111058

[195] Leylak C, Özdemir KS, Gurakan GC, Ogel ZB. Optimisation of spray drying parameters for Lactobacillus acidophilus encapsulation in whey and gum Arabic: Its application in yoghurt. International Dairy Journal. 2021;**112**:104865. DOI: 10.1016/j.idairyj.2020.104865

[196] Verma K, Tarafdar A, Kumar D, Kumar Y, Rana JS, Badgujar PC. Formulation and characterization of nano-curcumin fortified milk cream powder through microfluidization and spray drying. Food Research International. 2022;**160**:111705. DOI: 10.1016/j.foodres.2022.111705

[197] Obradović N, Volić M, Nedović V, Rakin M, Bugarski B. Microencapsulation of probiotic starter culture in protein–carbohydrate carriers using spray and freeze-drying processes: Implementation in whey-based beverages. Journal of Food Engineering. 2022;**321**:110948. DOI: 10.1016/j.jfoodeng.2022.110948

[198] FAO/WHO. Codex Alimentarius: Food Labelling. 5th ed. Rome: Food and Agriculture Organization of the United Nations; 2007

[199] Radosevich J, McGee N, Rawson NE. Functional ingredients in the pet food industry. In: Bagchi D, editor. Nutraceutical and Functional Food Regulations in the United States and around the World. 3rd ed. London: Academic Press; 2019. pp. 611-616. DOI: 10.1016/B978-0-12-816467-9.00038-1

[200] Schleicher M, Cash SB, Freeman LM. Determinants of pet food purchasing decisions. The Canadian Veterinary Journal. 2019;**60**(6):644-650

[201] Dima C, Dima S. Bioaccessibility study of calcium and vitamin D3 co-microencapsulated in water-in-oil-in-water double emulsions. Food Chemistry. 2020;**303**:125416. DOI: 10.1016/j.foodchem.2019.125416

[202] Boonlao N, Ruktanonchai UR, Anal AK. Enhancing bioaccessibility and bioavailability of carotenoids using emulsion-based delivery systems. Colloids Surfaces B Biointerfaces. 2022;**209**:112211. DOI: 10.1016/j.colsurfb.2021.112211

[203] Xu W, Lv K, Mu W, Zhou S, Yang Y. Encapsulation of α-tocopherol in whey protein isolate/chitosan particles using oil-in-water emulsion with optimal stability and bioaccessibility. LWT. 2021;**148**:111724. DOI: 10.1016/j.lwt.2021.111724

[204] Aditya NP, Aditya S, Yang H, Kim HW, Park SO, Ko S. Co-delivery of hydrophobic curcumin and hydrophilic catechin by a water-in-oil-in-water double emulsion. Food Chemistry. 2015;**173**:7-13. DOI: 10.1016/j.foodchem.2014.09.131

[205] Yu H, Wang H, Su W, Song Y, Zaky AA, Abd El-Aty AM, et al. Co-delivery of hydrophobic astaxanthin and hydrophilic phycocyanin by a pH-sensitive water-in-oil-in-water double emulsion-filled gellan gum hydrogel. Food Hydrocolloids. 2022;**131**:107810. DOI: 10.1016/j.foodhyd.2022.107810

[206] Aditya NP, Macedo AS, Doktorovova S, Souto EB, Kim S, Chang PS, et al. Development and evaluation of lipid nanocarriers for quercetin delivery: A comparative study of solid lipid nanoparticles (SLN), nanostructured lipid carriers (NLC), and lipid nanoemulsions (LNE). LWT. 2014;**59**(1):115-121. DOI: 10.1016/j.lwt.2014.04.058

[207] Muschiolik G, Dickinson E. Double emulsions relevant to food systems: Preparation, stability, and applications. Comprehensive Reviews in Food Science and Food Safety. 2017;**16**(3):532-555. DOI: 10.1111/1541-4337.12261

[208] Gravel A, Doyen A. The use of edible insect proteins in food: Challenges and issues related to their functional properties. Innovative Food Science and Emerging Technologies. 2020;**59**(3):102272. DOI: 10.1016/j.ifset.2019.102272

[209] Hall FG, Jones OG, O'Haire ME, Liceaga AM. Functional properties of tropical banded cricket (*Gryllodes sigillatus*) protein hydrolysates. Food Chemistry. 2017;**224**(11):414-422. DOI: 10.1016/j.foodchem.2016.11.138

[210] Ngo D-H, Kim S-K. Antioxidant effects of chitin, chitosan, and their

derivatives. Advances in Food and Nutrition Research. 2014;**73**:15-31. DOI: 10.1016/B978-0-12-800268-1.00002-0

[211] Mouithys-Mickalad A, Schmitt E, Dalim M, Franck T, Tome NM, van Spankeren M, et al. Black soldier fly (*Hermetia illucens*) larvae protein derivatives: Potential to promote animal health. Animals. 2020;**10**(6):941. DOI: 10.3390/ani10060941

[212] Liu Y, Wan S, Liu J, Zou Y, Liao S. Antioxidant activity and stability study of peptides from enzymatically hydrolyzed male silkmoth. Journal of Food Processing & Preservation. 2017;**41**(1):e13081. DOI: 10.1111/jfpp.13081

[213] Ma S, Li X, Sun Y, Mi R, Li Y, Wen Z, et al. Enzymatic hydrolysis of defatted *Antheraea pernyi* (Lepidoptera: Saturniidae) pupa protein by combined neutral protease yield peptides with antioxidant activity. Journal of Insect Science. 2021;**21**(2):276-279. DOI: 10.1093/jisesa/ieab013

[214] Dutta P, Dey T, Manna P, Kalita J. Antioxidant potential of *Vespa affinis* L., a traditional edible insect species of north East India. PLoS One. 2016;**11**(5):e0156107. DOI: 10.1371/journal.pone.0156107

[215] Ai H, Wang F, Xia Y, Chen X, Lei C. Antioxidant, antifungal and antiviral activities of chitosan from the larvae of housefly, *Musca domestica* L. Food Chemistry. 2012;**132**(1):493-498. DOI: 10.1016/j.foodchem.2011.11.033

[216] Di Mattia C, Battista N, Sacchetti G, Serafini M. Antioxidant activities *in vitro* of water and liposoluble extracts obtained by different species of edible insects and invertebrates. Frontiers in Nutrition. 2019;**6**:1-7. DOI: 10.3389/fnut.2019.00106/full

[217] Dechakhamphu A, Wongchum N. Investigation of the kinetic properties of Phyllanthus chamaepeuce Ridl. Extracts for the inhibition of pancreatic lipase activity. Journal of Herbal Medicine. 2022;**32**:100508. DOI: 10.1016/j.hermed.2021.100508

[218] Yoon S, Wong NAK, Chae M, Auh J-H. Comparative characterization of protein hydrolysates from three edible insects: Mealworm larvae, adult crickets, and silkworm pupae. Food. 2019;**8**(11):563. DOI: 10.3390/foods8110563

[219] Sousa P, Borges S, Pintado M. Enzymatic hydrolysis of insect *Alphitobius diaperinus* towards the development of bioactive peptide hydrolysates. Food & Function. 2020;**11**(4):3539-3548. DOI: 10.1039/D0FO00188K

[220] Hall F, Johnson PE, Liceaga A. Effect of enzymatic hydrolysis on bioactive properties and allergenicity of cricket (*Gryllodes sigillatus*) protein. Food Chemistry. 2018;**262**(3):39-47. DOI: 10.1016/j.foodchem.2018.04.058

[221] Hall F, Reddivari L, Liceaga AM. Identification and characterization of edible cricket peptides on hypertensive and glycemic *in vitro* inhibition and their anti-inflammatory activity on RAW 264.7 macrophage cells. Nutrients. 2020;12(11):3588. DOI: 10.3390/nu12113588

[222] Ahn MY, Kim BJ, Kim HJ, Jin JM, Yoon HJ, Hwang JS, et al. Anti-diabetic activity of field cricket glycosaminoglycan by ameliorating oxidative stress. BMC Complementary Medicine and Therapies. 2020;**20**(1):232. DOI: 10.1186/s12906-020-03027-x

[223] Zhao J-G, Wang H-Y, Wei Z-G, Zhang Y-Q. Therapeutic effects of

ethanolic extract from the green cocoon shell of silkworm *Bombyx mori* on type 2 diabetic mice and its hypoglycaemic mechanism. Toxicology Research. 2019;**8**(3):407-420. DOI: 10.1039/c8tx00294k

[224] Leem JY, Nishimura C, Kurata S, Shimada I, Kobayashi A, Natori S. Purification and characterization of N-β-Alanyl-5-S-glutathionyl-3,4-dihydroxyphenylalanine, a novel antibacterial substance of *Sarcophaga peregrina* (flesh fly). The Journal of Biological Chemistry. 1996;**271**(23):13573-13577. DOI: 10.1074/jbc.271.23.13573

[225] Leem JY, Jeong IJ, Park KT, Park HY. Isolation of p-hydroxycinnamaldehyde as an antibacterial substance from the saw fly *Acantholyda parki* S. FEBS Letters. 1999;**442**(1):53-56. DOI: 10.1016/S0014-5793(98)01614-7

[226] Azmiera N, Krasilnikova A, Sahudin S, Al-Talib H, Heo CC. Antimicrobial peptides isolated from insects and their potential applications. Journal of Asia-Pacific Entomology. 2022;**25**(2):101892. DOI: 10.1016/j.aspen.2022.101892

[227] Xia J, Ge C, Yao H. Antimicrobial peptides from black soldier fly (*Hermetia illucens*) as potential antimicrobial factors representing an alternative to antibiotics in livestock farming. Animals. 2021;**11**(7):1937. DOI: 10.3390/ani11071937

[228] Hadj Saadoun J, Sogari G, Bernini V, Camorali C, Rossi F, Neviani E, et al. A critical review of intrinsic and extrinsic antimicrobial properties of insects. Trends in Food Science and Technology. 2022;**122**:40-48. DOI: 10.1016/j.tifs.2022.02.018

[229] Sahoo A, Swain SS, Behera A, Sahoo G, Mahapatra PK, Panda SK. Antimicrobial peptides derived from insects offer a novel therapeutic option to combat biofilm: A review. Frontiers in Microbiology. 2021;**12**:661195. DOI: 10.3389/fmicb.2021.661195/full

[230] Van Moll L, De Smet J, Paas A, Tegtmeier D, Vilcinskas A, Cos P, et al. *In vitro* evaluation of antimicrobial peptides from the black soldier fly (Hermetia illucens) against a selection of human pathogens. Microbiology Spectrum. 2022;**10**(1):e0166421. DOI: 10.1128/spectrum.01664-21

Chapter 6

Flavoring and Coating Technologies for Processing Methods, Packaging Materials, and Preservation of Food

Ahmed El Ghorab, Hamdy Shaaban, Ibrahim H. Alsohaimi, Khaled El-Massry, Amr Farouk, Mohamed Abdelgawad and Shaima M.N. Moustafa

Abstract

The food sector addresses perhaps the main business with regard to degree, speculation, and variety. In a forever-evolving society, dietary requirements and inclinations are broadly factors. Alongside offering extraordinary mechanical help for inventive and valued items, the ongoing food industry ought to likewise cover the essential necessities of a consistently expanding populace. Active food packaging strategies have experienced a tremendous push forward in the last two decades. It is a great opportunity to decide which bioactive component will be more appropriate for each specific application once the microbiological hazards for each type of food item are recognized and the microbial targets are clearly differentiated. In order to improve Flavor delivery and preservation, the food industry and the science of Flavor are constantly creating new ingredients, processing techniques, and packaging materials. This improves the quality and acceptability of food by boosting Flavor stability. As most Flavors can be influenced by interactions with other food ingredients in addition to being volatile and chemically unstable to air, light, moisture, and high temperatures. The food sector will succeed in the long run if new technologies are quickly adopted and effectively used to meet both current and future consumer expectations.

Keywords: Flavors, preservation, processing of foods, Packaging

1. Introduction

A Flavor is a chemical substance or a combination of multiple chemical substances, that has an odor. They are abundant in a variety of products, including food, wine, spices, cosmetics, perfume, and essential oils. Their use in the food industry is not simple, and their effectiveness is frequently the result of the presence of numerous volatile ingredients with different chemical and physicochemical properties. In addition, different processing techniques for food can have different sensorial effects depending on the characteristics of each compound [1].

Interest in the stability of Flavors and bioactive ingredients in food items has grown. Foods' overall Flavor and nutritional value can be altered by the production and storage procedures, packaging materials, and types of ingredients utilized. Despite the fact that Flavor technology is well established, new goods should be incorporated into existing products and other approaches should be applied in order to enhance the performance of food Flavoring. Since they can be used to enhance the esthetic and quality of foods as well as boost their safety, edible coatings are one of the alternatives to conventional techniques [2].

One of the most crucial qualities of food items is Flavor, which is frequently changed during processing, necessitating the use of Flavoring compounds in the formulation [3]. Flavors are among the most expensive components used in food items, and even in little amounts, they have a significant impact on the food products' quality and price [2]. Since they can also be effective antioxidants, antimicrobials, and nutraceuticals, they are some of the chemicals that are added to food products most frequently, not only to improve their Flavor but also to promote food preservation [1, 4].

In order to improve Flavor delivery and preservation, the food industry and the field of Flavor science are constantly developing new ingredients, processing techniques, and packaging materials. This will improve the quality and acceptability of foods by increasing their Flavor stability and preservation. The majority of Flavors are chemically and physically unstable to air, light, moisture, and high temperatures; in addition, interactions with other food ingredients may have an impact [2, 5]. The development of fresher materials and the consolidation of bioactive mixtures in packaging films have responded to a change in the food industry during this time, and they have done so concurrently with the advancement of cutting-edge techniques for the identification of emerging and safe food-borne microbes. The most popular way of Flavor compound encapsulation is the employment of a variety of approaches. Different Flavor protection and application strategies have been employed in the industry (e.g., spray drying and extrusion). Coatings can be used to encapsulate Flavors and other functional ingredients to improve food qualities both before and after processing. Coatings are frequently used to protect food from environmental aggressions and extend shelf life. This "active packaging" approach was created to make the most of the coating system's edibility, high compound retention, and controlled release characteristics. Businesses who opt to recycle the materials, however, must also deal with the classification issue because the majority of packages are composed of a blend of materials with different qualities. Recycling is today a difficult and expensive task due to the recovery, selecting, cleaning, and reprocessing of resources [6]. Due to this, research into renewable raw materials has intensified recently in an effort to reduce pollution issues by using alternative biodegradable packaging [7]. As a result, biodegradable packaging has emerged to take the place of traditional materials that cannot be recycled. The product can be protected by such biodegradable materials, which are also reasonably easy to create, recycle, and decay [8]. The majority of biodegradable packaging often involves the use of ecologically friendly polymeric materials with the goal of preserving quality and prolonging the shelf life of minimally processed goods, like fruits and vegetables [9, 10]. With a focus on the key ingredients (such as biopolymers, additives, bioactive ingredients, and probiotic components), manufacturing processes (for edible film or coatings), and their use in specific products, this chapter's objective is to provide an overview of the state-of-the-art in edible film and coating used in various foods. This analysis also describes the essential conditions that biodegradable packaging must meet in order to be employed in specific applications for the preservation and improvement of different food products. Films and coatings that are edible are among these uses.

In this chapter, after describing the main aspects of Flavoring science, the utilization of edible coatings in foods and the application of Flavors using coating technology are detailed.

2. Purpose of food preservation

The primary goals of food preservation are to provide value-added goods, promote diet variety, and combat improper agricultural planning [11]. The primary ingredients for food are produced by several sectors of the agriculture industry. Inadequate management or bad planning in agricultural production can be remedied by avoiding erroneous locations, timings, and quantities of raw food components as well as by prolonging storage life using simple preservation procedures. Value-added food products can provide higher-quality foods with more nutritive, practical, and sensory qualities. The desire from consumers for more convenient and healthier food options has an impact on food preservation procedures. Eating should not be boring for customers; it should be entertaining. Many different dishes with a range of Flavors and tastes are enjoyed by people. Diet diversification is essential, especially in less developed countries, to reduce reliance on a single type of grain (i.e. rice or wheat). In addition, stockpiling, appropriation, and food protection are important factors in achieving food security. In terms of food preservation, it' is important to keep in mind that desired quality, desired rack life, and desired consumers should all be taken into account.

3. Methods for food preservation

There are currently many methods for food preservation available to the food industry. According to the method of action, the significant food protection procedures can be divided into three categories: (i) preventing recontamination both before and after processing; (ii) directly inactivating bacteria, yeasts, molds, or enzymes; and (iii) slowing down or inhibiting chemical deterioration and microbial growth [12]. A few of the methods or processes from the aforementioned domains are shown in **Figure 1**. It would often be quite difficult to make a clear distinction between inhibition and inactivation. Think about the freezing and drying methods of preservation. Although the main objective of freezing and drying is to restrict the growth of germs during storage, some bacteria are also eliminated. When items are frozen, between 10 and 60% of the viable microbial population seems to perish, and this percentage increases over time. The sections that follow provide a list of several food preservation methods [11] as shown in **Figure 1**. The part that follows the topic of food preservation talks about the flavor of foods, which is affected by the nature of foods and how to preserve them.

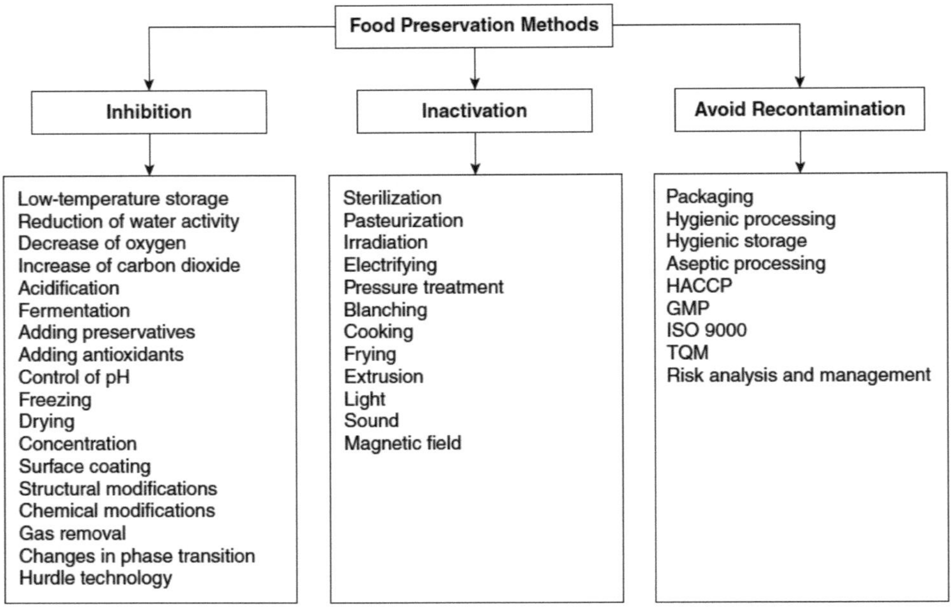

Figure 1.
Major food preservation techniques.

4. Flavoring of foods

Flavor is typically the result of the presence of volatile and nonvolatile components with a wide variety of physicochemical properties inside complicated matrices [6]. In reality, Flavors account for more than 25% of the global market for food additives, and the majority of Flavoring compounds are created through chemical synthesis or by extracting them from natural sources. The primary purpose of Flavors is to give food a certain taste or scent. The volatile chemicals affect both taste and scent, whereas the nonvolatile ones primarily affect taste [6, 7].

5. Flavors used in preservation and processing of foods

Flavor additives are used for specific purposes, such as improving the appeal of pharmaceuticals or nutraceuticals (such as vitamin C and multivitamin tablets that are coated with a sweet taste) [8], reducing the risk of foodborne illness, and phlegm contamination. Examples of these purposes include improving the appeal of low-flavor impact foods, giving a specific attribute to foods made from several component materials, restoring the integrity of Flavors that have been negatively impacted by processing conditions and more [9]. Flavors should work well with other components, withstand processing conditions while maintaining their qualities (such as high temperatures and pressures, irradiation, vacuum, and pH), and be stable after processing. Flavors must also adhere to all applicable safety regulations as well as other legal requirements. The consumer's acceptance is another crucial consideration because it has a significant impact on how the idea of employing Flavorings in food is perceived. A Flavor that meets most of the previous criteria will be more probably accepted by the consumer [8]. The most important requirements for Flavors used in foods are summarized in **Figure 2**.

Processing, packaging, storage, distribution, and retailing conditions will have an impact on how well additional Flavor characteristics are retained; as a result, it is crucial to choose the form and point of introduction of Flavorings in the process correctly.

To minimize exposure to harmful circumstances, Flavorings must be introduced near the conclusion of the processing activity. Flavorings should be included with the other ingredients wherever possible to prevent damage to them during processing. The majority of Flavorings are typically available in a variety of formats, including liquids [2], powders, and capsules [10, 11].

Food preservation can be achieved by promoting a longer shelf-life using techniques such as freezing, chilling, drying, curing, vacuum packing, modified atmosphere packaging, acidifying, fermenting, or adding chemical or natural preservatives (e.g., plant-derived antimicrobials) [12, 13]. In the last decade, the food industry has focused on procedures that deliver food providing a high level of microbial safety, good organoleptic quality, and nutritionally healthy, while minimizing the use of chemical preservatives [14, 15]. For example, spices and herbs, which are currently used as Flavoring and seasoning agents in foods, not only help preserving food due to their natural antimicrobial and antioxidant properties but also add Flavor [16].

Essential oils that are volatile have a lot of potential as food antimicrobials. These are primarily derived from herbs and spices and are in charge of giving food its Flavor [17–20]. Scientific research also indicates that these oils have potent antioxidant capabilities, which are advantageous for preventing free radicals' role in organoleptic deterioration. When added to food, these properties would delay microbial contamination and hence delay the start of rotting [16, 21, 22]. There are currently 3000 EOs documented, of which 300 are commercially available [23–25].

Despite some Eos having minimum inhibitory concentrations (MIC), they need to be two to ten times higher in food products than they were *in vivo* tests to have the same antibacterial action [25].

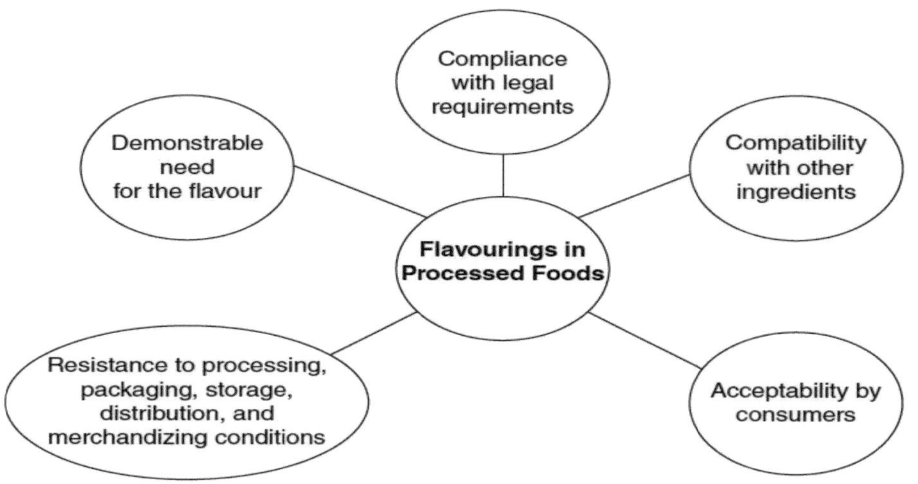

Figure 2.
Main requirements for Flavors used in foods.

In addition, Gram-positive bacteria appear to be more vulnerable to Eos than Gram-negative bacteria [26]. Despite all the EOs advantages, they also exhibit concerns associated with their use; for example, EOs exhibit an intense odor at unacceptable levels and inappropriate Flavors when used at effective doses [27]. Some of the studied Flavors that showed antibacterial and antifungal activities are presented in **Table 1**.

5.1 Application of EOs in the food industry: current trends

The creation of active packaging utilizing sustainable and eco-friendly materials is currently gaining popularity [28]. It has been demonstrated that cutting-edge active packaging can increase the shelf life of food items and slow down the growth of some germs. The creation of biocomposite systems based on synthetic or natural biopolymers can be used to attain these advantages. According to several sources, edible packaging has grown in favor among customers since edible films or coatings can be consumed along with food in addition to extending the shelf life of food products [29]. However, further research has shown that few films can actually be consumed (while others can just decay more quickly) and that customers will only accept edible films if they perceive them to be safe [30]. These packaging methods can include elements that are meant to be immobilized at the film, released into food, or able to absorb spoiled-food-causing chemicals [31]. Packaging shields food from dehydration and serves as a gas barrier from the environment, among other reasons. The active ingredients antimicrobials, antioxidants, texture enhancers, and essential nutrients, among others, can be transported using edible films. Additionally, these qualities can be improved by adding active ingredients to the films, such as EOs.

Plant from which EO is derived	Microorganisms	Activity
Callistemon lanceolatus	*Aspergillus flavus*	Antifungal
Onion	*Escherichia coli, Bacillus subtilis, Saccharomyces cerevisiae, Aspergillus niger*	Antibacterial, antifungal
Clove	*Escherichia coli, Bacillus subtilis, A. flavus, Mucor* sp.	Antibacterial, antifungal
Garlic	*B. subtilis, Staphyloccus aureus, E. coli A. flavus, Mucor* sp.	Antibacterial, antifungal
Cinnamon	*A. flavus*	Antifungal
Coriander, common myrrh, *Cananga odorata*	*A. flavus*	Antifungal, antibacterial
Mount Atlas mastic	*Salmonella typhimurium, E. coli, Staphylococus epidermidis, B. subtilis*	Antibacterial
Mint Thyme	*Micrococcus flavus, S. typhimurium*	Antibacterial
Zizyphus jujuba	*Staphylococcus aureus, B. subtilis, E, coli*	Antibacterial
Artemisia anomala S.	*E. coli, S. typhimurium, B. subtilis*	Antimicrobial
Ginger oil	*B. subtilis, Pseudomonas aeruginosa, Aspergillus niger*	Antifungal, antibacterial

Table 1.
Essential oils (EOs) and respective antimicrobial activity as potential food preservatives.

Food and beverage packaging systems (edible films) can contain EOs as an extra ingredient that is either applied directly to the edible films or encapsulated within the edible films. For instance, different amounts of EOs from *Origanum vulgare* and *Eugenia caryophyllata* were incorporated into cassava bagasse–polyvinyl alcohol-based trays. EOs were incorporated into the trays in two ways, directly by adding them into the mixture of ingredients at proportions 6.5, 8.5, and 10.0% (*w/w*) or by coating the surface with EOs at concentrations 2.5, 5.0, and 7.5 g oil/100 g tray. The greatest EOs concentrations produced the best results, particularly for *O. vulgare*, which completely inhibited molds, yeasts, and a small number of Gram-positive and negative bacteria, exhibiting entire suppression or a significant reduction in bacterial viability [32].

EOs from plants including jasmine, rosemary, peppermint, cinnamon, oregano, thyme, cumin, eucalyptus, rosewood, clove, tea tree, palmarosa, geranium, lavender, lemongrass, mandarin, bergamot, or lemon have been utilized in packaging systems rather frequently. Food items like fresh beef, butter, fresh octopus, ham, and fish can be found in the food matrices where packaging technologies with EOs have been used [3]. Regarding their major identified components, they belong to the hydrocarbon monoterpenes, such as _-pinene, _-pinene, _-selinene, and p-cymene, or the oxygenated monoterpenes group, such as thymol, carvacrol, geraniol, borneol, eugenol, linalool, terpineol-4-ol, 1,8-cineole, _-terpinyl acetate, and camphor [13–15, 33]. Another study analyzed the varying levels of certain significant hydrocarbons and oxygenated monoterpenes found in EOs. These substances had an antibacterial impact on *Listeria monocytogenes* and *E. coli* when used at a concentration of 0.2 g/mL

EOs	Film	Food	Main Results
Thyme EO	Curdlan-PVA	Chilled meat	Improvement of antioxidant activity and extension of the shelf life
Clove EO	Chitosan	Cooked pork sausages	Microbial growth inhibition, retarded lipid oxidation, and shelf-life extension when refrigerated storage
Oregano EO	Gelatin	Refrigerated Rainbow Trout Fillets	Lowering of total volatile basic nitrogen, peroxide value, thiobarbituric acid, and microbial growth
Rosemary EO	Whey protein isolate-cellulose nanofibers + TiO2 nanoparticles	Refrigerated lamb meat	Increased shelf life and antibacterial activity against *E. coli*, *Listeria monocytogenes*, *S. aureus*, and several Pseudomonas species as well as Enterobacteriaceae
Oregano EO + resveratrol nanoemulsion	Pectin	Fresh pork loin	Under high oxygen-modified environment packaging, the pH effect, color change, delayed lipid and protein oxidation, and prevention of microbial growth all occur
Cinnamon or lemongrass EO	Arabic gum-sodium caseinate	Guava fruit	Browning and related enzymes decrease, higher acceptability, antioxidant activity, and high content of phenolic compounds
Ginger EO nanoemulsion	Sodium caseinate	Chicken breast fillets	Growth inhibition of total aerobic psychrophilic bacteria

Table 2.
Recent illustrations of active films with EOs as the primary constituents demonstrate the benefits of the packaged food product.

and a pH of 4.0. The results of this study demonstrated that oxygenated monoterpenes had a stronger antibacterial impact than hydrocarbon ones. In actuality, these molecules' efficacy was tested in orange or apple juice together with heat treatments, and the results showed a synergistic fatal effect against *E. coli* [14].

Therefore, the objective of food packaging including biodegradable materials and EOs is to conduct antioxidant and antimicrobial assays to assess the final packaging system in contact with the food matrix and to produce reliable results. **Table 2** lists some of the most recent tests conducted on films containing EOs for novel food packaging solutions.

6. Edible coatings for food applications

6.1 Characteristics of edible films and coatings

Any substance with a thickness of less than 0.3 mm [34] that is created from a blend of biopolymers and other additives is considered to be an edible film or coating, in watery media, scattered [10, 35, 36]. Although some authors use the terms "edible film" and "coating" interchangeably, others contend that there is a distinction to be made because of how they are absorbed into the food product [37]. The edible coating is created directly on the food, as opposed to the edible film, which is manufactured in advance and then attaches to the product [37, 38]. However, in both cases, rigid matrices with similar properties are produced [39].

The primary traits edible films and coatings can exhibit:

i. protection against UV light [34];

ii. transport of solutes (e.g., salts, additives, and pigments), water vapor, organic vapors (e.g., aromas and solvents), and gases (e.g., oxygen, carbon dioxide, nitrogen, and ethylene) between food and the atmosphere [39];

iii. barrier against mechanical damage (e.g., dents or cuts) [37];

iv. increase the shelf-life of the product;

v. bioactive components (e.g., antioxidants) [40, 41];

vi. antimicrobial effect against bacterial reproduction and fungal contamination (e.g., silver nanoparticles) [42, 43];

vii. healthy microorganisms (e.g., probiotics) that confer benefits to the consumer; and

viii. biodegradable natural materials [10].

6.2 The components of edible films and coatings

Edible films and coatings are frequently assessed for their mechanical properties, such as their elasticity and rigidity, as well as the force required to break them, using terms like elongation at break (E), tensile strength (TS), and elasticity modulus (EM) [44, 45]. These terms also refer to similar characteristics, such as their elasticity and rigidity. Permeation, adsorption, and diffusion, which are associated with the flow of solutes between food and the atmosphere, are other mass transfer mechanisms they share [44, 46]. However, both mechanical properties and mass transfer phenomena are influenced by the type of material and manufacturing method that enables the formation of different biopolymeric matrix topologies [41–45]. The top biopolymers and additives used to make edible films and coatings are listed in **Table 3**, along with details on their properties and packing potential.

Starch is recognized as the universal biopolymer for bio-packaging and has been used extensively for decades [47] because of its characteristics and gelatinization abilities. Alginate is an important biopolymer that can also be used to make hydrogels and encapsulation barriers [48, 49]. However, chitosan has lately gained interest for the creation of edible films and coatings due to its capabilities as a gelling agent, chemical (it may establish hydrogen bonds and hydrophobic interactions), and biological (its biocompatibility, biodegradability, and bioactivity) [50–52]. While some writers have opted to use organic packaging materials such as proteins (such as collagen and protein isolates) [53, 54], lipids, and carbohydrates (such as canola oil and cinnamon bark oil) [55, 56], as well as other unusual materials (such as smooth-hound protein and papaya puree) [57, 58], to create bio-packaging with specific properties.

Materials	Examples	Properties	Function in Edible Films and Coatings
Biopolymers			
Polysaccharides	Starch Cellulose Pectin Gums Chitosan Agar Alginate Dextran	Thickeners Gellants Emulsifiers Stabilizers Coating	They form the base structure of a solid polymer matrix.
Proteins	Gelatin Casein Whey protein	Gellants Thickeners Stabilizers Foaming	They help in the transport of antimicrobials and antioxidants. They control the transport of gases (mainly oxygen).
Lipids	Waxes Paraffin Glycerides	Protectors Coatings	They help to avoid drying or dehydration of the edible film providing flexibility.
Additives			
Plasticizers	Glycerol Aloe	Viscosity Resistance	They decrease the in termolecular force and the melting temperature in the mixture.
Chaotropic agents	Resins Urea	Flexibility Destructuring agent	They also modify the viscosity and the rheological properties. They increase the solubility of polymers in water
Others	Polyphenols	Antioxidants Stabilizers Fungicides Herbicides Fertilizers	They work as stabilizers as well as protection for the products.

Table 3.
Main materials used and functionality in the manufacture of edible films and coatings.

However, the purpose of additives (such as plasticizers or stabilizers) in the formulation of edible films and coatings is to alter the mechanical characteristics (preferably increasing E and decreasing TS and EM) as well as mass transfer phenomena [37]. Additionally, adding anti-oxidant, fungicidal, or anti-microbial chemicals enables the creation of bioactive biopackaging [50, 59].

7. Food flavoring by coating

Edible coatings act as a physical barrier to prevent the widespread migration of contaminants from the environment into foods and vice versa. These barrier qualities are crucial for ensuring the safety of food. Consumers desire greater food safety, as well as improved nutritional and Flavor qualities. In recent years, active packaging has been developed to extend food shelf-life by improving and increasing coating properties and functionalities. For instance, adding active ingredients, such as Flavors, might give edible coatings more activity. To protect their activity and qualities, Flavors can either be encapsulated or added directly to the edible polymer matrix.

8. Influence of flavor incorporation on edible coating properties

Flavoring incorporation can influence the mechanical and barrier properties of edible coatings and films due to physical changes generated in the network structure. Edible coatings can incorporate a variety of active agents, such as Flavorings, to enhance food quality and safety. Changes that are induced depend on elements including molecular size, polarity, shape, Flavor affinity, and the polymer molecules of coatings and films [60].

9. Transfer of flavors and aromas in active food packaging

However, in the case of active packaging applied with EOs, they are not used for boosting food quality. Flavors and aromas are added to food to improve its Flavor and odor. In fact, when employed for food preservation, the potent odor of EOs may wind up imparting an undesirable Flavor to the packed food. Additionally, they have easily evaporated volatile components. As a result, the nanoencapsulation of Eos represents a practical method to hide their Flavor and aroma and stop them from evaporating. The encapsulation improves Flavor qualities and lessens the effect of EOs on the food's organoleptic features. Additionally, it increases the solubility of EOs in water-soluble polymers, offers efficient releasing properties, and promotes dispersion [61]. As a result, materials that are nanoencapsulated must have little affinity for EOs and little impact on the organoleptic properties of the packaged food. For instance, grain goods were packaged using eugenol that had been applied on cellulose. It enhanced the package's pesticide properties while maintaining the original product's organoleptic qualities [62–65]. Additionally, cinnamon essential oil (EO) was added to a polyvinyl alcohol nanofiber coating to disguise its potent Flavor while supplying antibacterial activity against Gram-positive and Gram-negative bacteria. This method works well for prolonging the shelf life of foods like strawberries that spoil quickly [66–71]. Different materials can be employed as well, depending on the characteristics of the chemicals or Eos that would be nanoencapsulated. Because of their biocompatibility, low toxicity, and biodegradability, biopolymers and biocomposites have recently received attention as viable building materials [72–76].

10. Flavor retention and release

It is essential that an edible coating be able to "catch" or hold on to Flavorings during the drying process. If Flavors are lost while being dried, the resulting Flavor will weaken and possibly lose its character balance. The possibility that a Flavor ingredient not preserved in the powder will be lost to the processing environment is a secondary worry about Flavor retention [77–80]. Modified food starches retain Flavor the best upon drying among the common edible coatings used in Flavor encapsulation. Excellent emulsifiers are modified food starches, and the quality of an emulsion greatly affects the Flavor preservation during spray drying [81]. Additionally, compared to 30–35% solids for acacia gum, modified food starches can often be employed at levels of 50–55% solids. Therefore, modified food starches and acacia gum should both result in good emulsions (and thus, good Flavor retention); however, modified food starches have the best Flavor retention since they may be utilized at a higher solid level, which also enhances Flavor retention [82]. High solid quantities of maltodextrins, corn syrup solids, or simple sugars (and their alcohols) are usually permissible, however, Flavor retention after drying is low due to these ingredients' poor emulsification abilities. The primary determining factors for Flavor preservation during drying are a suitable coating material's ability to emulsify and a high solid concentration at which they can be used.

11. Applications

To better safeguard their activity and qualities, Flavoring compounds can be encapsulated or added directly to edible coating matrices. Citric, malic, and tartaric acids are a few examples that are currently in use [83], along with many essential oils like oregano, thyme, cinnamon, lemongrass, and clove that can either enhance or hide the original Flavors of dishes. For instance, [84] assessed the sensory quality of coated fresh-cut 'Fuji' apples with edible coatings made of apple puree-alginate and Flavored with lemongrass, oregano, and vanillin. According to taste tests, coated fresh-cut apples with 0.3%vanillin inclusion were the most promising in terms of sensory quality. Carvacrol, a key ingredient in the essential oils of oregano and thyme, has reportedly attracted a lot of attention from researchers lately, despite the fact that the usage of all of these substances in food has been widely documented. Carvacrol is a Flavoring compound found in chewing gum, ice cream, baked products, and sweets [85]. Additionally, Laohakunjit and Kerdchoecuen [86] added sorbitol-rice starch coatings to milled rice that included 25%natural pandan leaf extract (*Pandanus amaryllifolius* Roxb.), enabling the creation of rice that was Flavored with jasmine after cooking.

Several items that use coating Flavoring technology are already available on the market. An example is a roasted peanut with a covering of curry Flavor that instantly dissolves in the tongue and imparts the taste of the Indian spice. A similar example intended for kids is a multi-layered sweet with varied tastes and Flavors in each layer, each layer being separated by Arabic gum or another hydrocolloid layer to stop the movement of scent components from one layer to the next. Volatile substances should have a very low diffusivity for this use and a high affinity for the coating, which should be very soluble in the mouth.

According to relevant research results, depending on the types of bioactive solutes added to edible films and coatings, the maturation of fruits and vegetables as well as the development of mold and microorganisms can be postponed, preserving certain qualities like texture, freshness, vitamin C content, and nutritional quality as well as bestowing new biological activities (e.g., antioxidant activity). In the case of dairy and animal goods, edible films and coatings allowed for the preservation of both the product's bioactive constituents and its sensory attributes. They improved these items' antioxidant, antifungal, and antibacterial properties as well as their shelf life.

The safety of the use of food Flavorings, both natural and synthetic, remains however a controversial topic and will likely elicit debate, motivate scientific studies, and entertain legislative actions in the near future.

12. Conclusion

The use of Flavoring agents has been documented in numerous studies and in a variety of food products; in addition to serving as Flavor enhancers, they can also be used as natural food preservatives because of their antibacterial and antioxidant properties, which guard food against known causes of food-borne illnesses and food spoilage. However, studies on their safety issues must be undertaken in order to further prove the safety of food Flavors. Food processing techniques that use coatings for Flavoring and preservation should also be investigated because they have been demonstrated to have no impact on food's nutritional value and offer an intriguing way to enhance functioning without altering the product's characteristics.

This chapter compiles and analyses the most recent studies on the use of edible films and coatings in various meals. Several types of materials have been employed in the manufacture of packaging for the preservation and improvement of food goods, with a focus on the bio-polymeric materials that have been used to produce innovative barriers to directly protect the product. Certain additives must also be added to improve the final packing's mechanical and physical characteristics. Bioactive ingredients and microorganisms (like probiotics) are now routinely incorporated into sustainable packaging to improve the functionality and nutrition of perishable and natural foods. Additionally, the main application techniques that distinguish edible films from edible coatings were shown. Along with the formation materials, these techniques also have an impact on pathogen inhibition, product quality, shelf life, maturation, and maturation effect.

In conclusion, bio-packaging has shown to be successful in preserving foods that have received minimal processing, and its use may save money by preventing food from naturally spoiling and extending the shelf life of the product. Depending on the biomaterials used and the types of biologically active compounds, certain characteristics of coated products, such as sensory, physicochemical, and nutritional aspects, can be improved. However, there are still a lot of biopolymers and additives (like zein) that have not been completely studied but have the potential to make edible films and coatings, which could provide encouraging insights into the safeguarding and preservation of food products.

More studies must be done to determine how Flavor retention and release function in edible coating matrices and whether nanotechnology and the use of nanostructured materials could result in better attributes than macro- and microstructures.

Funding sources

This research did not receive any specific grant from funding agencies in the public, commercial, or not-for-profit sectors.

Conflict of interest

The authors declare that there is no conflict of interest.

Author details

Ahmed El Ghorab[1*], Hamdy Shaaban[2], Ibrahim H. Alsohaimi[1], Khaled El-Massry[2], Amr Farouk[2], Mohamed Abdelgawad[3] and Shaima M.N. Moustafa[4]

1 Department of Chemistry, College of Science, Jouf University, Sakaka, Saudi Arabia

2 Department of Chemistry of Flavor and Aroma, National Research Center, Giza, Egypt

3 Department of Pharmaceutical Chemistry, College of Pharmacy, Jouf University, Sakaka, Saudi Arabi

4 Department of Biology, College of Science, Jouf University, Sakaka, Saudi Arabia

*Address all correspondence to: aghorab@ju.edu.sa

IntechOpen

© 2023 The Author(s). Licensee IntechOpen. This chapter is distributed under the terms of the Creative Commons Attribution License (http://creativecommons.org/licenses/by/3.0), which permits unrestricted use, distribution, and reproduction in any medium, provided the original work is properly cited.

References

[1] Rijk R. Chemical migration from active and intelligent packaging materials into food. In: Barnes KA, Sinclair CR, Watson DH, editors. Chemical Migration and Food Contact Materials. Cambridge, England: Woodhead Publishing Limited; 2007. pp. 371-391

[2] Madene A, Jacquot M, Scher J, Desobry S. Flavor encapsulation and controlled release – a review. International Journal of Food Science and Technology. 2006;**41**:1-21

[3] Grandison, A.S. (2006) Postharvest handling and preparation of foods for processing. In: Food Processing Handbook (Brennan, J.G., e.d). Wiley-VCH Verlag GmbH & Co., KGaA, Weinheim, Germany, pp. 1-30

[4] Pokorny J, Trojakova L. The use of natural antioxidants in food products of plant origin. In: Pokorny J, Yanishlieva N, Gordon M, editors. Antioxidants in Food: Practical Applications. CRC Press: Woodhead Publishing Ltd; 2001. pp. 355-372

[5] Gharsallaoui A, Roudaut G, Chambin O, Voilley A, Saurel R. Applications of spray-drying in microencapsulation of food ingredients: an overview. Food Research International. 2007;**40**:1107-1121

[6] European Parliament and Council. Directive 2008/98/EC of the European Parliament and of the Council of 19 November 2008 on Waste and Repealing Certain Directives (Waste Framework). Brussels, Belgium: European Parliament and Council; 2008. pp. 3-30

[7] Kehili M, Choura S, Zammel A, Allouche N, Sayadi S. Oxidative stability of refined olive and sunflower oils supplemented with lycopene-rich oleoresin from tomato peels industrial by-product, during accelerated shelf-life storage. Food Chemistry. 2018;**246**: 295-304 [Cross Ref] [PubMed]

[8] Putnik P, Bursa'c Kova˘cevi'c D, Režek Jambrak A, Barba FJ, Cravotto G, Binello A, et al. Innovative "Green" and Novel Strategies for the Extraction of Bioactive Added Value Compounds from Citrus Wastes—A Review. Molecules. 2017;**22**:680 [CrossRef] [PubMed]

[9] Settanni L, Palazzolo E, Guarrasi V, Aleo A, Mammina C, Moschetti G, et al. Inhibition of foodborne pathogen bacteria by essential oils extracted from citrus fruits cultivated in Sicily. Food Control. 2012;**26**:326-330 [CrossRef]

[10] Burt S. Essential oils: Their antibacterial properties and potential applications in foods—A review. International Journal of Food Microbiology. 2004;**94**:223-253 [CrossRef]

[11] Hyldgaard M, Mygind T, Meyer R. Essential Oils in Food Preservation: Mode of Action, Synergies, and Interactions with Food Matrix Components. Frontiers in Microbiology. 2012;**3**:12 [CrossRef] [PubMed]

[12] Falleh H, Ben Jemaa M, Saada M, Ksouri R. Essential oils: A promising eco-friendly food preservative. Food Chemistry. 2020;**330**:127268 [CrossRef]

[13] Rota MC, Herrera A, Martínez RM, Sotomayor JA, Jordán MJ. Antimicrobial activity and chemical composition of *Thymus vulgaris*, *Thymus zygis* and *Thymus hyemalis* essential oils. Food Control. 2008;**19**:681-687 [CrossRef]

[14] Ait-Ouazzou A, Cherrat L, Espina L, Lorán S, Rota C, Pagán R.

The antimicrobial activity of hydrophobic essential oil constituents acting alone or in combined processes of food preservation. Innovative Food Science and Emerging Technologies. 2011;**12**: 320-329 [CrossRef]

[15] Mohamed SAA, El-Sakhawy M, El-Sakhawy MAM. Polysaccharides, Protein and Lipid-Based Natural Edible Films in Food Packaging: A Review. Carbohydrate Polymers. 2020;**238**:116178 [CrossRef]

[16] Hassan B, Chatha SAS, Hussain AI, Zia KM, Akhtar N. Recent advances on polysaccharides, lipids and protein based edible films and coatings: A review. International Journal of Biological Macromolecules. 2018;**109**:1095-1107 [CrossRef] [PubMed]

[17] Kouhi M, Prabhakaran MP, Ramakrishna S. Edible polymers: An insight into its application in food, biomedicine and cosmetics. Trends in Food Science and Technology. 2020;**103**:248-263 [CrossRef]

[18] Longo MA, Sanroman MA. Vegetable lavors from cell culture. In: Hui YH, editor. Handbook of Fruit and Vegetable Flavors. Hoboken, New Jersey, Canada: John Wiley & Sons, Inc.; 2010. pp. 663-680

[19] Abegaz EG, Tandon KS, Scott JW, Baldwin EA, Shewfelt RL. Partitioning taste from aromatic lavour notes of fresh tomato (*Lycopersicon esculentum*, Mill) to develop predictive models as a function of volatile and monovolatile components. Postharvest Biology and Technology. 2004;**34**:227-235

[20] Omobuwajo TO. Food lavourings: principles of applications. In: Hui YH, editor. Handbook of Food Products Manufacturing. Hoboken, New Jersey, USA: JohnWiley & Sons, Inc.; 2007. pp. 117-130

[21] Turgis M, Vu KD, Dupont C, Lacroix M. Combined antimicrobial effect of essential oils and bacteriocins against foodborne pathogens and food spoilage bacteria. Food Research International. 2012;**48**:696-702

[22] Bonvehi JS. Investigation of aromatic compounds in roasted cocoa powder. European Food Research and Technology. 2005;**221**:19-29

[23] de Roos KB. Effect of texture and microstructure on lavour retention and release. International Dairy Journal. 2003;**13**:593-605

[24] Neetoo, H. and Chen, H. Application of high hydrostatic pressure technology for processing and preservation of foods. In: Bhat R, Alias AK, Paliyath G, editors. Progress in Food Preservation. United States: Wiley-Blackwell; 2012. pp. 247-276

[25] Gould GW. Preservation: past, present and future. British Medical Bulletin. 2000;**56**:84-96

[26] Cleveland J, Montville TJ, Nes IF, Chikindas ML. Bacteriocins: safe, natural antimicrobials for food preservation. International Journal of Food Microbiology. 2001;**71**:1-20

[27] Dorman HJD, Deans SG. Antimicrobial agents from plants: antibacterial activity of plant volatile oils. Journal of Applied Microbiology. 2000;**88**:308-316

[28] Gupta V, Nair J. Application of botanicals as natural preservatives in food. In: Bhat R, Alias AK, Paliyath G, editors. Progress in Food Preservation. United States: Wiley-Blackwell; 2012. pp. 513-530

[29] Regnier T, Combrinck S, Du Plooy W. Essential oils and other plant

extracts as food preservatives. In: Bhat R, Alias AK, Paliyath G, editors. Progress in Food Preservation. United States: Wiley-Blackwell; 2012. pp. 539-579

[30] Juneja VK, Dwivedi HP, Yan X. Novel natural food antimicrobials. Annual Review of Food Science and Technology. 2012;**3**:381-403

[31] Hamad SH. Factors affecting the growth of microorganisms in food. In: Bhat R, Alias AK, Paliyath G, editors. Progress in Food Preservation. United States: Wiley-Blackwell; 2012. pp. 405-427

[32] Albarracin W, Alfonso C, Sanchez IC. Application of essential oils as a preservative to improve the shelf life of Nile tilapia (*Oreochoromisniloticus*). Vitae, Revista da Faculdade de Química Farmacêutica. 2012;**19**:34-40

[33] Yanishlieva NV, Marinova E, Pokorný J. Natural antioxidants from herbs and spices. European Journal of Lipid Science and Technology. 2006;**108**:776-793

[34] Cutter CN. Antimicrobial effect of herb extracts against *Escherichia coli* O157:H7, *Listeria monocytogenes*, and *Salmonella typhimurium* associated with beef. Journal of Food Protection. 2000;**63**:601-607

[35] Hyldgaard M, Mygind T, Meyer RL. Essential oils in food preservation: mode of action, synergies, and interactions with food matrix components. Frontiers in Microbiology. 2012;**3**:1-24

[36] Bakkali F, Averbeck S, Averbeck D, Idaomar M. Biological effects of essential oils – a review. Food and Chemical Toxicology. 2008;**46**:446-475

[37] Inouye S, Takizawa T, Yamaguchi H. Antibacterial activity of essential oils and their major constituents against respiratory tract pathogens by gaseous contact. Journal of Antimicrobial Chemotherapy. 2001;**47**:565-573

[38] Belletti N, Kamdem SS, Tabanelli G, Lanciotti R, Gardini F. Modeling of combined effects of citral, linalool and -pinene used against *Saccharomyces cerevisiae* in citrus-based beverages subjected to a mild heat treatment. International Journal of Food Microbiology. 2010;**136**:283-289

[39] Coelho PM, Corona B, ten Klooster R, Worrell E. Sustainability of reusable packaging—Current situation and trends. Resources, Conservation and Recycling X. 2020;**6**:100037 [CrossRef]

[40] Jugreet BS, Suroowan S, Rengasamy RRK, Mahomoodally MF. Chemistry, bioactivities, mode of action and industrial applications of essential oils. Trends in Food Science and Technology. 2020;**101**:89-105 [CrossRef]

[41] Jeya Jeevahan J, Chandrasekaran M, Venkatesan SP, Sriram V, Britto Joseph G, Mageshwaran G, et al. Scaling up difficulties and commercial aspects of edible films for food packaging: A review. Trends in Food Science and Technology. 2020;**100**:210-222 [CrossRef]

[42] European Commission. Regulation (EC) No 450/2009 of 29 May 2009 on active and intelligent materials and articles intended to come into contact with food. Official Journal of European Communication. 2009;**135**:3-11

[43] Debiagi F, Kobayashi RKT, Nakazato G, Panagio LA, Mali S. Biodegradable active packaging based on cassava bagasse, polyvinyl alcohol and essential oils. Industrial Crops and Products. 2014;**52**:664-670 [CrossRef]

[44] Wen P, Zhu D-H, Wu H, Zong M-H, Jing Y-R, Han S-Y. Encapsulation of

cinnamon essential oil in electrospun nanofibrous film for active food packaging. Food Control. 2016;**59**:366-376 [CrossRef]

[45] Embuscado ME, Huber KC. Edible Films and Coatings for Food Applications. New York, NY, USA: Springer Science+Business Media; 2009 ISBN 9780387928234

[46] Montalvo C, López Malo A, Palou E. Películas comestibles de proteína: Características, propiedades y aplicaciones. Temas selectos de ingeniería de alimentos. 2012;**6**:32-46

[47] Castro-Muñoz R, González-Valdez J. New trends in biopolymer-based membranes for pervaporation. Molecules. 2019;**24**:3584 [CrossRef] [PubMed]

[48] Morales-Jiménez M, Gouveia L, Yañez-Fernandez J, Castro-Muñoz J, Barragan-Huerta BE. Microalgae-based biopolymer as a potential bioactive film. Coatings. 2020;**10**:120 [CrossRef]

[49] Guimarães A, Abrunhosa L, Pastrana LM, Cerqueira MA. Edible films and coatings as carriers of living microorganisms: A new strategy towards biopreservation and healthier foods. Comprehensive Reviews in Food Science and Food Safety. 2018;**17**:594-614 [CrossRef]

[50] Yai H. Edible films and coatings: Characteristics and properties. International Food Research Journal. 2008;**15**:237-248

[51] Guerrero Legarreta I, Rosmini MR, Armenta RE. Tecnología de Productos de Origen Acuático. Mexico City, Mexico: LIMUSA; 2009 ISBN 9786070500862

[52] Falguera V, Quintero JP, Jiménez A, Muñoz JA, Ibarz A. Edible films and coatings: Structures, active functions and trends in their use. Trends in Food Science and Technology. 2011;**22**:292-303 [CrossRef]

[53] Salvia-Trujillo L, Soliva-Fortuny R, Rojas-Graü MA, McClements DJ, Martín-Belloso O. Edible nanoemulsions as carriers of active ingredients: A review. Annual Review of Food Science and Technology. 2017;**8**:439-466 [CrossRef]

[54] Sánchez Aldana D, Contreras-Esquivel JC, Nevárez-Moorillón GV, Aguilar CN. Caracterización de películas comestibles a base de extractos pécticos y aceite esencial de limón Mexicano. CyTA Journal of Food. 2015;**13**:17-25 [CrossRef]

[55] Kra'sniewska K, Galus S, Gniewosz M. Biopolymers-based materials containing silver nanoparticles as active packaging for food applications—A review. International Journal of Molecular Sciences. 2020;**21**:698 [CrossRef] [PubMed]

[56] Zhang Y, Han J, Liu Z. Starch-based edible films. In: Environmentally Compatible Food Packaging. Sawston, UK: Woodhead Publishing; 2008. pp. 108-136

[57] Castro-Muñoz R, Buera-González J, de la Iglesia Ó, Galiano F, Fíla V, Malankowska M, et al. Towards the dehydration of ethanol using pervaporation cross-linked poly(vinyl alcohol)/graphene oxide membranes. Journal of Membrane Science. 2019;**582**:423-434 [CrossRef]

[58] Castro-Muñoz R, Galiano F, Fíla V, Drioli E, Figoli A. Matrimid®5218 dense membrane for the separation of azeotropic MeOH-MTBE mixtures by pervaporation. Separation and Purification Technology. 2018;**199**:27-36 [CrossRef]

[59] Kim YT, Min B, Kim KW. General characteristics of packaging materials for food system. In: Han JH, editor. Innovations in Food Packaging. San Diego, CA, USA: Elsevier Ltd.; 2014. pp. 13-35 ISBN 9780123946010

[60] Salama HE, Abdel Aziz MS, Sabaa MW. Novel biodegradable and antibacterial edible films based on alginate and chitosan biguanidine hydrochloride. International Journal of Biological Macromolecules. 2018;**116**:443-450 [CrossRef]

[61] Khazaei N, Esmaiili M, Djomeh ZE, Ghasemlou M, Jouki M. Characterization of new biodegradable edible film made from basil seed (Ocimum basilicum L.) gum. Carbohydrate Polymers. 2014;**102**: 199-206 [CrossRef] [PubMed]

[62] Martins JT, Cerqueira MA, Bourbon AI, Pinheiro AC, Souza BWS, Vicente AA. Synergistic effects between _-carrageenan and locust bean gum on physicochemical properties of edible films made thereof. Food Hydrocolloids. 2012;**29**:280-289 [CrossRef]

[63] Pérez-Guzmán CJ, Castro-Muñoz R. A review of zein as a potential biopolymer for tissue engineering and nanotechnological applications. PRO. 2020;**8**:1376 [CrossRef]

[64] Veiga-Santos P, Oliveira LM, Cereda MP, Alves AJ, Scamparini ARP. Mechanical properties, hydrophilicity and water activity of starch-gum films: Effect of additives and deacetylated xanthan gum. Food Hydrocolloids. 2005;**19**:341-349 [CrossRef]

[65] Xiao Q, Lim LT, Tong Q. Properties of pullulan-based blend films as affected by alginate content and relative humidity. Carbohydrate Polymers. 2012;**87**:227-234 [CrossRef]

[66] Lee KY, Mooney DJ. Alginate: Properties and biomedical applications. Progress in Polymer Science. 2012;**37**:106-126 [CrossRef] [PubMed]

[67] Liu J, Liu S, Chen Y, Zhang L, Kan J, Jin C. Physical, mechanical and antioxidant properties of chitosan films grafted with different hydroxybenzoic acids. Food Hydrocolloids. 2017;**71**:176-186 [CrossRef]

[68] Bautista-Baños S, Romanazzi G, Jiménez-Aparicio A. Chitosan in the Preservation of Agritultural Commodities. Oxford, UK: Elsevier Inc.; 2016 ISBN 9780128027356

[69] Castro-Muñoz R, Gonzalez-Valdez J, Ahmad Z. High-performance pervaporation chitosan-based membranes: New insights and perspectives. Reviews in Chemical Engineering. 2020;**8**:1-16 [CrossRef]

[70] Ahmad M, Nirmal NP, Danish M, Chuprom J, Jafarzedeh S. Characterisation of composite films fabricated from collagen/chitosan and collagen/soy protein isolate for food packaging applications. RSC Advances. 2016;**6**:82191-82204 [CrossRef]

[71] Slimane EB, Sadok S. Collagen from cartilaginous fish by-products for a potential application in bioactive film composite. Marine Drugs. 2018;**16**:211 [CrossRef]

[72] Moncayo-Martínez DC, Buitrago-Hurtado G, Néstor Y, Algecira-Enciso A. Películas comestibles a base de un biopolímero tipo dextrana Edible films based of dextran biopolymer. Agronomia Colombiana. 2016;**34**:107-109

[73] Beak S, Kim H, Song K. Bin sea squirt shell protein and polylactic acid laminated films containing cinnamon bark essential oil. Journal of Food Science. 2018;**83**:1896-1903 [CrossRef]

[74] Abdelhedi O, Nasri R, Jridi M, Kchaou H, Nasreddine B, Karbowiak T, et al. Composite bioactive films based on smooth-hound viscera proteins and gelatin: Physicochemical characterization and antioxidant properties. Food Hydrocolloids. 2017;**74**:176-186 [CrossRef]

[75] Rangel-Marrón M, Mani-López E, Palou E, López-Malo A. Effects of alginate-glycerol-citric acid concentrations on selected physical, mechanical, and barrier properties of papaya puree-based edible films and coatings, as evaluated by response surface methodology. Lwt. 2019;**101**:83-91 [CrossRef]

[76] Ghani S, Barzegar H, Noshad M, Hojjati M. The preparation, characterization and in vitro application evaluation of soluble soybean polysaccharide films incorporated with cinnamon essential oil nanoemulsions. International Journal of Biological Macromolecules. 2018;**112**:197-202 [CrossRef] [PubMed]

[77] Han J. Protein-based edible ilms and coatings carrying antimicrobial agents. In: Gennadios A, editor. Protein-Based Films and Coatings. Boca Raton, Florida, USA: CRC; 2002. pp. 485-498

[78] Alizadeh Sani M, Ehsani A, Hashemi M. Whey protein isolate/ cellulose nanofibre/TiO2 nanoparticle/ rosemary essential oil nanocomposite film: Its effect on microbial and sensory quality of lamb meat and growth of common foodborne pathogenic bacteria during refrigeration. International Journal of Food Microbiology. 2017;**251**:8-14 [CrossRef]

[79] Reineccius GA. Edible ilms and coatings for lavor encapsulation. In: Embuscado ME, Huber KC, editors. Edible Films and Coatings for Food Applications. New York, USA: Springer Science Business Media; 2009. pp. 269-294

[80] Baranauskiene R, Bylaite E, Zukauskaite J, Venskutonis RP. Flavor retention of peppermint (*Mentha piperita* L.) essential oil spray-dried in modiied starches during encapsulation and storage. Journal of Agricultural and Food Chemistry. 2007;**55**:3027-3036

[81] Reineccius GA. Flavor encapsulation. Food Reviews International. 1989;**5**(2):147-176

[82] Eswaranandam S, Hettiarachchy NS, Johnson MG. Antimicrobial activity of citric, lactic, malic, or tartaric acids and nisin-incorporated soy protein ilm against *Listeria monocytogenes, Escherichia coli* O157:H7, and *Salmonella gaminara*. Journal of Food Science. 2006a;**69**:79-84

[83] Eswaranandam S, Hettiarachchy NS, Meullenet JF. Effect of malic and lactic acid incorporated soy protein coatings on the sensory attributes of whole apple and fresh-cut cantaloupe. Journal of Food Science. 2006b;**71**:307-313

[84] Rojas-Grau MA, Avena-Bustillos RJ, Olsen C, Friedman M, Henika PR, Martin-Belloso O, et al. Effects of plant essential oils and oil compounds on mechanical, barrier and antimicrobial properties of alginate–apple puree edible ilms. Journal of Food Engineering. 2007a;**81**:634-641

[85] Fenaroli G, editor. Fenaroli's Handbook of Flavor Ingredients. 6th ed. Boca Raton, Florida, USA: CRC Press; 2009

[86] Laohakunjit N, Kerdchoecuen O. Aroma enrichment and the change during storage of non-aromatic milled rice coated with extracted natural lavor. Food Chemistry. 2007;**101**:339-344

Chapter 7

Drying Technology Evolution and Global Concerns Related to Food Security and Sustainability

Ayman Ibrahim, Tiziana M.P. Cattaneo, Alia Amer and Lajos Helyes

Abstract

Undoubtedly, rapid population growth has sharply increased global food demand. Although the green revolution, accompanied by food industrialization practices, helped a lot in meeting this demand, the food gap is still huge. Regardless of COVID-19, due to that 14% of the world's food is lost before even reaching the market, and thus the food insecurity prevalence by rate (9.7%), where the food losses are valued at $400 billion annually according to FAO. In the face of such issues related to food insecurity and food losses, drying technology since its inception has been known as the most common operation in food processing and preservation. However, the excessive use of the drying process and exposure to heat for long periods led to a severe deterioration in the physicochemical quality characteristics of these products. At the same time, growing attention on human health through monitoring the quality and safety of food to avoid chronic diseases led to increasing awareness of the consumer to obtaining products with high nutritional value. Therefore, there has been a great and rapid evolution in drying technology to preserve food with high quality. Hence, this chapter aims to shed light on the drying technology evolution in food processing and preservation as one of the most important post-harvest treatments in the agriculture field.

Keywords: drying technology, drying methods, food processing, quality, food security

1. Introduction

Nobody denies that the world is inclining toward alternative sources of energy: a world where human habitats are safe, resilient, and sustainable and where there is universal access to affordable, clean, reliable, and sustainable energy. One of the most important applications of energy is food preservation using drying technology. Preserving food by drying is the oldest method used at home and by food transformation industry. Dried food became an important result, allowing humans to be independent from food supply in any environmental conditions [1]. Food drying techniques have long been applied since ancient times in conventional ways, such as drying in the sun.

From the traditional to the most innovative, there are several drying solutions. Depending on the drying method used, the processing time, the product's final quality, or the efficiency of the process can vary considerably. Currently, drying methods have been developed with the latest technology to reduce the damages caused by biochemical changes, which decrease nutritional value during the drying process. Conserving energy and achieving the best quality of dried products have become the most important factors that determine the usefulness and success of operating any drying unit, where heat is transmitted through the drying unit in three forms conduction, convection, and radiation. In this regard, when the temperature of the fresh material increases, the molecular motion gains more energy; as a result, it causes changes in the structure of the material as well as its chemical properties to increase the shelf life and improve the quality [2]. As indicated by [3], literature data reported more than 400 different types of plants. Among them, only 100 different types are applied in practice, design, heat input, operating temperature and pressure, and quality specifications of the final dried product are the main variables considered for building dryers. However, such technologies are expensive, high-energy demanding, and greenhouse gases producing. In the past, food products were dried, exploiting sun energy. In several countries, sun drying is still used for domestic food transformation. At the industry level, the spoilage reduction, in order to improve the stability of dried products during their storage, is an important input to stimulate a great and rapid evolution in drying technology [4]. A lot of published literature in the areas of drying of food products is already available, but often each paper has been focalized on selected types of drying plants. Furthermore, this chapter will shed light more on the most common industrial drying techniques applied in the food processing process, independently from the basic principles on which they are based to build them. In this way, the reader could choose the methodology most adequate to his objective in an easy and fast manner.

2. Solar drying methods

In the past, the first method used to dry food was sun. Solar energy directly hit the food, left exposed to the air or placed on the ground. It should be considered that this type of drying is influenced by external contamination, the presence of insects and other small animals, the available area, the poor standardization of the process, and the potential development of bad smells [5]. Therefore, this methodology is a most economical solution. Fruits and vegetables that grow and are cultivated in remote areas lend themselves well to conservation by solar drying, while losing some of their peculiarities and nutritional value. But in the past, this method required large space and time and offered poor process control. This natural method takes place when materials are dried with unheated forced air, taking advantage of its natural drying potential. A natural evolution of sun drying has been realized by planning the use of solar dryers, as an efficient system for solar energy use [6]. Countries that enjoy large amount of natural sun exposure have, as the obvious option for drying, the solar dryers based on natural convection [7]. In this way, is possible to save costs and energy. However, uncontrollable climate changes are the main negatively influenced factors [8]. On the basis of climate and environment conditions or if a better quality is needed, the use of mechanical air drying can be the best solution. In any case, sun drying is the cheapest method to dry food. Nowadays, both sun drying and mechanical air driers are available on the market.

More than 250,000,000 tons of horticultural products and cereals are dried every year using natural drying [8]. Especially in developing countries, the products are placed on the ground and turned periodically until completely dry. The newly developed solar drying allows the use of renewable energy sources, minimizing the defects of traditional techniques.. The drying process influences the amount and the organization of water molecules in the food cells producing changes in the product matrix [9]. Modern solar drying techniques try to combine aspects of the natural process with industrial needs, applying the concept of drying for maintenance of the characteristics of the raw materials and the environment in which they are grown (developing countries). The devices designed to effectively carry out this type of drying use mainly the energy of the sun for their functioning, shielding their deleterious effects and creating a hygienically suitable treatment environment, in tune with the microclimatic conditions of the environment. In African countries, for example, during the last 10 years, a great effort has been made to improve the horticultural drying process, with the introduction of more efficient solar drying systems [10–14].

The management of the production cycle is influenced both by intrinsic factors linked to the nature of the material to be transformed and by external factors, therefore determined by the context in which the production takes place. The final quality of the product strongly depends on the choices of technological parameters regarding processing and packaging. These choices can favor the conservation of compounds with high nutritional value present in raw materials that can still be found in final products, improving their nutritional quality. Important, especially in this pandemic era [15–17]. Solar drying has been proven to be adequate to be applied to several crops [18]. It allows the production of products with desirable quality together with minimal environmental impact. Since the process is slow and weather-dependent, so a wise strategy "fan off-fan on" must be devised considering the following conditions: air temperature, relative humidity, moisture content, and temperature of the material being dried. Solar dryers are systems capable of exploiting solar radiant energy to heat a flow of air used to dry products. There are types with natural ventilation (which exploit the "chimney" effect) or with forced ventilation, which can be powered by a photovoltaic system or other electrical source [6]. Recently, a very comprehensive review of several types of natural convective and direct-type (NCDT) solar dryers has been brought [19] with the aim of collecting the most interesting practices of solar drying technology for the benefit of one and all. Authors have classified in a precise way the different types of dryers. They explain the main characteristics of the "direct types" (i.e., (i) direct solar cabinet dryer, (ii) modified solar cabinet dryer with natural convection, (iii) direct solar drying using the chimney, (iv) foldable solar crop dryer; and the "natural convection type," then particular attention has been made at modeling the process and the evaluation of the performances. Furthermore, regarding indirect solar drying technologies, the most exhaustive review has been reported in Ref. [20] and has been written some years ago. Efforts have been made all over the world around to increase efficiency and drying time; thus, modern dryers are equipped with fans, efficient collectors, different thermal storage materials, reflectors, auxiliary heat sources, and sun-tracking systems. A recent paper [21] has reported an interesting development of closed, indirect heating, forced convection solar dehydrator with desiccant, which is economical, hygienic, and works off the grid. The whole experimental design has been discussed, paying attention to (i) energy balance, (ii) quantity of air required, (iii) flow of air required, (iv) area of the solar heater/collector, and (v) rates of dehydration.

3. Industrial drying techniques

Generally, drying involves the coexistence of complex processes, which may be operated simultaneously. Initially, the energy is transferred from hot drying agent to fresh product. Then, the unbounded moisture evaporation (free water) occurs, and eventually, water particles bound inside the cellular structure. These particles exposed to diffusion and thus migration, and transferred to the product surface, where the water is lastly evaporated [22]. The attributes of dried products can be effectively influenced by the parameters of process, such as pressure, temperature, gas feed rate, and relative humidity. In addition to protein formulations characteristics, such as composition and excipients type, and solutes concentration, viscosity as well as solvent type [23]. A large assortment of dryers has been developed to dehydrate and preserve these products to meet different quality and cost requirements. As indicated by [3], over 400 dryer kinds have been described in the reviews, although only about 100 types are commonly use. Differences in dryer design are due to the product physical characteristics, modes of heat input, temperatures and pressure operating, dried product qualifications, etc. Conventionally, food stuffs are dried by open sun drying system. Though this system is still common in many places for non-commercial use, there have been several efforts to improve advanced drying systems for products on a commercial scale. It is obligatory to enhance the drying systems to reduce waste [4]. Regarding this concern, this section will shed light more on the most common industrial drying techniques applied in the food processing process.

3.1 Rotary drying

Rotary drying is also known as tumbling dryer. It is considered the oldest continuous and the most common high-volume dryer used in industry. It has developed more adaptations of the technology than any other dryer classification. Rotary dryers stand out for their flexibility and applicability to a large number of materials, as well as their high processing capacity [24]. Large amounts of granular substantial with particles of 10 mm or larger that are not too fragile or heat-sensitive or cause any other solids handling problems are dried in rotary dryers [25]. Conventional rotary dryers have flights, which lift solids and make them cascade across the dryer section (active phase) when there is effective fluid-particle contact [26]. It is one of the many drying methods existing in unit operations of chemical engineering. The drying takes place in rotary dryers, which consist of a cylindrical casing usually made of steel plates rotating on bearings, lightly inclined horizontally. Usually, its length is 5–90 m, while its diameter is 0.3–5 m, and rotates at 1–5 [25]. It is regularly worked at a negative interior pressure to stop dust get-away. Wet feedstuff is presented into the system upper edge and the feed proceeds through it by force of rotation, head influences and shell inclines. Then the dried stuff is withdrawn at lower edge. The direct heat rotary dryer diagram is illustrated in **Figure 1**. Gas flow direction throughout the cylinder in the case of the solids is dictated by the processed substance characteristics. Current flow is utilized for heat-susceptible substances even with high entrance gas temperature because of the gas's fast cooling through initial surface moisture evaporation. While for other substances, the countercurrent flow is desired for higher thermal efficiency. In the first status, the solid flow rate is boosted by gas flow, whereas it retarded in the second status [27–30].

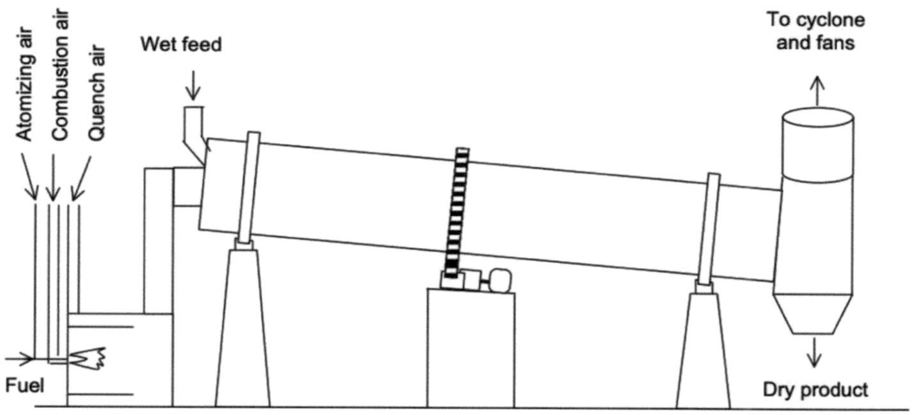

Figure 1.
Diagram of direct-heat rotary dryer.

3.1.1 Types of rotary dryers

A. Direct rotary dryer; it comprises a nude metal cylinder attached with or without flights. It is appropriate for low or medium-temperature processes, which is limited by the of metal characteristics strength;

B. Direct rotary kiln; it comprises a metal cylinder padded in the inner with isolate bulk or refractory brick, to be appropriate for process under maximum temperatures;

C. Indirect steam tube dryer; it comprises a bare metal cylindrical shell together with one or more rows of metal tubes composite longitudinally within it. It is appropriate for the process up to the obtainable steam temperature or in processes that required cooling for the water tubes;

D. Indirect rotary calciner; it comprises a nude metal cylinder encompassed by an electrically or fired heated furnace. It is suitable for procedure at temperatures up to the highest, which could be tolerated by the cylinder metal that being generally from 800 to1025 K for stainless steel, when from 650 to700 K for carbon steel; and

E. Direct Roto-Louvre dryer; it is the most important type, as the product proceeds in a crosscurrent movement to the gas. It is appropriate for medium and low - temperature processes.

However, the period that molecules spend in the active step is low, consequently affecting the process energy efficiency. The prospect of the rotary and crossflow for the wood chips drying was acheived by Cairo et al. [31] when they reported that the best execution was pointed out by the crossflow dryer.

3.1.2 Combined conduction-convection heating rotary dryer

For drying the high-moisture paddy in the field conditions, a combined conduction and convection heating rotary dryer for 0.5 t/h capacity was designed and developed

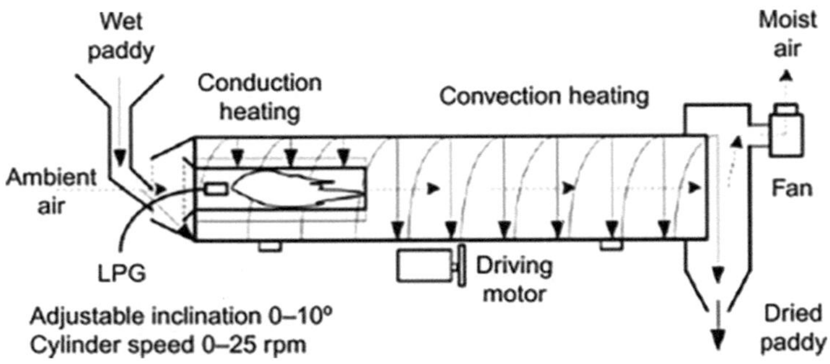

Figure 2.
A schematic drawing of combined conduction and convection-type rotary dryer.

by Likitrattanaporn et al. [32] by using liquefied petroleum gas (LPG) as the heat source. Moreover, a trial rotary dryer proposed with a system of concurrent flow comprising two main parts, a discharge cover, and a double cylinder is illustrated in **Figure 2**. The forward motion of the paddy happens by rotary motion of the cylinder and inclination angle. The air is inflated within the cylinder by a suction fan positioned on the discharge cover top. A 1-hp motor with a 1–60 reduction gear was utilized for driving the rotary dryer. LPG bulb on the entry edge heats the air. The heated air stirs to another end by the suction fan. Through the forward movement, the paddy contacts the external surface for the interior cylinder. Where, the conduction heating occurs followed by a cascading activity over the interior of the outer cylinder, which causes a convection heating. After this, paddy is placed into the emptying cover and got out of the dryer. When the suction fan absorbs the humid air, comparatively less humidity is taken away through the last or third pass at 100°C and 110°C, meaning 1.5% and 1.7%, respectively. At 120°C, the humidity content of 2.1% could be extracted. Clearly, this is because there was less free water available at the third pass of drying [33].

3.2 Fluidized-bed dryers

Fluidized-bed dryers (FBDs) are used extensively for the drying of wet particulate and granular materials with sizes ranging from 50 μm to 5 mm that can be fluidized, and even slurries, pastes, and suspensions that can be fluidized in beds of inert solids [34]. Here, the product is held aloft in a high-velocity hot air stream, thus promoting good mixing and heat transfer for uniform and rapid drying (**Figure 3**). FBDs are commonly used in processing many products such as chemicals, carbohydrates, foodstuff, biomaterials, beverage products, ceramics, medicines in agglomerated or powder form, healthcare stuffs, fertilizers, agrochemicals and pesticides, detergents, pigments, and surface-active agents, tannins, polymer and resins, as well as products for combustion, calcination, incineration, waste management, and environmental protection processes. Fluidized-bed operation offers imperative benefits such as good materials mixing, high amounts of heat and mass transfer, as well as easy material moving [34].

The air passages across a cribriform tray from below the foodstuff in addition to check it. The fragments are consumed in one end and from above. This support pushing along pieces formerly in a dryer, which they leave at the other termination. This is fortified by the lower-humidity fragments having lesser density and mass. This

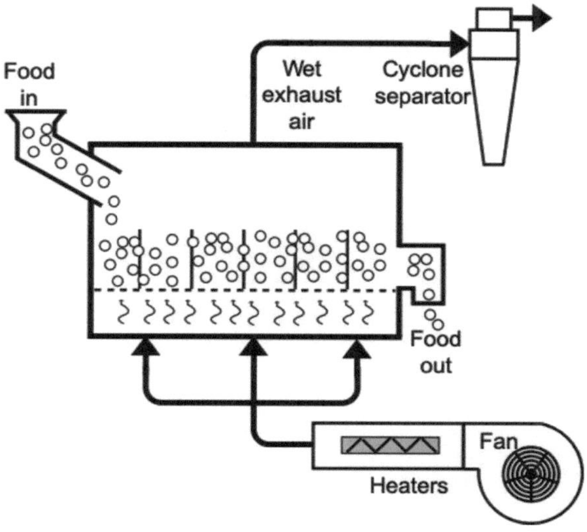

Figure 3.
Fluidized-bed dryer.

system has a suitable thermal efficiency and reduces individual fragments over warming [35].

The traditional fluidized bed is made by passaging a gas stream from bottom of bed for special materials. With little gas speeds, the bed of units is packed, which rests on a gas distributor plate. The passages of fluidizing gas is done over wholesaler, which regularly spread crossways bed [35]. The pressure droplet through the expansion of bed as the velocity of fluidizing gas is improved. At a particular gas speed, a fluidized bed is formed when the whole bed weight is completely supported by gas stream. This state is identified as minimum fluidization,and the corresponding gas velocity is named minimum fluidization velocity. Pressure drop through the bed remains near the same as pressure drop at minimum fluidization even if the gas velocity is increased further. **Figure 4** demonstrates numerous particulate bed regimes from filled to bubbly bed when the gas velocity is boosted. The graphs express the bed pressure drops and bed void age under many regimes [35].

By superficial gas velocities greater than the lowest fluidization velocity, typically from 2 to 4 umf, the fluidized bed is controlled. The minimum fluidization velocity could be assessed experimentally and using numerous correlations [36]. It should be noted that these correlations have limitations such as particle size, column dimensions, and operating parameters. Thus, they are valid in a certain range of criteria and operating conditions.

3.2.1 Fluidized-bed dryer's benefits and limitations

Commonly fluidized-bed drying advantages included a removal of moisture high rate, easy material transport inside the dryer, high thermal efficiency, control easing, as well as low cost of maintenance. The fluidized-bed dryer's limitations comprise high electrical power consumption, high pressure drop, poor quality of some particulate stuffs, unsimilarity product quality for certain categories of such dryers, pipes and vessels erosion, fine particles entrainment, and particle pulverization or attrition, as well as fine particle agglomeration, etc. For detailed discussion, see [37].

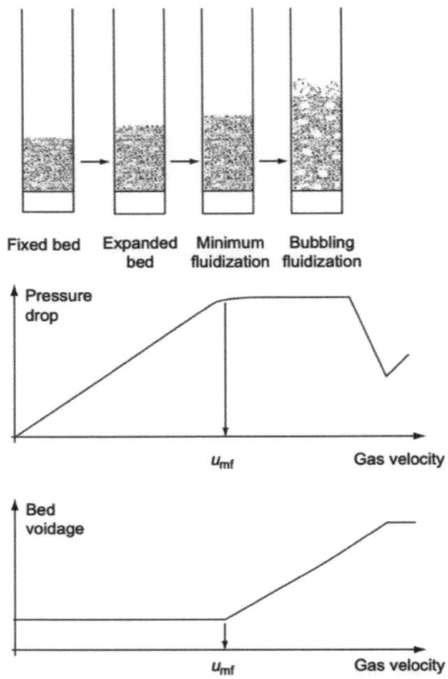

Figure 4.
Various regimes of a bed of particles at different gas velocities.

Similar to drying, the fluidized bed has located a varied range of industrial purposes in many industries for dedusting, granulation, mixing, coating, agglomeration, incineration, chemical reactions, gasification, combustion as well as cooling, etc. Many procedures could be combined to fluidized bed drying in the same processor for achieving more than two procedures in one unit. Practices that could be favorably integrated to fluidized-bed drying are explained briefly in the following. The mixing influence in a fluidized bed is commonly suitable for particles size within 50 and 2000 mm. In the case of fine particles, which are less than 50 mm, or for the case of particles that are hard to fluidize when moistened, vibration is generally applied to enhance the mixing effect and fluidization characteristic [35–37]. For large particles, the supplement of internals or the usage of the spouting mode can assist to develop the operation. Good particle mixing is essential for fluidized-bed drying. So, the awareness of particle characteristics and properties is required to ensure good fluidized-bed dryer performance. Moreover, the particles bed can be fluidized by an energetic flow or by bed fluidizing sections regularly such as the entire bed is fluidized in order once over a cycle. Evidently, this process results in saving drying air and therefore electrical power, but it leads to an elongated operating time due to the intermittent heat input mode. In addition, intermittent fluidization can decrease the mechanical damage problem to particles because of continual vigorous particles clash and corrosion-induced dusting [35–37].

Spray drying, coating, agglomeration, and granulation share similar basic operative standard. The fine spray of the solution paste slurry suspension is sprayed in a fluidized bed of inert particles or the drying substance oneself, which is already overloaded in the room drying. Solid particle development and formation take place in the room as evaporation and drying enthuse moisture [35]. In granulation, the growth of solid particles is acheived by succeeding wetting and liquid feed coating onto the solid particles and coated layer solidification by hot drying air. In the coating, a layer of the priceless active agent could be coated on a less expensive substrate or adding a surface agent on solid particles, which is necessary for downstream processing. Through spraying a proper binder onto the solid particles bed, granulated or agglomerated solid big-size particles are formed [35–37]. During most situations, sole spray drying is not power efficient to remove the whole wet content in solids. It is because a significant amount of heat and time are necessary for the elimination of internal water, which is stuck in the solids internally. The drying system of the fluidized bed could be combined as the second phase of drying for removing interior moisture. This could be followed by the third phase of fluidized bed cooling for preventing condensation troubles during the packaging in various applications [35–37].

3.3 Drum dryers

The drum dryer is normally used to dry slurries, pastes, concentrated solutions, or viscous on rotating steam-heated drums [38, 39]. This is because of the moisture boiling off and flashing or of irreversible thermochemical transformations of their content, which take place on their first interaction with the hot drum surface [40–42]. The viscous slurry or paste is automatically spread by spreading action of both counter rotating drums into a fine sheet, which follows to warmer drum within single-drum dryers or splitting sheets on both hot cylinders in double-drum dryers. The adhering fine paste sheet is then speedily dried conductively by excessive heat flux of steam condensation in the drum. In the case of humid slurries, which generate wet slabs, the wet thin slab drying could be improved by blowing heated dry air on the sheet outward. The fine slab has heat-sensible supplies, which could be dried also at a minor temperature in the vacuum [43].

It can be used in the irreversible thermochemical transformations during the slurry's first contact with the hot drum. This is to impart the certain required qualities together of the dried product [44]. In the case of starch slurries, to produce pregelatinized starch, the starch can be gelatinized before the sheet is dried. When the thin sheet is exposed to the high heat flux and high temperature for a short period, the pored structure is imparted to the dried slab, because of the speedy formation of vapor bubbles in the slab during drying. The pored goods are premium in immediate food constructions because they are more readily wetted and can be simply rehydrated. For this, a drum dryer is broadly utilized over the universe in pregelatinized starch fabrication for immediate food construction.

3.3.1 Types of drum dryers

The drum dryer was first patented for use in the manufacture of pregelatinized starch in Germany [45]. Number and configuration of drums, heating systems, and product removal have been considered in many experiments. The drum diameter varies from 0.45 to 1.5 m, and its length varies from 1 to 3 m. The drum wall thickness is between 2 and 4 cm. The drum dryer is classified according to the steam-heated drums number and configuration as well as the atmosphere pressure around the drying sheet.

3.3.1.1 Atmospheric double-drum dryer

This dryer type has a higher rate of production, it can handle a wide range of goods as well as it is more efficient [38–40, 46]. Across the pendulum nozzle or a header, there are several nozzles, the paste or slurry is nourished onto the tweak of two

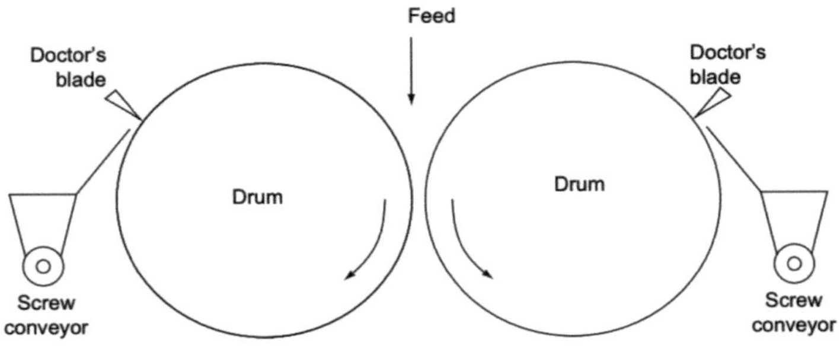

Figure 5.
Double-drum dryer with nip feed.

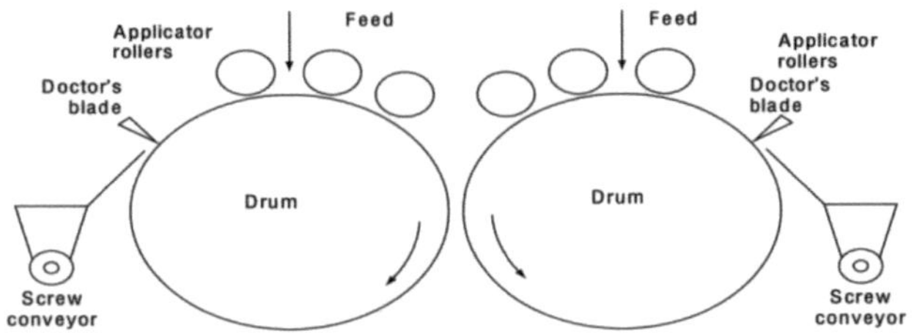

Figure 6.
Twin-drum dryer with applicator roller feeds.

steam-heated drums counter-rotating toward each other, making a boiling pool at the tweak as illustrated in **Figure 5**. The feedstuff could be nourished in the drum tweaks and applicator cylinder as shown in **Figure 6**. Starch slurries turn into gelatine in boiling pool to form pastes, which be extra sticky. The counter-rotation of the drums spread the slurry or paste into two soft slabs on both drums that subsequently dry conductively.

3.3.1.2 Atmospheric single-drum dryers

The paste or slurry is nourished throughout a header that has various nozzles or across a pendulum nozzle onto the tweak of a steam-heated drum and highly cooler applicator cylinder counter-rotating toward each other, then making a boiling pool at the tweak (**Figure 7**) [38–40, 46]. Starch slurries turn into gelatine in boiling pool, making pastes extra sticky. The drum counter rotation and applicator cylinder diffuse the paste/slurry into a soft slab on the warm drum, which subsequently dries

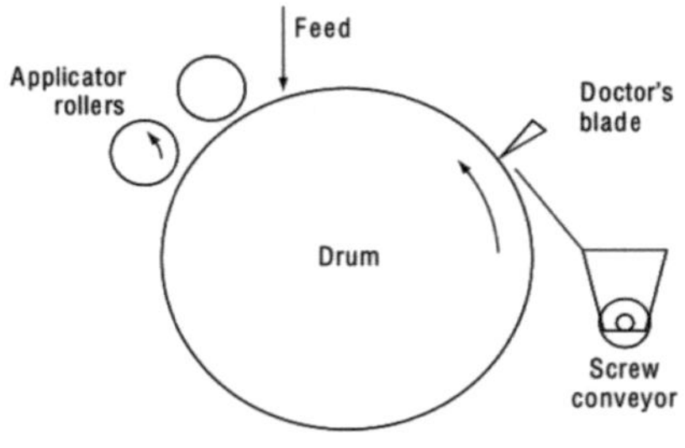

Figure 7.
Single-drum dryer with applicator roller feed.

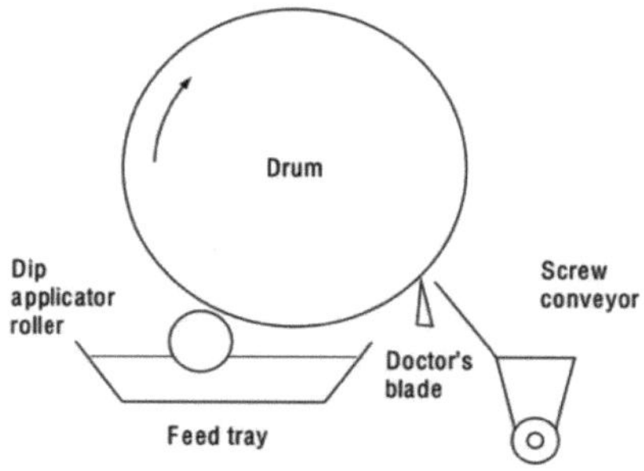

Figure 8.
Single-drum dryer with dip roller feed.

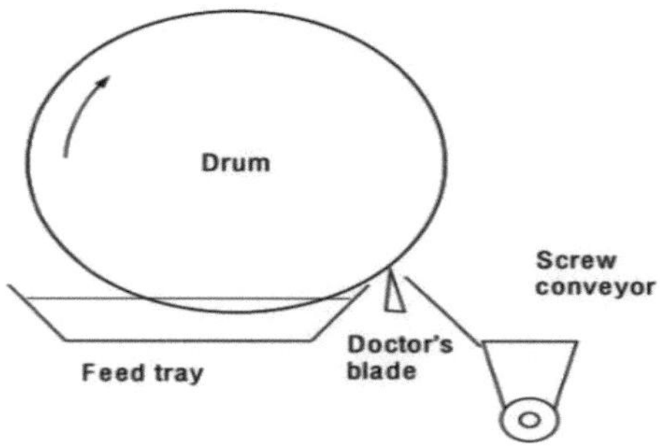

Figure 9.
Single-drum dryer with dip feed.

conductively. Otherwise, the slurry could be nourished by dip coating a dip or applicator cylinder in a feed plate at the dryer bottommost, then cylinder coated on the drum as illustrated in **Figure 8**. Also, the slurry could be nourished by dip coating the drum directly in the feed tray (see **Figure 9**) or scattered or marked from a feed plate as illustrated in **Figure 10**.

3.3.1.3 Atmospheric twin-drum dryers

The applied of slurry is by direct dip coating of the twin drums in the feed tray at the dryer bottom (**Figure 11**), or by splash, or spray feeders from a feed tank at the

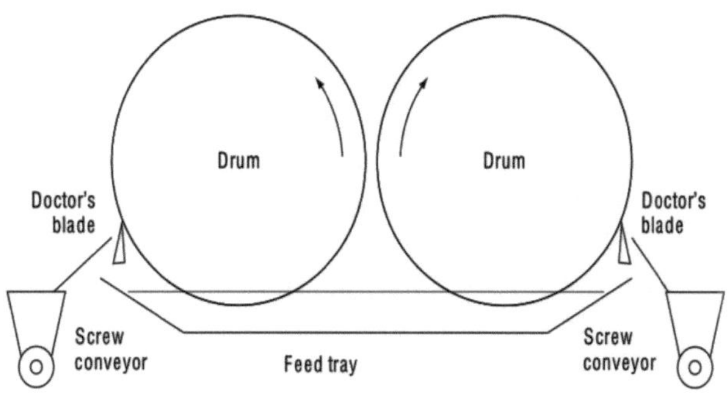

Figure 10.
Twin-drum dryer with dip feed.

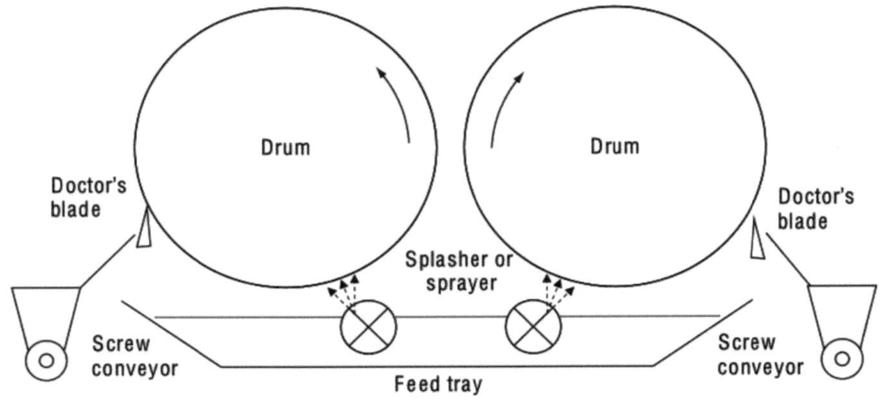

Figure 11.
Twin-drum dryer with splasher or sprayer feed.

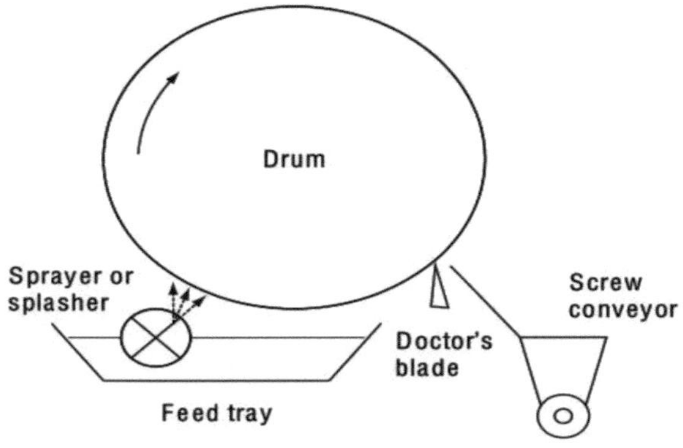

Figure 12.
Single-drum dryer with splasher or sprayer feed.

dryer bottommost (**Figure 12**) onto the top of the both steam-heated drums, which are counter-rotating faraway from each other [38–40, 46]. Sheets are made by cohesion onto the drum top and are held up versus gravity by their surface tension, then dry conductively. This kind of dryer is appropriate for mixtures that produce dusty stuffs.

3.3.1.4 Enclosed drum dryers

If the solvent steam other than humid discharged in the course of the drying by the drum requires to be improved, or if dried foodstuffs produce much powder, the atmospheric twin-drum dryers could be bounded in vapor or in dust-tight enclosure [38–40, 46]. The steam vapor could be retrieved by an appropriate condenser, and the dust could be eliminated using a wet scrubber.

3.3.1.5 Vacuum double-drum dryer

Heat-sensitive materials can be dried in a vacuum double-drum dryer where the dryer is enclosed in an airtight enclosure under a vacuum (**Figure 13**) [38–40, 46].

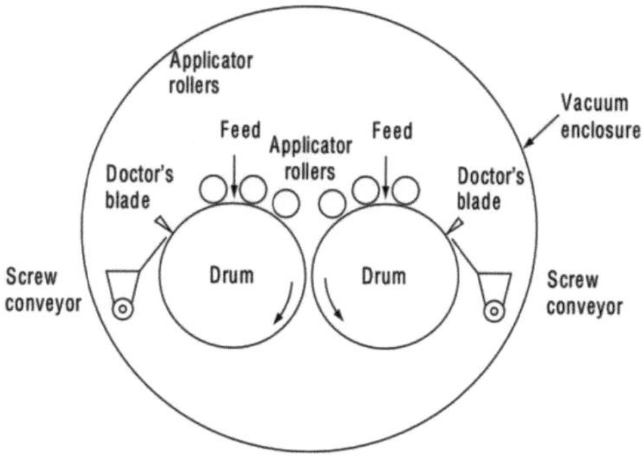

Figure 13.
Vacuum double drum dryer.

This dryer type is also attached to a vacuum pump, a scrubber, and a condenser. Dryer operation is like its atmospheric version excluding that there are two product basins: one is for vacuum breaking and the other is for product discharge.

3.4 Industrial spray drying systems

The development of spray drying equipment and techniques evolved over several decades. With the sudden need to minimize the food materials transport weight, spray drying was established during World War II. This method enables feed transformation from a fluid state into a dried particulate shape by feed spraying into hot drying media. It is a continual particle processing drying operation. The feed could be an emulsion, suspension, solution, or dispersion. The dried product could be in the shape of granules, agglomerates, or powders dependent on feed's chemical and physical characteristics, and the dryer scheme, as well as the final powder properties, desired [47]. Spray drying is a suspended particle processing (SPP) operation. This method uses fluid atomization for creating droplets. The droplets are dried into single particles when transferred to a hot gaseous drying media, generally it is air. It is considered a one-step continual unit processing procedure. Nowadays, over 25,000 spray dryers are used commercially to dry products such as fine and heavy chemicals, dyestuffs, dairy products, agrochemicals, and biotechnology products as well as mineral concentrates used in pharmaceuticals for preparing with evaporation capacities ranging from a few kg/h to over 50 tons/h.

Spray drying advantages:

- Biological products, pharmaceuticals, and heat-sensitive foods could be dried at atmospheric pressure with minimum temperatures. Sometimes the inactive atmosphere is used.
- Product characteristics are considerably further effectively controlled.
- Spray drying authorizes high-capacity production within the continual process and comparatively modest tools.
- Goodstuff links into contact together with the equipment surfaces in an anhydrous condition.

- A comparatively regular, spherical particle with about the same attribution of nonvolatile compounds as in the liquid feed is produced by spray drying.
- The efficiency could be compared with that of other types of direct dryers because the operative gas temperature might range from 150 to 600°C.

Spray drying disadvantages:

- If the product has a high bulk density, spray drying fails.
- Generally, the system is not elastic. If the unit is designed for fine atomization, it might not be able to make a coarse product and vice versa.
- The feed has to be pumpable.
- Has a high initial investment compared with other continuous dryer types.
- Increasing the drying cost when the product is recovered and dust collection.

Spray drying involves of three process phases: (1) atomizing, (2) moisture evaporation and spray-air mixing, and (3) separation of dry product from the exit air. Every phase is conducted according to the operation of the dryer as well as the design or chemical and physical properties of the feed, determining the features of the final product. A spray drying procedure model is illustrated in **Figure 14**.

3.4.1 New developments in spray drying

Numerous innovative designs and operational modifications have been proposed in the literature [48–51] although very few are readily available commercially due to incomplete knowledge about the new systems.

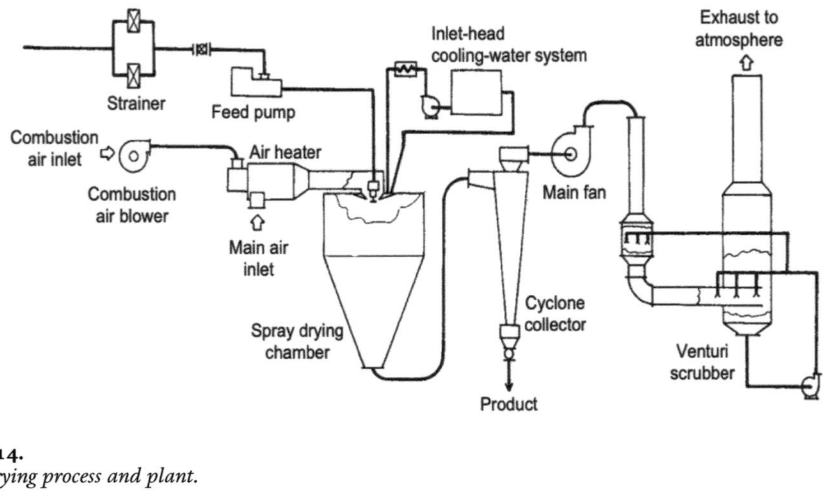

Figure 14.
Spray drying process and plant.

3.4.1.1 Superheated steam spray drying

Although superheated steam drying was proposed one century ago, the potential of superheated steam like a drying medium was not exploited industrially for half a century. The advantages of this type of drying are reviewed by [52] as follows:

- No oxidative damage
- No explosion hazards
- Recovery of potential heat supplied for evaporation
- Ability to operate at vacuum and high operating pressure conditions
- Minimizing air pollution by closed-system operation
- Producing a product with better quality under appointed conditions

When the constraints could brief to the following:

- The higher temperature of product
- The capital costs is higher than the hot-air drying
- The possibility of air infiltration making recovery of heat from exhaust steam is difficult using compression or condensation

The initial scheme of a superheated steam spray dryer was made by Gauvin [53, 54]. The patent on superheated steam spray drying was issued by Raehse and Bauer [55]. This drying has been examined using the CFD system by Frydman et al. [56].

3.4.1.2 Two-stage horizontal spray dryer

A horizontal spray dryer has been suggested as an alternative to the conventional vertical types [57, 58] for pharmaceutical materials and heat-sensitive food [59, 60]. The low temperatures of drying permit flavor reservation, controlled porosity and density, high solubility, as well as fine goodness agglomerated products. Essential energy provisions relative to the conventional vertical spray dryer with a lower electrical loading for a specific capacity.

A suggested design is described in **Figure 15**. The high-pressure pump is used for feed pushing to the spray nozzles. These nozzles are organized in several configurations. Then, the spray travels horizontally from the nozzles, and due to gravity, it goes down. After that, transporting of the dried powders using a conveying strap in the bottom and carried to bag filters or cyclones. A model for the horizontal spray dryer (6.0 m $\times$ 3.0 m $\times$ 3.0 m) has been presented by Cakaloz et al. [61].

In the single-stage horizontal spray dryer process (**Figure 16**), it has been shown that the dwell period might be overly short for allowing the droplets for drying completely. Therefore, to overcome this constraint, a new two-stage, two-dimensional horizontal spray dryer idea is suggested by Mujumdar [58] and shown in **Figure 15**,

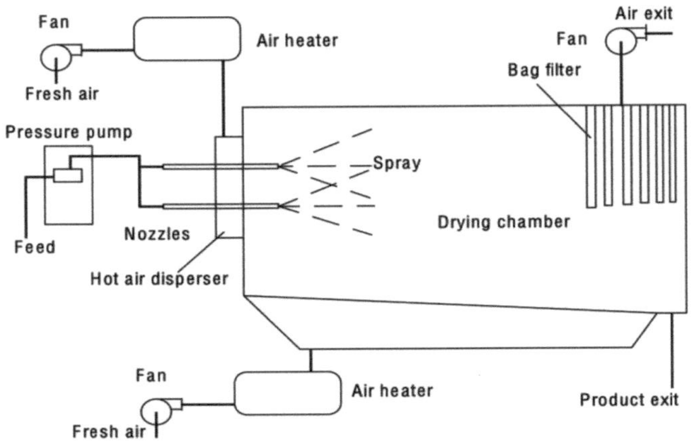

Figure 15.
Proposed layout of a two-stage horizontal spray dryer.

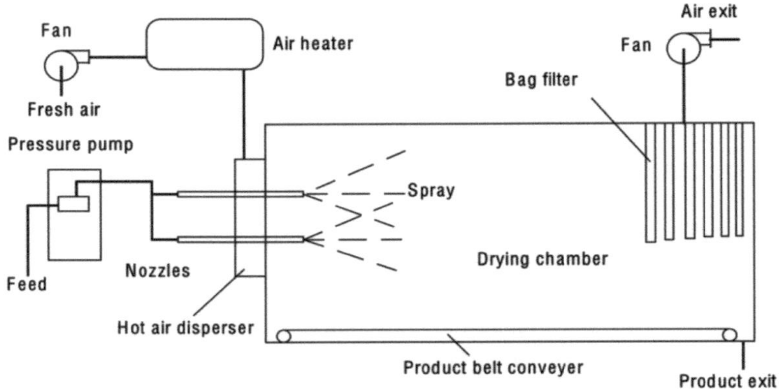

Figure 16.
Schematic layout of the horizontal spray dryer.

which is prepared for permitting long drying periods. This is necessary for big droplet sizes and heat-sensitive products. The variance between single and two-stage horizontal spray driers is that in the single spray drier, there is a fluid-bed dryer, which is installed at the horizontal chamber bottom. Whereas, the two-stage horizontal spray drier is not commercialized yet. The CFD process could be utilized to emulate the horizontal spray dryer, which should be connected to a fluid-bed drying system. The droplet's surface moisture is taken away rapidly from the chamber of spraying. While the inner water occupies longer, which could be extracted in a thin layer, out of a deliberation dryer in the bottom of the room.

3.4.1.3 Low-humidity spray drying

On occasion, it has to choose freeze dryers instead of spray dryers, particularly for pharmaceuticals and biochemicals drying. This is because spray drying utilizes high-temperature gas like the medium of drying and the sediments on the wall cannot be averted in feasible spray dryers. The products are subject to degradation by

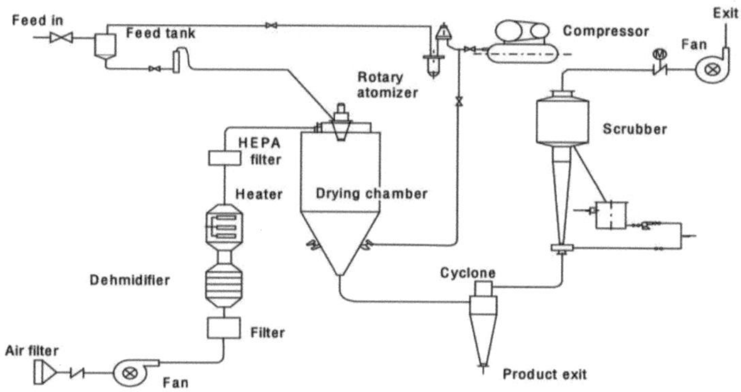

Figure 17.
Schematic layout of the low-dew-point spray dryer.

overheating [48]. Yet, the freeze dryer has high operating, and capital costs, as well as high energy consumption, compared with the spray dryer [62, 63].

To decrease the active compound degradation in the dried powders, a novel kind of spray dryer was proposed, such as low dew point (LDP) spray dryer. This type utilizes air at a nearby surrounding temperature of 80°C and quite minimum humidity (**Figure 17**). It comprises four units: the preprocessing system of drying air including heating and dehumidification; the preparation system of feed; the drying system and feed atomization including drying chamber, air disperser, and atomizer, etc., in addition to the system of product collection.

3.4.1.4 Spray freeze drying

The freeze drying (FD) is a significant drying machinery for pharmaceuticals, foods, and biochemicals, which could preserve the biological activity, aroma or flavor, etc. Freeze drying is a drying method, in which the solvent in the material is solidified at low temperatures, then sublimates directly from the solid into the vapor state under vacuum. The freeze-drying process of an aqueous solution contains three phases: product freezing, ice sublimation, and removal of solvent vapor. Nevertheless, compared with spray drying, the freeze drying is an order-of-magnitude costlier drying method because of its need for vacuum, refrigeration, and long operating times [64]. In a recent combination of both drying processes, called spray freeze drying (SFD), this process was carried out as a batch method as follows: The liquid nitrogen is atomized in a cryogenic medium to spray the droplets to freeze. Frozen droplets are scuttled in a cryogenic medium and processed out and dried in the freeze-dryer under a vacuum. A model flow slab is illustrated in **Figure 18**. This method is conducted in batches instead of continuously, because of the required long freeze-drying time. In addition, it is appropriate for high-significant products with low tonnages [65].

A new method called a combined atmospheric spray and fluidized-bed freeze drying (ASFBFD) process was examined. The planned schematic flow graph is described in **Figure 19**. It mainly comprises internal fluidized bed, drying or freezing room, liquid nitrogen cooler, internal bag filter, fan, nozzles, valves, pipes, pump, etc. [65].

The process is briefly explained as follows: The feed is transported into the nozzle from the top through a pump, where it is atomized using a nozzle. The fine spray is

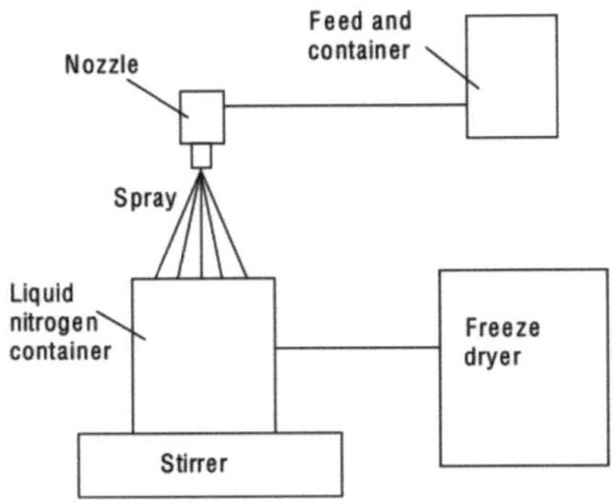

Figure 18.
A schematic flowchart of the conventional spray freeze drying.

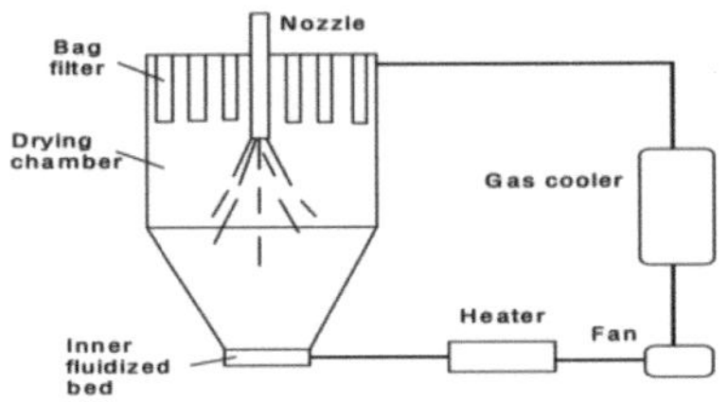

Figure 19.
Batch-type atmospheric pressure combined with spray and fluidized-bed dryer.

connected with freeze nitrogen or air immediately. Upon the material of feed, the drying medium is frozen to about −90°C using liquid nitrogen and inflated by a fan within the fluidized-bed cribriform tray. The fluidized bed is placed at the bottom of the chamber. As the temperature of the cool air is lower enough for spray freezing, the frozen atoms conserve their main conditions and go to the bottom or to the fluidized bed through gravity. Nearly a few frozen atoms might be restricted by air or nitrogen; nonetheless, they are detached from the air using the interior bag filter. Throughout the phase, the spray freezing drying is investigated continuously until the exhaustion of feed in the batch procedure. After this phase is completed, suitable drying terms are chosen, such as liquid nitrogen, which is regulated to meet freeze-drying terms. Drying actual operation and its time terms should be specified. Many investigators have already investigated many studies on atmospheric freeze drying [66–72]. These reviews showed that the atmospheric spray and fluidized-bed freeze drying grouping is a viable new procedure that needs additional R&D. Lastly, a summary of the comparison between the four drying processes, such as SD, FD, SFD, and ASFBFD, is assumed in **Table 1**.

3.4.1.5 Encapsulation

Spray drying and fluid-bed drying lead to another common food application of these technologies, that is, microencapsulation. It is defined as a process by which one material or more is entrapped within alternative material [73]. This method is generally used to prevent the core material from degradation and to control releasing or separate reactive constituents in a formulation. Because the spray dryer is generally fast, available, economical and produces good-quality products [74], it enhances the

Drying method	Spray drying (SD)	Freeze drying (FD)	Spray freeze drying (SFD)	Atmospheric spray and fluidized-bed freeze drying (ASFBFD)
Drying period	Short	Long	Long	Medium
Final shape	Irregular or agglomerated	Cake	Porous and spherical, mono	Porous and spherical
Power requirments	Low	High	Highest	Medium
Quality dried product	Medium	Good	Good	Good
Operation	Continuous	Batch	Batch	Semicontinuous
Productivity	High	Medium	Low	Medium
Cost	Low	High	High	High

Table 1.
Drying method properties [65].

most public means of encapsulation process. This procedure is simple and comparable with the one-phase spray drying technique. The coated substance is named active or core substance, and the coating substance is entitled a shell, while the wall substance is called a carrier or encapsulant [75]. The encapsulated active substance is scattered in the hydrocolloid transporter, such as maltodextrin, dextrin, Arabic gum, gelatin, and modified starch. Subsequently, the emulsifier is supplemented, and the mix should be homogenized to compose an oil-in-water emulsion, then it's consumed by the atomizer for the spray dry. In the drying room, the aqueous stage dries, and the active substance is captured as atoms through the protein film or hydrocolloid, discharging of the active substance from capsules under specific terms. The temperature, moisture, and pressure are the main controlling factors of its release [76].

3.4.2 Energy efficiency enhancement

Spray drying is an energy-intensive procedure. With the increased energy costs with overall production stage changes, spray dryer consumers have to look for techniques to enhance the spray dryer system's thermal efficiency. For typical single-stage drying, the best method to control energy practice is to raise the inlet temperature and keep the outlet temperature low, as well as take full benefit of the energy introduced. Nevertheless, the weakness of this procedure is the potential product degradation in food spray drying [52, 77, 78].

3.5 Freeze drying

Freeze drying is used for high-quality foodstuffs stabilization, certain biological materials, and pharmaceuticals such as proteins, vaccines, bacteria, and mammal cells. These substances are freeze-dried; thus, the quality of the dried product is retained [63, 79]. Freeze drying is a process in which the water or another solvent is sublimated by the direct transition of water from solid (ice) to vapor, thus omitting the liquid state, and then desorbing water from the "dry" layer [80–84].

As a rule, freeze drying produces the highest-quality food product obtainable by any drying method. This is due to the fact that freezing water in the material prior to lyophilization inhibits chemical and biochemical such as non-enzymatic browning, protein denaturation,, microbiological, and enzymatic reactions processes. Consequently, the content of various nutrients, smell, and taste do not change. Raw materials comprise water, ranging from 80% to 95%. The water is removed by sublimation resulting in extremely porous structure creation of the freeze-dried products, and the finalized rehydration or lyophilizing happens immediately when water is added to the substance at a later time [85, 86]. However, the water in foodstuffs could be free or bound to the solution using different powers. The free water is freezed, but the bound does not. In the freeze-drying method, the iced and some bound water should be detached. Consequently, lyophilization is an extremely complex and multistep procedure that comprises (a) freezing under atmospheric pressure phase, (b) main drying; appropriate freeze drying; ice sublimation, at decreased pressure. For example, iced water is managed, then sublimation at 0°C and absolute pressure of 4.58 mmHg could happen. Nevertheless, subsequently the water usually exists in a solution or a joint state, the substance should be cooled below 0°C to preserve water in a frozen phase. So, during the main drying step, the frozen layer temperature as showed in **Figure 20** is at −10°C or lower at absolute pressures of about 2 mmHg or less. As the ice sublimes, the sublimation interface, which began at the external surface (**Figure 20**), dried material retreats and porous shell residues. The latent heat (2840 kJ/kg ice) could be operated using the dried substance and frozen layers, as illustrated in **Figure 20**. The vaporized water is conveyed within the dried substance porous sheet. Through the main drying step, in the dried layer, a large amount of nonfrozen water might be desorbed. The desorption method might affect the amount of heat that reaches the sublimation interface, and subsequently, it might affect the velocity of the sublimation front moving. The period when there are no frozen sheets is possessed to perform the end of the primary drying phase. (c) Secondary drying phase; desorption drying; drying the product to the required final humidity, this stage starts at the end of the primary drying stage, and the desorbed water vapor is transported through the pores of the material that is dried [63, 80, 87]. During the above three phases, six main physical phenomena could be distinct, which have a significant impact on the process path, the quality of the obtained substance, and

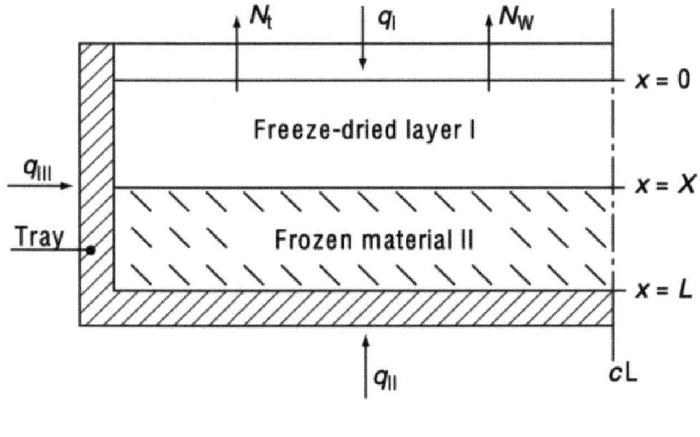

Figure 20.
Diagram of a material on a tray during freeze drying. The variable X denotes the position of the sublimation interface (front) between the freeze-dried layer (layer I) and the frozen material (layer II).

its overall costs [88]. Those phenomena are as follows: the water is transitioned into ice, then the ice to a vapor phase, the water particles are desorpted from substance structures, the obtainment of sufficiently low pressure, the re-sublimation of water vapor removed from the substance on the condenser surface, and the removal of a layer of ice from the capacitor surface.

In this context, the conditions of the freeze-drying process should be selected in a way that does not melt the water. The presence of water during freeze drying may result in many changes in the composition, morphology, and physical properties of foods (e.g., shrinkage), then reducing product quality during storage [89]. The effect of freeze-drying conditions on the nutritional properties, antioxidant activities, and glass transition characteristics of different food materials can be found in the literature [90–95]. Despite the long processing time and high cost, it is preferred for extreme-quality products. Though some damages in bio compounds could be set after freeze drying, this method is preferable to maintain nutritional qualities, particularly when operated under a vacuum. Moreover, the quality parameters such as freeze-dried products' rehydration and porosity are favorable for manufacturing foods. Newly, the freeze-drying process condensation with innovative technologies or pretreatments permit overcoming some of these drying processing challenges [96].

3.5.1 Microwave freeze drying

The limitations on heat transfer amounts in conventionally managed freeze-drying operations have led to providing internal heat generation with the use of microwave energy [97, 98]. Hypothetically, the use of microwaves must result in an instant drying rate, because the transferring of heat does not need internal temperature slopes and the temperature of ice could be preserved near to the maximum allowable temperature for the frozen layer exclusive of the need for extreme surface temperatures. For example, if it is allowable to keep the iced layer at −12°C, after that it has been assessed that the drying time for an ideal process using microwaves for a hypothetical 1-in. slab would be 1.37 h [99]. It should be notable that this drying time compares very approvingly with 8.75 h needed for the heat input within the dry layer, while 13.5 h for heat input within the iced layer without the removal of dry layer, and even with relatively short time of 4 h of drying for continuous removal of dry layer. Examining the freeze drying of a 1-in.-thick slab beef, the actual drying time of slightly over was 2 h, compared with 15 h for traditionally dried slabs [100]. Despite these benefits, the microwave application has not been successful [99, 101–105]. Because of the following reasons: (A) Supplied energy in the microwave form is too expensive [101]. (B) The tendency to glow discharge, which could cause gases ionization in the room and food deleterious changes, and losses of useful energy. The tendency to glow discharge is larger in the pressure ranged from 0.1 to 5 mmHg and could be decreased by operating the freeze dryers at pressures under 50 mm. The operation at low pressures has a double drawback: (1) it is much expensive, because of the demand for condensers operating at a very low temperature and (2) At these low pressures, the drying rate is much slower. (C) The process control is very difficult. Meanwhile, water has an inherently greater dielectric loss factor than ice, any localized melting produces a rapid chain reaction, which results in runaway overheating. (D) The economical equipment suitable for a large continuous scale is not yet available. In spite of these constraints, microwave freeze drying is considered a potential development [101].

3.5.2 Industrial freeze dryers

The main types of industrial freeze dryers include the following:

3.5.2.1 Tray and pharmaceutical freeze dryers

The hugest amount of industrial freeze dryers in process are of the vacuum batch style with freeze drying of the food stuff in trays. There are two main styles, depending on the type of condenser used. The first style showed the condenser plates in the same chamber and near the tray-heater assembly. The second style is representing the condenser in a separate chamber linked to the first by a wide, in over-all, butterfly valve. This latter type of the factory is permanently used in pharmacological industries, but it can also be used for freeze-drying foods. To reduce product contamination risk, especially in the pharmaceutical industries, a new freeze-dryer plant concept has been developed. The system, as illustrated in **Figure 21,** has two doors: a little one is for charging the stuff before drying, when the full door, which is opposed to the little door, is for discharging the stuff after drying. The condenser is positioned on the floor, which is under the first floor, where the drying chamber is placed. The freeze dryer shelves are lowered to the drying chamber bottom and then lifted one by one to a location in line with the filling machine. The charging of foodstuff is prepared under a laminar flow of sterilized air; the little door is unlocked only for each plate filling and then is directly closed [87].

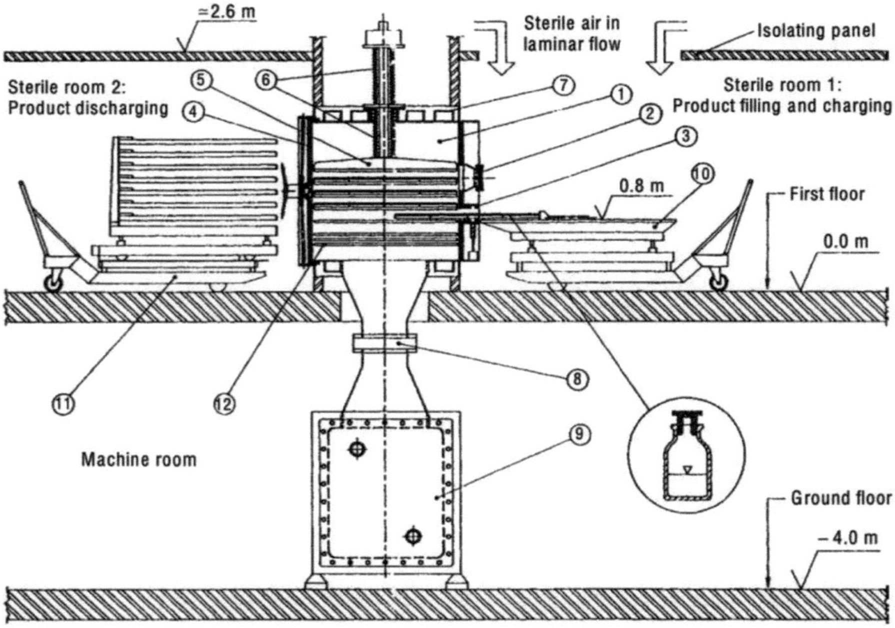

Figure 21.
Industrial freeze dryer design with stoppering device (Criofarma model C300-7): 1, drying chamber; 2, inspection window; 3, automatic small door opening; 4, full door; 5, hydraulic press for stoppering the bottles after drying; 6, PTFE elbows for double sterile condition inside the stoppering plug; 7, reenforcing member and cooling coils after steam sterilization; 8, isolation butterfly valve; 9, ice condenser chamber; 10, loading device; 11, discharging device; 12, unloaded shelves.

3.5.2.2 Multi-batch freeze dryers

The freeze-drying system in a batch factory is usually programmed and organized to minimalize the drying time and enlarge the factory production. With a single-batch factory, the load on the several systems will be very variable during the drying rotation. The product flow and handling operations will be alternating because of the batch process specifically. This means that optimum use of supplies will not be probable in a sole-cabinet batch system. To extent this weakness, an industrial freeze-drying factory is built with numerous batch cabinets instructed to operate with staggered and overlying drying rotations. Each cabinet can be filled with materials from the same system, and they are assisted by the similar central system for vacuum pumping, tray heating, and condenser refrigeration. Although, the procedure is individually organized in each cabinet from a unattached control panel. This builds probably the simultaneous different foodstuffs production, which rises the operation flexibility of the factory. With only two cabinets in operation, an important part of the batch weakness may be excluded; for example, with four cabinets, an excellent leveling of loads will be accomplished. A huge amount of industrial freeze-drying factories operate today in this style as multi-cabinet batch factories [81, 99, 106].

3.5.2.3 Tunnel freeze dryers

The process in the tunnel freeze dryer (**Figure 22**) takes place in a large vacuum cabinet into which the tray-carrying trolleys are loaded at intervals through a large vacuum lock at one end of the tunnel and discharged similarly at the other end. The drying conditions are carefully controlled in a number of sections of the tunnel by temperature-pneumatic controllers [81]. The plates of vapor constriction fit closely inside the channel walls yet permitting the trolleys to passage through two locations in the tunnel main unit, and gate valves turn off the locks from the main unit. Thus, the tunnel is divided into five autonomous process zones. When trolley is not moving during the period, a tray-lifting scheme causes all trays to sit in each trolley on heaters below the top. The heaters consume flat-top surfaces and ribs under which vacuum

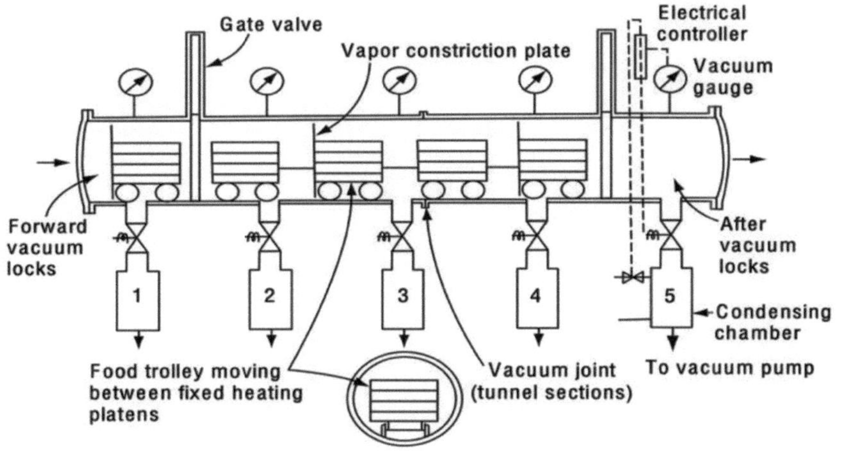

Figure 22.
Typical tunnel freeze dryer schematic diagram.

steam passes. They are cantilevered in couples from both sides of the tunnel. Vacuum steam heating has numerous benefits, including high latent heat of condensation and temperature control operating pressure. The cooling system comprises a great aqua ammonia absorption freezer instead of a compression factory. Because of easiness with the refrigeration, load can be diverse by controlling the feed of oil to the boiler that heats the absorber.

The whole capacity of the tunnel freeze dryer can be boosted as it increases volume of business. Large business factories for cottage cheese and coffee processing have been set up in this way. The tunnel freeze dryers have the same benefits of factory capacity operation that can be attained as in multi-batch factories. On the other hand, the flexibility for simultaneous production of materials or in swapping between products is missing.

3.5.2.4 Vacuum spray freeze dryers

The vacuum spray freeze dryer system is illustrated in **Figure 23**. It has been industrialized for tea infusion, coffee extract, or milk. The product is applied by spraying from a sole jet upward or downward in a cylindrical tower of about 3.7-m diameter by about 5.5-m high [81, 107]. The solutions are solidified into little particles by evaporative freezing. In the tower, a frozen helical condenser is coiled between the internal wall and central hopper, the latter accumulating the partially dry powder as it drops freely to the tower bottommost. This is in turn associated with a tunnel where the drying technique is accomplished on a stainless-steel belt migrant among radiant heaters. The dried material passes into a hopper that feeds a vacuum padlock, allowing alternating product removal for packing. The whole system operates under a vacuum of about 67 Pa. In the initial evaporation, the diameter of frozen particles obtained by spraying into a vacuum is about 150 mm and loses about 15% moisture, and there is no sticking of these particles.

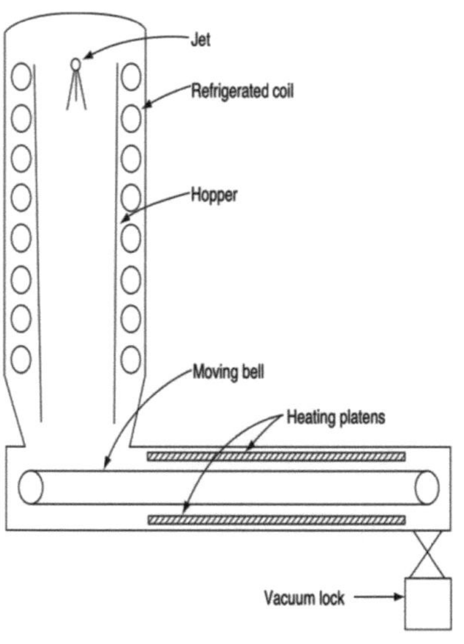

Figure 23.
Vacuum spray freeze dryer layout.

3.5.2.5 Continuous freeze dryers

Continuous freeze dryers are used for products drying in trays and for agitating bulk material drying. When the product is handled in trays, then the most delicate treatment of the product is attained. The food material is placed in the tray. Hence, it is not exposed to scratch, and it falls in contact with surfaces only that completely meet required hygiene values. While agitating a crushed product, more efficient heat transfer to sole product particles can be realized. Thus, a significant decrease in the heating surface is achievable. Although, scratch of the product by agitation increased the production of water vapor per unit, the heating surface tends to bring little product particles with vapor stream away from the bulk bed of product and cause product loss. Any system problems for water vapor elimination to retrieve the product loss may more than counterbalance the advantage of the higher heater surface load [108]. The heat transfer to the product and the trays is by radiation, which is the easiest mode to safeguard a correct as well as consistently distributed heat transfer to the material during the practice. The radiant heat is formed by horizontal heater saucers that are gathered within temperature zones. Each tray remains for a fixed time in each temperature zone for drying time minimized [108].

4. Electromagnetic waves dryers

Electromagnetic spectrum ((EMS) is the range of all wave types of EM radiation represented in Radio, Microwave, Infrared, Visible light, Ultraviolet, X-ray, and Gamma-ray. It is known that the Sun is a source of energy across the full spectrum, and its electromagnetic radiation covers the atmosphere constantly. The EM waves already have several applications in the agriculture field such as imaging, remote sensing, quality sensing, and dielectric heating in both pre-harvest and post-harvest treatments as shown in **Figure 24**. Agricultural products are considered dielectric materials and thus can store electric energy and convert it into heat. The converted heat is different from one plant to other according to its permittivity (ε) in general. This value of permittivity (ε) is noticeably frequency-dependent. Therefore, the dielectric constant parameter for agricultural products varies with frequency. For instance, the permittivity (ε) of water has an absorption peak in 24 GHz frequency. As a result, the temperature of the water stored in the agricultural products rises and evaporates, its moisture content decreases, and the drying process takes place. As a result, the drying processes of agricultural products using electromagnetic waves took a great place compared with traditional methods of drying, especially for agricultural products that are highly sensitive to heat, as one of the best modern technological solutions to maintain the quality of agricultural products and the production of dried products. Therefore, in the following, the most famous electromagnetic spectrum bands used in the drying process will be presented, which are both infrared and microwaves.

4.1 Infrared dryers

Conserving energy and achieving the best quality of dried products have become the most important factors that determine the usefulness and success of operating any drying unit. Where heat is transmitted through the drying unit in three forms

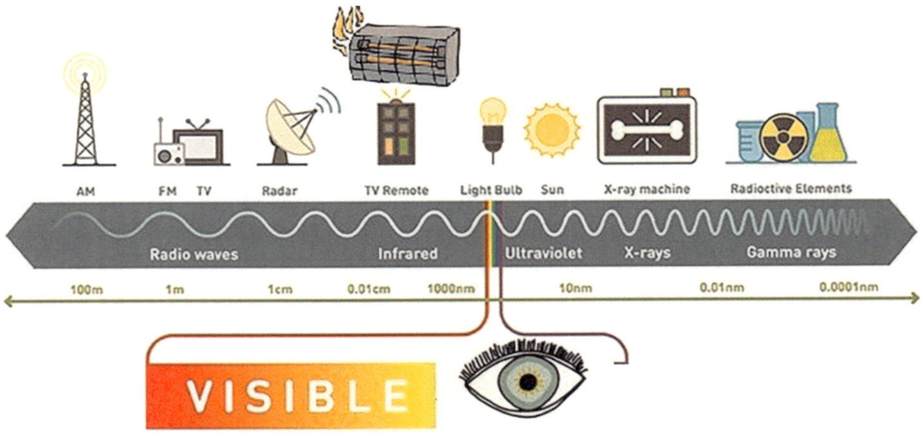

Figure 24.
Shows some applications of the EM spectrum.

conduction, convection, and radiation. In this regard, when the temperature of the fresh material increases, the molecular motion gains more energy; as a result, it causes changes in the structure of the material as well as its chemical properties to increase the shelf life and improve the quality [2]. Drying agricultural products, one of the most important processes of food processing and preservation, is a high priority in achieving food security. Where, drying process aims to improve the stability of food products, by reducing water activity, which leads to a decrease in microbial activity and thus limits physical and chemical changes during the storage of dried products [109]. The common conventional approach to the drying process is using convection heat transfer by transferring heat from hot air to the target product by convection, and as a result, the water evaporates into the air also by convection. However, this conventional drying method was characterized by many related disadvantages such as time consumption, low efficiency, unfair exposure to high temperatures, quality variation, in addition to high energy consumption [110]. These disadvantages of traditional drying by convective led to the innovation of other drying technologies to overcome these drawbacks such as drying by microwave, infrared (IR), osmotic drying, fluidized bed, and hybrid drying methods (integrating two or more drying methods). IR drying technique is one of the most important modern drying technologies compared with traditional drying because it is characterized by short process time, uniform temperature, high heat transfer coefficient, good quality of final products, improved energy efficiency, and safe as mentioned [111–113]. Therefore, [114–119] concluded that drying by IR heating is a promising method to achieve high-quality dried products and is suitable for fruits, vegetables, grains, and other high-value products. IR heat derives from the IR radiation that lies between the visible light (Vis) spectrum and the microwave band along the electromagnetic spectrum as depicted in **Figure 25**. IR drying technology idea is based on making changes in the electronic, rotational, and vibrational states of the atoms and molecules of the fresh material when exposed to IR radiation within the wavelength range of 780–106 nm.

IR band is divided into three regions, near-IR (NIR) is the first region whose wavelength ranges from 750 to 2500 nm to 0.75–1.4 μm at temperatures between 400 and 1000°C, the second region called mid-IR (MIR) with wavelength ranges from 2500 to 25,000 nm to 1.4–3 μm at temperatures between 400 and 1000°C.

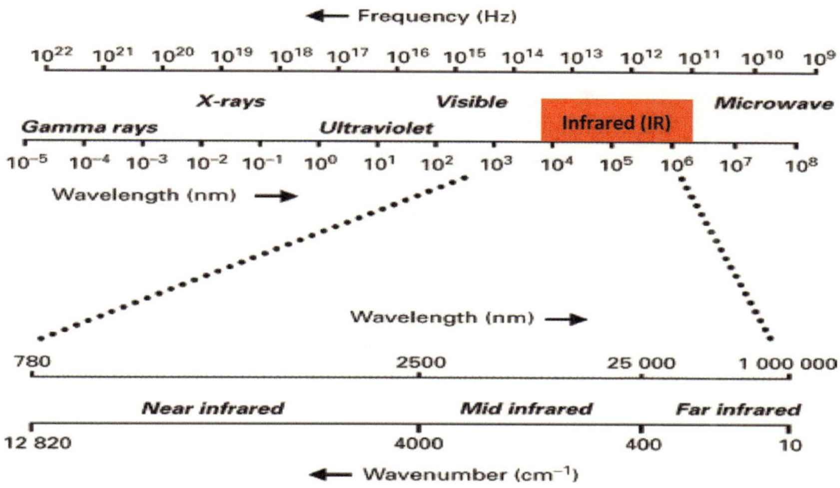

Figure 25.
Depict highlighting infrared (IR) radiation region.

While the wavelength of the third region (far-IR, FIR) ranged from 25 × 103 to 106 nm ≃3–1000 μm at temperatures above 1000°C [120, 121]. After IR radiation penetrates the fresh material surface, IR rays move and vibrate the constituent molecules of the fresh material through the frequency of the IR band within a frequency of 60 × 103–150 × 103 MHz, which results in friction between the molecules and leads to rapid internal heating [114, 122–124]. There are many sources of IR radiation according to their operational power, whether they are electrically heated or gas-fired generators to generate IR energy. The most common electrical heated sources are reflector-type IR incandescent lamps (incandescent vacuum lamp, gas-filled lamp, and tungsten halogen lamp), IR emitter-type quartz tubes, ceramic, and radiant panels as shown in **Figure 26**. While the traditional IR sources used for heating are electric heaters that produce infrared radiation in a range of 1100–2200°C and gas-fired generators, which consist of a perforated metal plate heated by gas flames until the temperature of the metal plate rises and infrared energy is radiated at a temperature range of 343–1100°C [110, 125, 126]. IR radiation sources can provide different wavelengths ranging from short to long wavelengths according to the voltage value applied to the IR emitters. The efficiency is another significant factor in the evaluation of both electric IR heaters and IR gas heaters, where the efficiency of electric IR heaters ranges from 40 to 70%, more than the efficiency of IR gas heaters, which ranges from 30 to 50%, and emit medium to long IR. In addition, Sheridan and Shilton, 1999, mentioned that the appropriate IR wavelength for industrial heating ranges from 1.17 to 5.4 μm, which corresponds to temperature values from 260 to 2200°C, and found that the best efficiency of heat transfer by IR radiation source occurs when the heater turns red. The idea of electric infrared emitters is based on passing an electric current to an electric heater through a high-resistance wire such as nichrome wire, iron chromium wire, and tungsten filament. When the metal wire is heated to the glowing stage and the temperature rises to 2200 K, it will emit NIR with a wavelength between 0.7 and 1.4 μm. Accordingly, the Incandescent lamp type for producing IR radiation is classified as a short-wave IR emitter, while the quartz tube type is classified as a medium-IR wave emitter [126].

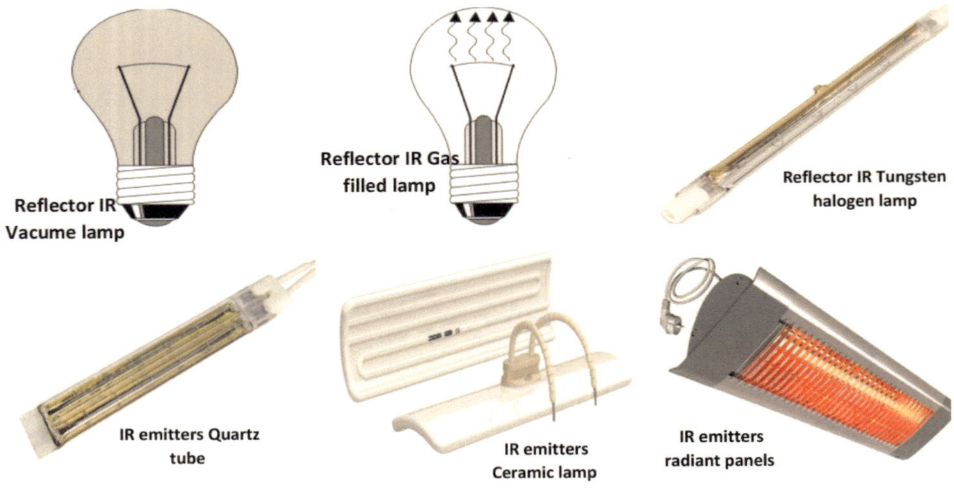

Figure 26.
Shows the types of infrared sources.

Many studies have reported that organic matter and water are the main constituents of fresh material or foodstuffs, which is based on the absorption of IR radiation significantly, especially at Mid and Far IR [113, 127, 128]. Where, the water absorption spectral coefficient was noted at 3 (MIR) and 6 (FIR) μm different wavelengths as shown in **Figure 27** and concluded that these wavelengths are considered suitable to be fixed in large-scale IR dryers for food products that generally contain 90% water. Confirming this, both [129, 130] mentioned that food efficiently absorbs IR radiation at wavelengths greater than 2.5 μm through a change in the state of vibration of the vibrating mechanism, which leads to a high temperature of the product being dried.

Several studies have proved that IR heating as a non-traditional drying method has many advantages and benefits such as uniform heating, short processing time, high heat transfer rate, high efficiency (80–90%), low energy consumption, low cost

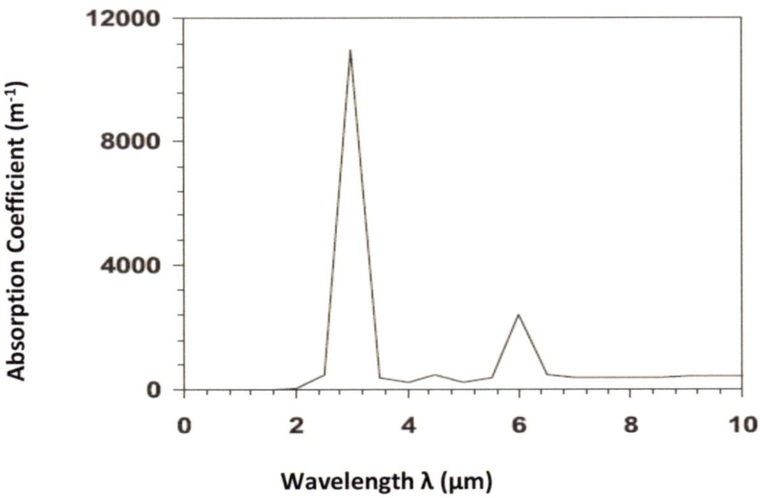

Figure 27.
Depicts the water absorption coefficient at different IR wavelength.

characterized, and improving final product quality in addition to it could be used as an application to measure water content in food products. Nowak et al. [113, 131–133] observed that far-infrared drying helps retain sensory quality in products such as sweet potatoes, grapes, Cordyceps militaris, and mangoes. Also, [134] investigated the effects of both hot air temperature and the IR drying method on the kinetics of persimmon slices, and the results found that the logarithmic model was the best model fitted to the experimental IR drying. In this regard, [135] studied the combined hot air at temperature levels (of 55, 65, and 75°C) and the IR drying method at radiation lamp power levels of 150, 250, and 375 W, on the persimmon fruits' moisture loss kinetics. The study finds that the drying time was reduced by 36% when increasing the drying hot air temperature from 55 to 75°C, while the drying time was reduced by 68.4% with increasing IR radiation power from 150 to 375 W. Nowak and Lewicki [113] dried the apple slices with IR radiation and by convection under equivalent conditions and reported that the heat-irradiated apple slices evaporated much more water than that not heated by IR energy until 80% of water is removed. Sun et al. [136] mentioned that the IR drying method combined with hot air as a pre-drying method can save 20% of drying time as compared with the IR drying alone throughout the drying of a thin layer of apple pomace. Chen et al. [117] conducted a comparative study between traditional hot-air (HA) and innovative drying methods of short- and medium-wave infrared radiation (SMIR) for drying jujube slices. The results find that the jujube slices dried by SMIR were of better color, higher retentions of vitamin C, total flavonoids content (TFC), and cyclic adenosine monophosphate (cAMP) content than the HA drying method, in addition to shorter drying time and higher drying efficiency. Also, the effects of the IR drying method on carrots were studied by [137]; the results pointed out that increasing IR drying time caused dramatic changes in the water state in dried carrots. Moreover, [138] used a combined drying method of IR and freeze drying to produce high-quality dried bananas at reduced cost, and the results showed that the dried banana samples were of a better color, higher crispness, and higher shrinkage compared with those produced by using regular freeze drying. As well, [139] achieved a considerable moisture reduction and higher drying rates in drying bananas with IR drying compared with hot air drying in the early stage.

4.2 Microwaves dryers

Microwaves (MW) are a form of electromagnetic beam, as are radio waves, ultraviolet radiation, X-rays, and gamma-rays. MV is located in the electromagnetic beam range between radio and infrared bands as shown in **Figure 28**. MW has wavelengths of about 30 cm to 1 mm, and frequencies ranging from about 1 GHz to nearly 300 GHz [140]. Microwaves have many applications such as communications, radar, astronomy, and remote sensing, and the most famous application for most people is cooking. Where the MV rays are absorbed by water at certain frequencies.

This property of MV is useful in cooking. Water in the food absorbs microwaves, which cause the water to heat up, then cook the food. Therefore, MV was used as a heating system in several industrial applications such as food, chemical, and materials processing, for example, cooking food and drying fruits and vegetables in both batch and continuous operations. MW radiation as a drying technique is based on the passage of microwaves through the material causing a molecule oscillation [114, 141], which leads to the volumetric heating of the material. MW volumetric heating (MWVH) is a way of using MW to penetrate uniformly throughout the volume of the product, thus delivering energy evenly into the body of the material. Hence, equally

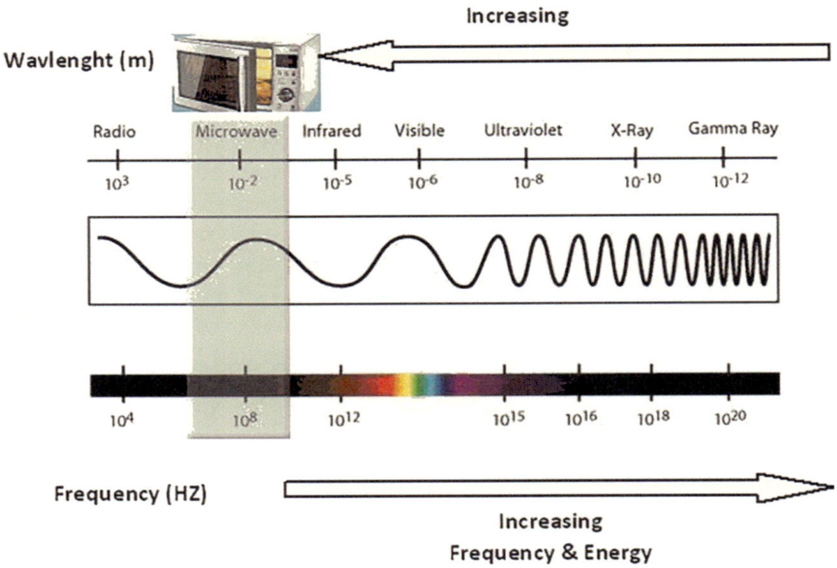

Figure 28.
Shows microwave band on electromagnetic beam.

heated the entire volume of a flowing liquid, suspension, or semisolid. Conversely, the traditional thermal processing methods rely on conduction and convection from hot surfaces to deliver energy into the product. A comparison study between MW and convective drying methods is shown in **Figure 29**. Alibas and Yilmaz [142] studied the effects of both these two drying methods on the drying kinetics, thermal properties, and quality parameters of orange slices. The results show that the MW drying

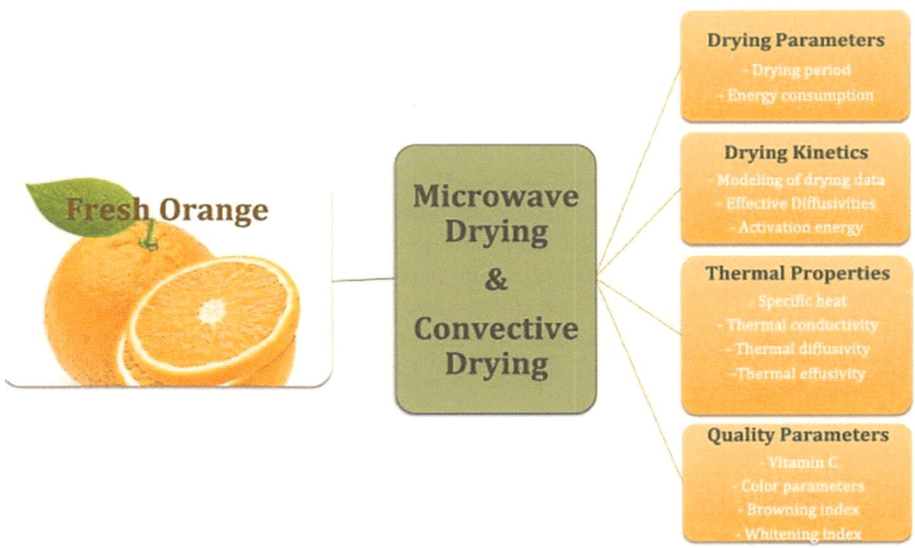

Figure 29.
Graphical comparison steps of the MW and convective drying methods.

processes were completed between 16 and 136 min depending on eight different microwave output power levels between 90 and 1000 W. On the other hand, in convective drying processes completed within the range 460–3120 min, at four different drying temperatures of 50, 75, 100, and 125°C. As well, the energy consumption was measured, and it observed that the MW drying method's energy consumption was very low at high and low powers. Finally, it is concluded that the most suitable drying method is MW drying at medium powers of 350 and 500 W by considering both drying and quality parameters.

Microwave drying technology is characterized by low energy consumption and short processing time, more uniform and energy efficient, making it an attractive source of thermal energy. However, some results indicated that microwave radiation alone is not sufficient to complete a drying process with high quality. Therefore, it is recommended to combine techniques, such as forced air or vacuum, to further improve the efficiency of the MW process [114, 141, 143–148]. Accordingly, several studies on drying using MW of various fruits and vegetables attached with an auxiliary system such as convective and vacuum methods reported that is more efficient than both MW and conventional drying techniques individually. As well, both [141, 148–153] mentioned that the MW drying technique is widely used in incorporation with hot air-drying systems. Where, the hot air removes water in a free state from the product surface, while the MW radiation removes water from the inner of the product. Furthermore, it is concluded that the MW-hot air combination drying systems not only increase the drying rates but also better retain the quality of the final products. In this regard, Sharma and Prasad [154], modified and developed a 600 W, 2450 MHz MW oven into an MW-hot air drier as shown in **Figure 30**. In order to explore the possibility of using a combined microwave-convective drying technique for processing garlic and assessment of the quality of the finished product. The results showed that the combined MW-hot air drying resulted in a reduction in the drying time and an extent of 80–90% in comparison to conventional hot air drying and a superior-quality final product. Alibas [155] studied the chard leaves quality characteristics during drying by MW, convective, and combined MW-convective. The results showed that the drying periods lasted 5–9.5, 22–195, and 1.5–7.5 min for MW,

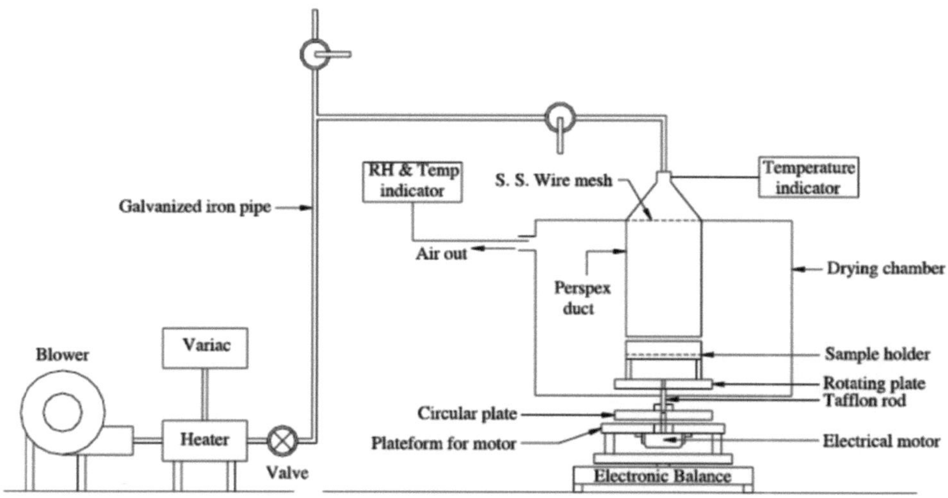

Figure 30.
Shows the microwave dryer attached to hot air.

convective, and combined MW-convective drying methods, respectively. Furthermore, the optimum drying period, color, and energy consumption were obtained for the combined MW-convective drying method. The optimum combination level was 500 W MW applications at 75°C.

Also, [156] studied the influence of MW-convective drying on chlorophyll and the color of herbs. The findings proved that the MW with auxiliary convective is a promising technique permitting the obtainment of dried material of high quality, additionally processing short time that can be an economic factor and incentive for the application of that method of drying on an industrial scale. In this regard [157] used different drying methods for Pistacia Atlantica seeds to study the impact on drying kinetics and quality properties. The results indicated that the MW drying method has afforded higher moisture removal in a shorter period compared with the traditional drying methods. Moreover, it was found that the essential oil composition was not considerably influenced by the MW drying method, and the texture quality is appropriate. Furthermore, [153] studied the effect of convective and vacuum MW drying on the bioactive compounds, color, and antioxidant capacity of sour cherries. It is found that in case of an increase in air temperature during convective drying as well as the increase in material temperature during VMWD deteriorated all the quality parameters of dried product. However, VMWD turned out to be much better than convective drying and competitive with a freeze-drying method. As it turned out that the best quality of the dried product and its more attractive color were found at VMWD at 480 W, followed by drying at MW power reduced to 120 W, which corresponds to anthocyanins content. As well, [158] dried pumpkin slices using convective and vacuum-MW drying methods to determine the drying kinetics, drying shrinkage, and bulk density, as well as to measure the color and carotenoid content of pumpkin slices dehydrated. They find that the vacuum-MW method has approximately tenfold shortened the time of pumpkin slice drying as compared with the convective method. Considering the viewpoint of color and carotene content, the vacuum-MW drying method was more effective than the convective method. Moreover, when use was made of vacuum-MW method, the dried products had a more attractive color. Sutar and Prasad [159], used microwave vacuum drying as shown in **Figure 31** to study the effect of vacuum in microwave drying operation and track the kinetics and moisture diffusivity of carrot slices. The results find that with the

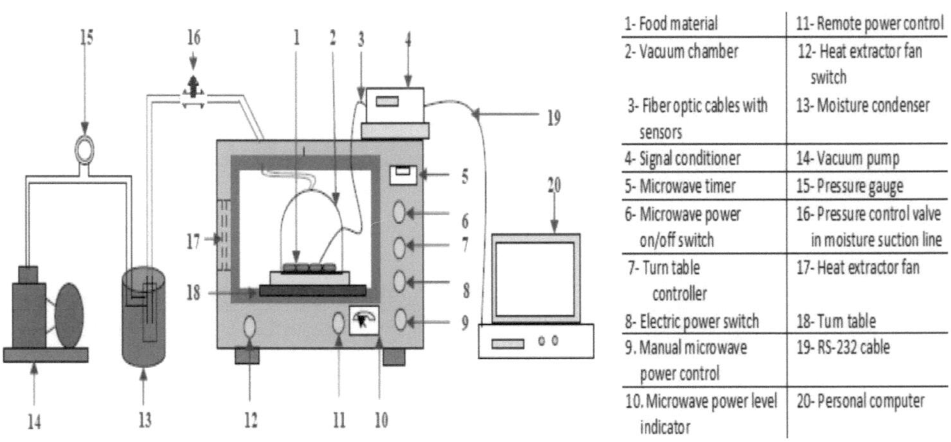

Figure 31.
Schematic diagram of the laboratory microwave vacuum dryer.

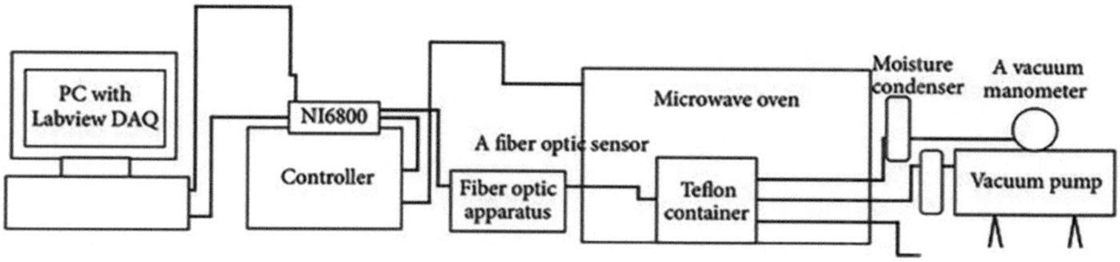

Figure 32.
Schematic diagram of the microwave vacuum drying system.

increase in microwave power density, the drying rates were increased and proved that the optimum model to predict the drying behavior of carrot slices' overall process conditions was the Page model. The combination of convective and vacuum-microwave (VMW) methods for drying kinetics and quality of beetroots was investigated by Figiel [160]. Where, convective drying with 60°C hot air and integration between convective pre-drying (CPD) and VMW drying method at 240, 360, and 480 W were used to dehydrate the Beetroot cubes.

The results showed that the VMFD method significantly reduced the total time of drying and decreased drying shrinkage in comparison with the convective method. Furthermore, the VMM-treated samples exhibited lower compressive strength, better rehydration potential, and higher antioxidant activity than those dehydrated in convection. Also, it is found that increasing the MW wattage and decreasing the time of CPD improved the quality of beetroot cubes dried by the combined method. Additionally, [161] optimized the drying process of Polygonum cuspidatum by using MW-vacuum drying and pretreatment methods as shown in **Figure 32**.

Where, a microwave vacuum drying system is designed and built that consists of a microwave drying unit, a power and temperature control unit, a moisture condenser, a vacuum pump, a vacuum manometer, and a PC-based data acquisition unit. The pretreatment methods were blanching for 30 s at 100°C, drying at 60°C, microwave pretreatment methods, and followed by microwave vacuum for 200 mbar at 50°C. Finally, it is concluded that it can be used to scale up the microwave vacuum drying system to a commercial scale.

5. Advanced and original drying methods

5.1 Instant controlled pressure drop method (détente instantanée controlee, DIC)

Major technical innovations in fruit and vegetable dehydration include pre or post-treatment under high or low pressure. At present, the instant controlled pressure drop method known by its French abbreviation DIC (détente instantanée contrôlée) was innovated as a solution to overcome the issues of drying shrinkage/collapse, to obtain a better quality of the dried plants. The DIC is characterized by its ability to handle a broad range of food products and is considered one of the newest innovative drying methods, regardless of its sensitivity to heat. The DIC processing method was developed, defined, and studied by [162]. DIC as a drying method is focused on exposing the product to high-pressure saturated steam for a few seconds and then followed by an abrupt pressure drop toward a vacuum. Then hot air-drying processing took place after DIC treatment as a conventional drying method. As well, [163] mentioned that

DIC processing is based on fundamental studies concerning the thermodynamics of instantaneity. Where, DIC system consists of six main parts processing vessel, vacuum tank, pneumatic valve, steam generator, air compressor, and vacuum pump as shown in **Figure 33**. The DIC work idea depends on exposing a partially dried product (usually the humidity is close to 30% db) to a high temperature/short time (HTST) such as a vapor pressure ($P < 1.0$ MPa) at high temperature (below 180°C) for a short time (less than a minute). Then this HTST stage was treated with a sharp pressure drop to a vacuum (within the pressure range of 3–5 kPa, and Δt from 10 to 60 ms). This leads to a mechanical influence represented in a severe pressure drop in a very short time. As a result, automatic evaporation of a part of water inside the product and at the same time a cooling operation were stimulated, which stopped their thermal degradation and gives a controlled expansion of the product [164–168].

Hence, **Figure 34** accurately describes the different stages of the work idea of the DIC processing unit as follows:

A. An elementary vacuum is implemented to minimize air resistance and simplify the steam diffusion within the product, then rapid heating is happen.

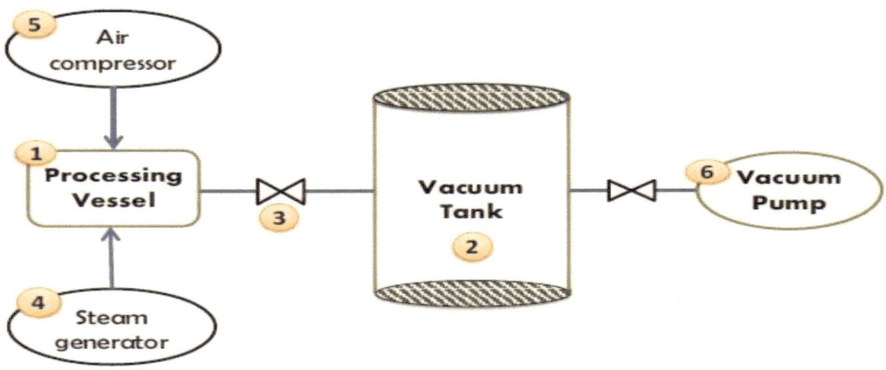

Figure 33.
Schematic of the DIC technology.

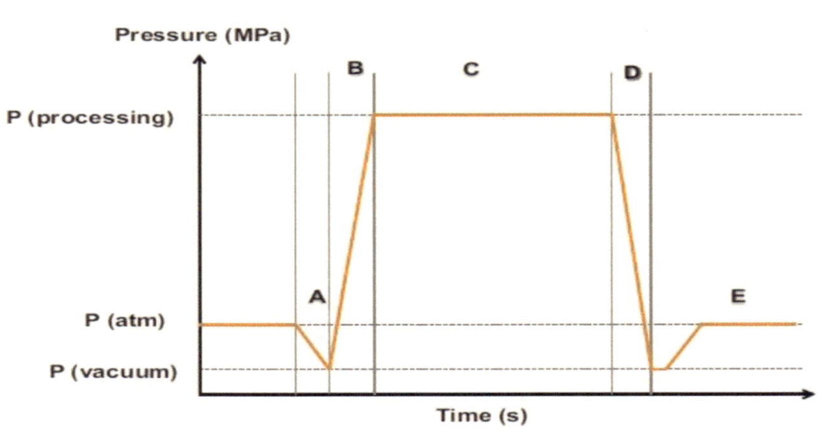

Figure 34.
Describes the work steps of the DIC system.

B. Secondly, injected saturated steam into the vessel until it reaches the highest pressure, then it is fixed.

C. Pressure and time parameters are given depending on the product treatment.

D. The pressurization needs to be followed by a sharp decompression till reached the vacuum.

E. Atmospheric air is injected to come again to atmospheric pressure, after the vacuum point, for sample recovery.

According to, [169], over the past 33 years, this technology has continued to expand its food applications and improve its characteristics on an industrial scale. DIC technology has continued over the past 33 years, to expand its food applications and improve its characteristics on an industrial scale. Since the DIC implementation has shown a quantum leap in quality enhancement and reducing the cost in the food industry generally, this is achieved by reducing the drying time of fruits and vegetables and enhancing the essential oils extraction, vegetable oils, and antioxidant elements. Additionally, it eliminates plant microorganisms and germs, which cause contamination, and reduces non-food ingredients and allergens. As well as it provides strong decontamination eliminating vegetative microorganisms and spores and reducing non-nutritional and allergenic components. It is known that convective air drying is one of the most industrial food drying operations applied. However, it has some disadvantages such as shrinkage and collapse of the product structure, low kinetics, nutritional losses, long drying periods, and microbial contamination. To overcome these disadvantages, the convection drying method has been enhanced by combining it with the instant controlled pressure drop technology (DIC), which has obtained excellent results, to improve the drying process and the total quality of dried products [170]. This integration has been termed "Swell-drying," which involves a hot airflow for the pre-drying stage coupled with a DIC texturing stage. A swell-drying operation generally consists of a first hot airflow pre-drying stage until a specific water content (between 0.20 g and 0.50 g H2O/g of solid material), then followed by a DIC texturing stage (100–900 kPa during a few seconds, followed by an abrupt controlled pressure drop), and a final air-drying stage until the weight is stable. In this regard, [171] applied the DIC technology to the quality characteristics of cell wall polysaccharides of apple slices and studied their relationship to the texture. The results showed that it is possible to get apple chips with a crisp texture and excellent honeycomb-like structure by coupling convective air drying with the DIC technology (swell-drying). Also, the samples showed an excellent rehydration ratio to a homogeneous porous structure and a large specific surface area. As well, [172] applied the swell-drying technology on fresh banana pieces and studied the effect of dehydration kinetics, water and oil holding capacity, and nutritional characteristics. Drying kinetics showed that DIC technology increased the effective water diffusivity by 23%. Moreover, the water holding capacity increased by 290% under high-pressure conditions under DIC treatments, reaching 7.8 against 2.0 g H2O/g db, while the oil holding capacity was 0.60 against 1.30 g oil/g db for non-textured samples. Finally, it is noted that the DIC treatment inhibited the transformation of banana starch to reduced sugar. Additionally, concerning the strawberry fruits, the swell-drying technology is an excellent processing method to produce strawberry snacks with a high crispness behavior and a high rehydration capacity in less time than convect air-drying conventional method [173]. As pointed out by [174, 175], the DIC treatment has a significant impact on strawberries' drying and rehydration kinetics, increasing the effective diffusivity, as well obtain the highest levels of phenols, flavonoids, and anthocyanins and antioxidant activities were achieved at 350 kPa for 10 s.

6. Conclusions

Human habitats are safe, resilient, sustainable, and a universal access to affordable, reliable, and sustainable energy, It's a world everyone dreams about. The world's eyes are turning toward alternative sources of traditional energy. Although wind and hydro power are also making progress, solar energy is becoming preferred due to its increasing affordability. Food preservation by drying is the time-honored and most common method used by humans. Food dehydration is one of the most important achievements in human history, it has achieved food security for humans even under the adverse environmental conditions that make it difficult to supply fresh food daily. Food drying techniques have long been applied since ancient times in conventional ways, such as drying in the sun. From the traditional to the most innovative, there are several drying solutions. Successive technological developments that have been going on for a long time had a significant impact on the growing consumer desire to obtain high-quality and safe agricultural products, whether the quality of the external appearance or the quality of the nutritional value of these products, whether fresh or dried. Therefore, a huge development has been achieved in drying technology for agricultural products. Starting with solar drying, traditional methods by convection, and drying using electromagnetic spectra such as microwave drying and infrared dryers, and newly invented methods. This continuous, renewable, and accelerated innovation in the field of food drying technology is to fulfill the consumer's increasing desire to achieve food security not only by providing food but with high quality, healthy and safe food. So far, no comprehensive research on industrial solar drying has been performed, and hopefully, further development in solar dryers based on the three "A" principles will take the shape in future, that is, Affordability, Availability, and Accessibility.

Food for all, no poverty, renewable energy, and employment generation are some of the key features. The future scope of solar dehydrators is mostly to increase usability for small-scale farmers and users by decreasing costs, increasing reliability and versatility, and making it easier to obtain. This in turn will increase production and lower the crisis of food for the human population.

Acknowledgements

The work is supported by the 2020-1.1.2-PIACI-KFI-2021-00328 project. All thanks and appreciation to both the Agricultural Engineering Research Institute (AEnRI), Horticulture Research Institute, Agricultural Research Center (ARC), Egypt; Research Centre for Engineering and Food Processing, CREA, Via G. Venezian 26, 20133 Milan, Italy; and the Hungarian University of Agriculture and Life Sciences, Gödöllő, Hungary, for their support.

Conflict of interest

The authors declare no conflict of interest.

Author details

Ayman Ibrahim[1*], Tiziana M.P. Cattaneo[2], Alia Amer[3] and Lajos Helyes[4]

1 Agricultural Engineering Research Institute (AEnRI), Agricultural Research Center (ARC), Giza, Egypt

2 Research Centre for Engineering and Food Processing, CREA, Milan, Italy

3 Medicinal and Aromatic Plants Research Department, Horticulture Research Institute, Agricultural Research Center, Giza, Egypt

4 Horticultural Institute, Hungarian University of Agriculture and Life Sciences, Gödöllő, Hungary

*Address all correspondence to: aymanelgizawee@gmail.com

IntechOpen

© 2023 The Author(s). Licensee IntechOpen. This chapter is distributed under the terms of the Creative Commons Attribution License (http://creativecommons.org/licenses/by/3.0), which permits unrestricted use, distribution, and reproduction in any medium, provided the original work is properly cited.

References

[1] Guiné RPF. Food Drying and Dehydration – Technology and Effect on Food Properties. Saarsbruken: LAP LAMBERT Academic Publishing; 2015. ISBN: 978-3-659-75392-3

[2] Lee SC, Jeong SM, Kim SY, Park HR, Nam KC, Ahn DU. Effect of far-infrared radiation and heat treatment on the antioxidant activity of water extracts from peanut hulls. Food Chemistry. 2006;**94**(4):489-493

[3] Hii CL, Law CL. Product quality evolution during drying of foods, vegetables and fruits. In: Jangam SV, Law CL, Mujumdar AS, editors. Drying of Foods, Vegetables and Fruits. Vol. 1. 2010. pp. 125-144. ISBN: 978-981-08-6759-1

[4] Mujumdar AS, editor. Principles, Classification, and Selection of Dryers in Handbook of Industrial Drying. 3rd ed. CRC Press; 2006

[5] Araya-Farias M, Ratti C. Dehydration of foods: General concepts advances in food dehydration. In: Ratti C, editor. Advances in Food Dehydration. Boca Raton: CRC Press Taylor & Francis Group LLC; 2009. pp. 1-36. ISBN: 9780367386368

[6] Bala BK, Debnath N. Solar drying technology: Potentials and developments. Journal of Fundamentals of Renewable Energy and Applications. 2012;**2**. Article ID R120302. DOI: 10.4303/jfrea/R120302

[7] Udomkun P, Romuli S, Schock S, Mahayothee B, Sartas M, Wossen T, et al. Review of solar dryers for agricultural products in Asia and Africa: An innovation landscape approach. Journal of Environmental Management. 2020;**268**:110730. DOI: 10.1016/j.jenvman.2020.110730

[8] Rossi I, Asyifaa AH, Adiningsih FCA, Aulia GA, Achmad SR. Conventional and advanced food-drying technology: A current review. International Journal of Scientific & Technology Research. 2021;**10**(01):99-107

[9] Muncan J, Tsenkova R. Review: Aquaphotomics—from innovative knowledge to integrative platform in science and technology. Molecules. Open Access. 2019;**24**:2742. DOI: 10.3390/molecules24152742

[10] Jalata MW. Solar drying of fruits and windows of opportunities in Ethiopia. African Journal of Food Science. 2010;**4**(13):790-802. Available from: http://www.academicjournals.org/ajfs

[11] Bennamoun L. Reviewing the experience of solar drying in Algeria with presentation of the different design aspects of solar dryers. Renewable and Sustainable Energy Reviews. 2011;**15**(7):3371-3379. DOI: 10.1016/j.rser.2011.04.027

[12] Ndukwu MC, Bennamoun L, Abam FI. Experience of solar drying in Africa: Presentation of designs, operations, and models. Food Engineering Reviews. 2018;**10**(4):211-244. DOI: 10.1007/s12393-018-9181-2

[13] Ssemwanga M, Makule EE, Kayondo SI. The effect of traditional and improved solar drying methods on the sensory quality and nutritional composition of fruits: A case of mangoes and pineapples. Heliyon. 2020;**6**(6):e04163. DOI: 10.1016/j.heliyon.2020.e04163

[14] Imre LL. Solar drying. In: Mujumdar AS, editor. Handbook of

Industrial Drying. New York: Marcel Dekker; 1987. p. 357. DOI: 10.1080/07373938808916399

[15] Galanakis CM, Aldawoud TMS, Rizou M, Rowan NJ, Ibrahim SA. Food ingredients and active compounds against the coronavirus disease (COVID-19) pandemic: A comprehensive review. Food. 2020;**9**: 1701. DOI: 10.3390/foods 91117 01

[16] Galanakis CM, Rizou M, Aldawoud TMS, Ucak I, Rowan NJ. Innovations and technology disruptions in the food sector within the COVID-19 pandemic and post-lockdown era. Trends in Food Science and Technology. 2021;**110**:193-200. DOI: 10.1016/j.tifs.2021.02.002

[17] Galanakis CM. The food systems in the era of the coronavirus (COVID-19) pandemic crisis. Food. 2020;**9**:523. DOI: 10.3390/foods 9040523

[18] Mathlouthi M. In: Caballero B, Finglas P, Toldrá F editors. Encyclopedia of Food Sciences and Nutrition, 2nd ed. Academic Press, New York. DOI: 10.1016/B0-12-227055-X/ 00369-2.

[19] Chavan A, Vitankar V, Mujumdar A, Thorat B. Natural convection and direct type (NCDT) solar dryers: A review. Drying Technology. 2021;**39**(13):1-22. DOI: 10.1080/07373937.2020.1753065

[20] Shrivastava V, Kumar A, Baredar P. Developments in indirect solar dryer: A review. International Journal of Wind and Renewable Energy. 2014;**3**(4):67-74

[21] Siddiqui AM, Thange S, Thite R, Todkar A. Forced-convection indirect solar dryer. International Research Journal of Engineering and Technology. 2021;**8**(5):2872-2877. e-ISSN: 2395-0056, p-ISSN: 2395-0072

[22] Pydi Setty Y, Ramana Murthy JV. Development of a model for drying of solids in a continuous fluidized bed dryer. Indian Journal of Chemical Technology. 2003;**10**: 477-482

[23] Guerrero M, Albet C, Palomer A, Guglietta A. Drying in pharmaceutical and biotechnological industries. Food Science and Technology International. 2003;**9**:237-243

[24] Barrozo MAS, Mujundar A, Freire JT. Air-drying of seeds: A review. Drying Technology. 2014;**32**:1127-1141. DOI: 10.1080/07373937.2014.915220

[25] Yliniemi L. Advanced control of a rotary dryer [Ph.D. thesis]. Oulu, Finland: University of Oulu; 1999

[26] Santos DA, Duarte CR, Barrozo MAS. Segregation phenomenon in a rotary drum: Experimental study and CFD simulation. Powder Technology. 2016;**294**:1-10. DOI: 10.1016/j.powtec.2016.02.015

[27] Baker CGJ. Drying Technology. 1988;**6**(4):631-653, 754

[28] Kelly JJ, Tech. Ireland. 1969;**1**(2):25

[29] Kelly JJ, O'Donnell JP. Dynamics of granular material in rotary dryers and coolers. ICE Symposium Series. 1968;**29**: 38

[30] Poersch W. Verfahrenstechnik. 1971;**5**(5):186

[31] Cairo N, Colangelo G, Starace G. Performance analysis of two industrial dryers (cross flow and rotary) for ligno-cellulosic biomass desiccation. Renewable Energy & Power Quality Journal (RE&PQJ). 2012;**1**:274-280. DOI: 10.24084/repqj10.292

[32] Likitrattanaporn C, Ahmad I, Sirisoontaralak P, Noomhorm A, et al. Performance evaluation of a mobile rotary dryer for high moisture paddy. In: Proceedings of the Third Asia-Pacific Drying Conference, 1-3 September 2003; Asian Institute of Technology, Bangkok, Thailand. pp. 199-207

[33] Imran A, Noomhorm A. Grain process engineering. In: Kutz M, editor. Chapter 12 in Handbook of Farm, Dairy and Food Machinery Engineering. 3rd ed. Academic Press. ISBN 9780128148037; 2019. pp. 267-298. DOI: 10.1016/B978-0-e12-814803-7.00012-9

[34] Kerr WL. Food drying and evaporation processing operations. In: Handbook of Farm, Dairy and Food Machinery Engineering. 2019. pp. 353-387. DOI: 10.1016/b978-0-12-814803-7.00014-2

[35] Law CL, Mujumdar AS. Fluidized bed dryers. In: Handbook of Industrial Drying. 2006. p. 3

[36] Gupta CK, Sathiyamoorthy D. Fluid Bed Technology in Material Processing. New York: CRC Press; 1999. (Chap. 1)

[37] Mujumdar AS, Devahastin S. Applications for fluidized bed drying. In: Yang WC, editor. Handbook of Fluidization and Fluid Systems. New York: Marcel Dekker; 2003. (Chap. 18)

[38] Moore JG. Drum dryers. In: Mujumdar AS, editor. Handbook of Industrial Drying. 1st ed. New York: Marcel Dekker; 1987. p. 227

[39] Moore JG. Drum dryers. In: Mujumdar AS, editor. Handbook of Industrial Drying. 2nd ed. New York: Marcel Dekker; 1995. p. 249

[40] Okos MR, Narsimhan G, Singh RK, Weitnauer AC. Food dehydration. In: Heldman DR, Lund DB, editors. Handbook of Food Engineering. New York: Marcel Dekker; 1992. p. 437

[41] Baumann R. Efficiency of roller dryers. Chem. Ing. Technol. 1953;**25**:607

[42] Daud WRW, Armstrong WD. Pilot plant study of the drum dryer. In: Mujumdar AS, editor. Drying '87. New York: Hemisphere Publishing Corporation; 1987. p. 101

[43] Daud WRW. Drum dryers. In: Handbook of Industrial Drying. 4th ed. Boca Raton: CRC Press; 2006

[44] Fritze H. Dry gelatinized starch produced on different types of drum dryers. Die Strake. 1973;**25**:244

[45] Mahler S. Verfahren zur Verhinderung von Klumpenbildung bei der Aufloesung kalt quellender Staeund Staerkepraeparate. Germ. Pat. 1921; DE389023 C

[46] Tang J, Feng H, Shen GQ. Drum drying. In: Encyclopedia of Agricultural, Food, and Biological Engineering. New York: Marcel Dekker; 2003. p. 211

[47] Michael JK. Spray drying and spray congealing of pharmaceuticals. In: Encyclopedia of Pharmaceutical Technology. Vol. 14. NY: Marcel Dekker INC; 1993. pp. 207-221

[48] Abuaf N, Staub FW. Drying of liquid–solid slurry droplets. In: Mujumdar AS, editor. Drying '86. Vol. 1. New York: Hemisphere/McGraw-Hill; 1986. pp. 277-284

[49] Amelot MP, Gauvin WH. Spray drying with plasma-heated water vapor. In: Mujumdar AS, editor. Drying '86.

Vol. 1. New York: Hemisphere/McGraw-Hill; 1986. pp. 285-290

[50] Anderson BS. Low ox spray dryers. Chemical Age of India. 1979;**30**(11)

[51] Strumillo C, Markowski A, Kaminski W. Modern developments in drying of paste-like materials. In: Mujumdar AS, editor. Advances in Drying. Vol. 2. 1983. pp. 193-232

[52] Mujumdar AS. Superheated steam drying. In: Mujumdar AS, editor. Handbook of Industrial Drying. 2nd ed. New York: Marcel Dekker; 1995. pp. 1071-1086

[53] Gauvin WH, Costin MH. Spray drying in superheated steam. In: Mujumdar AS, editor. Drying '80. Vol. 1. New York: Hemisphere/McGraw-Hill; 1980. pp. 320-331

[54] Gauvin WH. Novel approach to spray drying using plasmas of water vapor. Canadian Journal of Chemical Engineering. 1981;**59**(6): 697-704

[55] Raehse W, Bauer V. Process for spray drying materisals and mixtures thereof using superheated steam, 1995; USA Patent 5431780

[56] Frydman A, Vasseur J, Ducept F, Moureh J. Simulation of spray drying in superheated steam using computational fluid dynamics. Drying Technology. 1999;**17**(7):1313-1326

[57] Kwamya, Mujumdar AS. M. Eg. Project Report. Canada: Dept. Chem. Eng., McGill University; 1984

[58] Mujumdar AS. Dryers for particulate solids, slurries and sheet form materials. In: Devahastion S, editor. Mujumdar Practical Guide to Industrial Drying. 2000. pp. 37-71

[59] Available from: http://www.ceroge rs.com/html/horizontal_dryer.html

[60] Available from: http://www.fesintl.com/htmfil.fld/sprydryh.html

[61] Cakaloz T, Akbaba H, Yesugey ET, Periz A. Drying model for a-amylase in a horizontal spray dryer. Journal of Food Engineering. 1997;**31**:499-510

[62] Flink JM. Energy analysis in dehydration processes. Food Technology. 1977;**31**(3):77-79

[63] Ratti C. Hot air and freeze drying of high-value foods: A review. Journal of Food Engineering. 2001;**49**: 311-319

[64] Bruttini, Ligus. Freeze drying. In: Mujumdar AS, editor. Handbook of Industrial Drying. 2nd ed. New York: Dekker; 1995. p. 1423

[65] Dunoyer JM, Larouse J. Experience nouvelles sur la lyophilization. In: Trans. of Eighth Vacuum Symposium and Second International Congress. Vol. 2. 1961. p. 1059

[66] Tadeusz K, Mujumdar AS. Advanced Drying Technologies. New York: Marcel Dekker; 2001

[67] Wolff E, Gibert H. Atmospheric freeze drying, part I: Design, experimental investigation and energy saving advantages; part II: Modeling drying kinetics using adsorption isotherms. Drying Technology. 1990;**8**(2):385-428

[68] Meryman HT. Sublimation freeze drying without vacuum. Science. 1959; **130**:628

[69] Woodard HT. Freeze-drying without vacuum. Food Engineering. 1963;**35**(6):9

[70] Boeh-Gcansey O. Freeze drying in a fluidized-bed atmospheric dryer and in a vacuum dryer: Evaluation of external transfer coefficients. Journal of Food Engineering. 1988;**7**:127-146

[71] Mumenthaler M, Leuenberger H. Atmospheric spray-freeze drying: A suitable alternative in freeze-drying technology. International Journal of Pharmaceutics. 1991;**72**:97-110

[72] Lombrana J, Villaran MC. The influence of pressure and temperature on freeze drying in an adsorbent medium and establishment of drying strategies. Food Research International. 1997;**30**(3–4):213-222

[73] Risch SJ. Encapsulation: Overview of uses and techniques, encapsulation and controlled release of food ingredients. In: Risch SJ, Reineccius GA, editors. ACS Symposium Series 590. Washington, DC: American Chemical Society; 1995. pp. 2-7

[74] King CJ. In: Toei R, Mujumdar AS, editors. Control of Food-Quality Factors in Spray Drying, Drying '85. Hemisphere, New York; 1985. pp. 59-66

[75] Dziezak JD. Micro-encapsulation and encapsulated ingredients. Food Technology. 1988;**42**(4):136-148, 151

[76] Arshady R, George MH. Suspension, dispersion, and interfacial polycondensation—A methodological survey. Polymer Engineering and Science. 1993;**33**:865-876

[77] Huang LX, Wang Z, Tang J. Recent progress of spray drying in China (in Chinese). Chemical Engineering. (China). 2001;**29**(2):51-55

[78] Filková I, Huang LX, Mujumdar AS. Industrial spray drying systems. In: Mujumdar AS, editor. Handbook of Industrial Drying. 3rd ed. Boca Raton, Fl: CRC Press Taylor and Francis Group; 2006. pp. 215-256

[79] Assegehegn G, Brito-de la Fuente E, Franco JM, Gallegos C. The importance of understanding the freezing step and its impact on freeze-drying process performance. Journal of Pharmaceutical Sciences. 2019;**108**:1378-1395

[80] Haseley P, Oetjen GW. Freeze-Drying. Veinheim, Germany: Wiley-VCH; 2018. p. 421

[81] Mellor JD. Fundamentals of Freeze-Drying. London, UK: Academic Press Inc.; 1978. p. 386

[82] Adams GDJ, Cook I, Ward KR. The principles of freeze-drying. In: Wolkers WF, Oldenhof H, editors. Cryopreservation and Freeze-Drying Protocols. 3rd ed. Vol. 1257. Totowa, NJ, USA: Humana Press Inc.; 2015. pp. 121-143

[83] Franks F, Auffret T. Freeze-Drying of Pharmaceuticals and Biopharmaceuticals. Cambridge, UK: RSC Publishing; 2008. p. 218

[84] Liu YZ, Zhao YF, Feng X. Exergy analysis for a freeze-drying process. Applied Thermal Engineering. 2008;**28**: 675-690

[85] Meda L, Ratti C. Rehydration of freeze-dried strawberries at varying temperatures. Journal of Food Process Engineering. 2005;**28**:233-246

[86] Jia Y, Khalifa I, Hu L, Zhu W, Li J, Li K, et al. Influence of three different drying techniques on persimmon chips' characteristics: A comparison study among hot-air, combined hot-air-microwave, and vacuum-freeze drying techniques. Food and Bioproducts Processing. 2019;**118**:67-76

[87] Liapis AI, Bruttini R. Freeze drying. In: Mujumdar AS, editor. Handbook of Industrial Drying. 4th ed. Boca Raton, FL, USA: CRC Press; 2014. pp. 259-282

[88] Nowak D, Jakubczyk E. The freeze-drying of foods—The characteristic of the process course and the effect of its parameters on the physical properties of food materials. Food. 2020;**9**:1488. DOI: 10.3390/foods9101488

[89] Oikonomopoulou VP, Krokida MK, Karathanos VT. The influence of freeze drying conditions on microstructural changes of food products. Procedia Food Science. 2011;**1**:647-654

[90] Krzykowski A, Dziki D, Rudy S, Gawlik-Dziki U, Polak R, Biernacka B. Effect of pre-treatment conditions and freeze-drying temperature on the process kinetics and physicochemical properties of pepper. LWT. 2018;**98**: 25-30

[91] Martinez-Navarrete N, Salvador A, Oliva C, Camacho MM. Influence of biopolymers and freeze-drying shelf temperature on the quality of a mandarin snack. LWT. 2019;**99**:57-61

[92] Wu XF, Zhang M, Bhandari B. A novel infrared freeze drying (IRFD) technology to lower the energy consumption and keep the quality of Cordyceps militaris. Innovative Food Science and Emerging Technologies. 2019;**54**:34-42

[93] Silva-Espinoza MA, Ayed C, Foster T, Camacho MDM, Martinez-Navarrete N. The impact of freeze-drying conditions on the physico-chemical properties and bioactive compounds of a freeze-dried orange puree. Foods. 2019;**9**:32

[94] Khalloufi S, Ratti C. Quality deterioration of freeze-dried foods as explained by their glass transition temperature and internal structure. Journal of Food Science. 2003;**68**: 892-903

[95] Egas-Astudillo LA, Martínez-Navarrete N, Camacho MM. Impact of biopolymers added to a grapefruit puree and freeze-drying shelf temperature on process time reduction and product quality. Food and Bioproducts Processing. 2020;**120**:143-150

[96] Bhatta S, Stevanovic Janezic T, Ratti C. Freeze-drying of plant-based foods. Foods. 2020;**9**(1):87. DOI: 10.3390/foods9010087

[97] Burke RF, Decareau RV. Advances in Food Research. 1964;**13**:1-88

[98] Copson DA. Microwave Heating. Westport, CN: AVI Publishing; 1962

[99] Goldblith SA, Rey L, Rothmayr WW. Freeze Drying and Advanced Food Technology. London: Academic Press; 1975

[100] Hoover MW, Markantonatos A, Parker WN. Food Technology. 1966;**20**: 107-110

[101] Bouldoires JP, LeViet T. Microwave freeze-drying of granulated coffee. In: Second International Symposium on Drying. Montreal, Canada: McGill University; 1980

[102] Sunderland JE. An economic study of microwave freeze- drying. Food Technology. 1982;**36**(2):50-56

[103] Rey L, May JC, editors. Freeze Drying/Lyophilization of Pharmaceutical and Biological Products. New York: Marcel Dekker; 1999

[104] Wang ZH, Shi MH. Numerical study on sublimation-condensation

phenomena during microwave freeze drying. Chemical Engineering Science. 1998;**53**:3189-3197

[105] Wang ZH, Shi MH. The effects of sublimation-condensation region on heat and mass transfer during microwave freeze drying. Journal of Heat Transfer. 1998;**120**:654-660

[106] Snowman JW. Lyophilization techniques, equipment, and practice. In: Downstream Processes: Equipment and Techniques. New York: Alan R. Liss; 1988. p. 315

[107] Thuse E, Ginnette LF, Derby R. U.S. Patent 1968;**3**:362,835

[108] Bruttini R, Liapis AI. Freeze Drying. Handbook of Industrial Drying: CRC Press; 2007

[109] Mayor L, Sereno AM. Modelling shrinkage during convective drying of food materials: A review. Journal of Food Engineering. 2004;**61**(3):373-386

[110] Mujumdar AS. Handbook of Industrial Drying. 3rd ed. CRC Press; 2006. DOI: 10.1201/9781420017618

[111] Rastogi NK. Recent trends and developments in infrared heating in food processing. Critical Reviews in Food Science and Nutrition. 2012;**52**(9): 737-760

[112] Kocabiyik H, Tezer D. Drying of carrot slices using infrared radiation. International Journal of Food Science and Technology. 2009;**44**(5):953-959

[113] Nowak D, Lewicki PP. Infrared drying of apple slices. Innovative Food Science and Emerging Technologies. 2004;**5**(3):353-360

[114] Sakare P, Prasad N, Thombare N, Singh R, Sharma SC. Infrared drying of food materials: Recent advances. Food Engineering Reviews. 2020;**12**:381-398

[115] Yan JK, Wu LX, Qiao ZR, Cai WD, Ma H. Effect of different drying methods on the product quality and bioactive polysaccharides of bitter gourd (*Momordica charantia* L.) slices. Food Chemistry. 2019;**271**:588-596

[116] Venkitasamy C, Zhu C, Brandl MT, Niederholzer FJ, Zhang R, McHugh TH, et al. Feasibility of using sequential infrared and hot air for almond drying and inactivation of enterococcus faecium NRRL B-2354. LWT—Food Science and Technology. 2018;**95**(11):123-128

[117] Chen Q, Bi J, Wu X, Yi J, Zhou L, Zhou Y. Drying kinetics and quality attributes of jujube (*Zizyphus jujuba* Miller) slices dried by hot-air and short- and medium-wave infrared radiation. LWT—Food Science and Technology. 2015;**64**(2):759-766

[118] Zare D, Naderi H, Ranjbaran M. Energy and quality attributes of combined hot-air/infrared drying of paddy. Drying Technology. 2015;**33**(5): 570-582

[119] Bualuang O, Tirawanichakul Y, Tirawanichakul S. Comparative study between hot air and infrared drying of parboiled rice: Kinetics and qualities aspects. Journal of Food Processing & Preservation. 2013;**37**(6):1119-1132

[120] Sakai N, Hanzawa T. Applications and advances in farinfrared heating in Japan. Trends in Food Science and Technology. 1994;**5**:357-362

[121] Skjöldebrand C. Infrared heating. In: Richardson P, editor. Thermal Technologies in Food Processing. New York, NY, USA: CRC Press; 2001

[122] Calín-Sánchez Á, Lipan L, Cano-Lamadrid M, Kharaghani A, Masztalerz K, Carbonell-Barrachina ÁA, et al. Comparison of traditional and novel drying techniques and its effect on quality of fruits, vegetables and aromatic herbs. Foods. 2020;**9**:1261. DOI: 10.3390/foods9091261

[123] Boudhrioua N, Bahloul N, Ben Slimen I, Kechaou N. Comparison on the total phenol contents and the color of fresh and infrared dried olive leaves. Industrial Crops and Products. 2009;**29** (2–3):412-419

[124] Fasina O, Tyler B, Pickard M, Zheng GH, Wang N. Effect of infrared heating on the properties of legume seeds. International Journal of Food Science and Technology. 2008;**36**(1): 79-90

[125] Pan Z, Atungulu GG. Infrared Heating for Food and Agricultural Processing. 1st ed. CRC Press; 2010. DOI: 10.1201/9781420090994

[126] Hung JY, Wimberger RJ, Mujumdar AS. Drying of coated webs. In: Mujumdar AS, editor. Handbook of Industrial Drying. 2nd ed. New York, NY, USA: Marcel Dekker Inc; 1995. pp. 1007-1038

[127] Pawar SB, Pratape VM. Fundamentals of infrared heating and itsapplication in drying of food materials: A review. Journal of Food Process Engineering. 2017;**40**:1745-4530. DOI: 10.1111/jfpe.12308

[128] Sayed-Yagoobi J, Wirtz JW. An experimental study of gas-fired infrared drying of paper. Drying Technology. 2001;**19**:1099-1112

[129] Sun DW. Thermal Food Processing: New Technologies and Quality Issues. 2nd ed. Boca Raton, FL, USA: CRC Press; 2012. pp. 618-620

[130] Richardson P. Thermal Technologies in Food Processing. 1st ed. Boca Raton, FL, USA: CRC Press; 2001. pp. 292-294

[131] Salehi F. Recent applications and potential of infrared dryer systems for drying various agricultural products: A review. International Journal of Fruit Science. 2020;**20**(3):586-602

[132] Zhu Y, Pan Z. Processing and quality characteristics of apple slices under simultaneous infrared dry-blanching and dehydration with continuous heating. Journal of Food Engineering. 2009;**90**: 441-452

[133] Zartha Sossa JW, Orozco GL, García Murillo LM, Peña Osorio M, Sánchez SN. Infrared drying trends applied to fruit. Frontiers in Sustainable Food Systems. 2021;**5**:650690

[134] Mosavi Baygi SF, Farahmand A, Taghi Zadeh M, Zia FA. Modeling on hot air and infrared thin layers drying of persimmon slices. Food Science & Technology. 2015;**13**(53):161-171

[135] Salehi F, Kashaninejad M, Siahmansouri P, Moradi E. Moisture loss kinetics of persimmon during combined hot air-infrared drying process. Journal of Food Technology & Nutrition Sciences. 2017;**14**(2):39-48

[136] Sun J, Hu X, Zhao G, Wu J, Wang Z, Chen F, et al. Characteristics of thin-layer infrared drying of apple pomace with and without hot air pre-drying. Food Science and Technology International. 2007;**13**(2):91-97

[137] Xu C, Li Y, Yu H. Effect of far-infrared drying on the water state and

glass transition temperature in carrots. Journal of Food Engineering. 2014;**136**: 42-47

[138] Pan Z, Shih C, McHugh TH, Hirschberg E. Study of banana dehydration using sequential infrared radiation heating and freeze-drying. LWT - Food Science and Technology. 2008;**41**(10): 1944-1951

[139] Pekke MA, Pan Z, Atungulu GG, Smith G, Thompson JF. Drying characteristics and quality of bananas under infrared radiation heating. International Journal of Agricultural and Biological Engineering. 2013;**6**(3):58-70

[140] Stuchly SS, Stuchly MA. Microwave drying: Potential and limitations. In: Mujumdar AS, editor. Advances in Drying. New York: Hemisphere Publishing Corporation; 1980

[141] Feng H, Yin Y, Tang J. Microwave drying of food and agricultural materials: Basics and heat and mass transfer modeling. Food Engineering Reviews. 2012;**4**(2):1-18

[142] Alibas I, Yilmaz A. Microwave and convective drying kinetics and thermal properties of orange slices and effect of drying on some phytochemical parameters. Journal of Thermal Analysis and Calorimetry. 2021;**147**:8301-8321

[143] Ramaswamy HS, Pillet-Will T. Distribution and equalization of temperature in a microwave heated food model. Journal of Food Quality. 1992; **15**(16):435-448

[144] Tulasidas TN, Raghavan GSV, Mujumdar AS. Microwave drying of grapes in a single mode cavity at 2450 MHz –II: Quality and energy aspects. Drying Technology. 1995;**13**(8/9): 1973-1992

[145] Ruız Dıaz G, Martınez-Monzo J, Chiralt PFA. Modelling of dehydration-rehydration of orange slices in combined microwave/air drying. Innovative Food Science and Emerging Technologies. 2003;**4**:203-209

[146] Soysal Y. Microwave drying characteristics of parsley. Biosystems Engineering. 2004;**89**:167-173

[147] Orsat V, Changrue V, Vijaya RGS. Microwave drying of fruits and vegetables. Stewart Postharvest Review. 2006;**6**(4):1-7

[148] Joardder MUH, Karim A, Kumar C. Effect of temperature distribution on predicting quality of microwave dehydrated food. Journal of Mechanical Engineering Science. 2013;**5**:562-568. DOI: 10.15282/jmes.5.2013.2.0053

[149] Maskan M. Drying, shrinkage and rehydration characteristics of kiwi fruits during hot air and microwave drying. Journal of Food Engineering. 2001;**48**: 177-182

[150] Sham PWY, Scaman CH, Durance TD. Texture of vacuum microwave dehydrated apple chips as affected by calcium pretreatment, vacuum level, and apple variety. Journal of Food Science. 2001;**66**(9): 1341-1347

[151] Andres A, Bilbao C, Fito P. Drying kinetics of apple cylinders under combined hot air–microwave dehydration. Journal of Food Engineering. 2004;**63**:71-78

[152] Sunjka PS, Rennie TJ, Beaudry C, Raghavan GSV. Microwave–convective and microwave–vacuum drying of cranberries: A comparative study. Drying Technology. 2004;**22**(5): 1217-1231

[153] Wojdyło A, Figiel A, Lech K, Nowicka P, Oszmiański J. Effect of convective and vacuum–microwave drying on the bioactive compounds, color, and antioxidant capacity of sour cherries. Food and Bioprocess Technology. 2014;7:829-841

[154] Sharma GP, Prasad S. Drying of garlic (Allium sativum) cloves by microwave-hot air combination. Journal of Food Engineering. 2001;**50**:99-105

[155] Alibas I. Characteristics of chard leaves during microwave, convective, and combined microwave-convective drying. Drying Technology. 2006;**24**: 1425-1435

[156] Śledź M, Witrowa-Rajchert D. Influence of microwave-convective drying on chlorophyll content and colour of herbs. Acta Agrophysica. 2012;**19**(4): 865-876

[157] Yarahmadi N, Hojjatoleslamy M, Boroujeni LS. Different drying methods of Pistacia Atlantica seeds: Impact on drying kinetics and selected quality properties. Food Science & Nutrition. 2019;**8**:3225-3233

[158] Nawirska A, Figiel A, Kucharska AZ, Sokół-Łetowska A, Biesiada A. Drying kinetics and quality parameters of pumpkin slices dehydrated using different methods. Journal of Food Engineering. 2009;**94**: 14-20

[159] Sutar PP, Prasad S. Modeling microwave vacuum drying kinetics and moisture diffusivity of carrot slices. Drying Technology. 2007;**25**(10): 1695-1702

[160] Figiel A. Drying kinetics and quality of beetroots dehydrated by combination of convective and vacuum-microwave methods. Journal of Food Engineering. 2010;**98**:461-470

[161] Wanxiu X, Guanyu Z, Chunfang S, Shaogang H, Zhenfeng L. Optimization of microwave vacuum drying and pretreatment methods for polygonum cuspidatum. Mathematical Problems in Engineering. 2018;**2018**:1-11

[162] Allaf K, Vidal P. Feasibility study of a new process of drying/swelling by instantaneous decompression toward vacuum of in pieces vegetables in view of a rapid re-hydration. Gradient activity plotting university of technology of compiegne UTC N° CR/89/103, industrial SILVA-LAON partner. 1989

[163] Allaf K, Allaf T. Instant Controlled Pressure Drop (D.I.C.) in Food Processing: From Fundamental to Industrial Applications. New York: Springer; 2013

[164] Albitar N, Mounir S, Besombes C, Allaf K. Improving the drying of onion using the instant controlled pressure drop technology. Drying Technology. 2011;**29**:993-1001

[165] Ben Amor B, Allaf K. Impact of texturing using instant pressure drop treatment prior to solvent extraction of anthocyanins from malaysian roselle (hibiscus sabdariffa). Food Chemistry. 2009;**115**:820-825

[166] Haddad J, Allaf KA. A study of the impact of instantaneous controlled pressure drop on the trypsin inhibitors of soybean. Journal of Food Engineering. 2007;**79**(1):353-357

[167] Haddad J, Greiner R, Allaf K. Effect of instantaneous controlled pressure drop on the phytate content of lupin. LWT—Food Science and Technology. 2007;**40**(3):448-453

[168] Louka N, Allaf K. Expansion ratio and color improvement of dried vegetables texturized by a new process "controlled sudden decompression to the vacuum": Application to potatoes, carrots and onions. Journal of Food Engineering. 2004;**65**(2):233-243. DOI: 10.1016/j.jfoodeng.2004.01.020

[169] Pech-Almeida JL, Téllez-Pérez C, Alonzo-Macías M, Teresa-Martínez GD, Allaf K, Allaf T, et al. An overview on food applications of the instant controlled pressure-drop technology, an innovative high pressure-short time process. Molecules. 2021;**126**:6519, 1-37

[170] Mounir S, Allaf T, Mujumdar AS, Allaf K. Swell drying: Coupling instant controlled pressure drop DIC to standard convection drying processes to intensify transfer phenomena and improve quality—An overview. Drying Technology. 2012;**30**:1508-1531

[171] Xiao M, Yi J, Bi J, Zhao Y, Peng J, Hou C, et al. Modification of cell wall polysaccharides during drying process affects texture properties of apple chips. Journal of Food Quality. 2018;**2018**:1-11

[172] Setyopratomo P, Fatmawati A, Sutrisna PD, Savitri E, Allaf K. The dehydration kinetics, physical properties and nutritional content of banana textured by instantaneous controlled pressure drop. Asia-Pacific Journal of Chemical Engineering. 2012;**7**: 726-732

[173] Alonzo-Macías M, Montejano-Gaitán G, Allaf K. Impact of drying processes on strawberry (Fragaria var. Camarosa) texture: Identification of crispy and crunchy features by instrumental measurement. Journal of Texture Studies. 2014;**45**:246-259

[174] Alonzo-Macías M, Cardador-Martínez A, Mounir S, Montejano-Gaitán G, Allaf K. Comparative study of the effects of drying methods on antioxidant activity of dried strawberry (Fragaria Var. Camarosa). Journal of Food Research. 2013;**2**: 92-107

[175] Maritza AM, Sabah M, Anaberta CM, Montejano-Gaitán JG, Allaf K. Comparative study of various drying processes at physical and chemical properties of strawberries (Fragaria var camarosa). Procedia Engineering. 2012;**42**:297-312

Section 2

Food Packaging

Chapter 8

The Importance of Packaging in Certified Organic Food: A Matter of Conscience

Rosana Carvalho Esteves

Abstract

Organic food production offers many advantages and is more environmentally friendly than conventional food. However, despite the certification guaranteeing organic management, soil and water preservation, and social well-being, it generally does not assess the type of packaging the food is sold in. The legislation of some countries, such as Brazil, does not require the use of biodegradable packaging. Thus, it is common to find organic food sold in plastic or styrofoam packaging, making it ecologically incorrect throughout its life cycle. This study demonstrates the importance of consumer awareness and how it can act in a global paradigm shift, demanding greater environmental responsibilities from those who produce the food. Also, the role of consumers, fulfilling their part as non-polluting agents, is essential for ecological well-being. A qualitative meta-analysis showed that, although green consumers are gradually increasing, they are still in the minority. For the time being, they are currently insufficient to generate significant changes in the production chain. Because of this, it is important that public, technological, and environmental institutions talk about these issues more and start calling for specific laws about the right way to use packaging, especially for certified organic foods.

Keywords: organic food, circular economy, packaging, sustainable packaging, waste; eco-friendly packaging, eco-friendly consumers; green polymers

1. Introduction

Organic food is not simply food that has been grown without the use of pesticides. It is a complex manufacturing system with production units that must be legalized and certified by a third party. Furthermore, each unit must implement sustainable management practices, which must be documented in the field notebook and approved and verified regularly by the certifying entity. In terms of epistemology, it must also ensure the social development of the productive area and rural community where it is inserted [1, 2].

An organic production system is guided and supported by four main pillars: soil preservation, water preservation, social well-being development, and environmental biodiversity promotion. Organic food production is much better for the environment than monocultures or systems that use a large number of pesticides. The essence of this system is to provide much more than just kilograms of food per hectare (kg/ha). While it ensures long-term environmental conservation and mitigates the effects of climate change, the purpose of organic production as established in the laws of several countries is to offer products free of genetically modified organisms (GMO) and intentional contaminants. Likewise, it must guarantee the preservation of the biological diversity of natural ecosystems as well as the restoration or increase in the life multiplicity of modified ecosystems [3].

In accordance with a standard definition, the soil, air, and water must be conserved and kept healthy, fruitful, and free of toxins during the entire process [4]. In Brazil and the European Community, the certification requirements established in the legislation are quite congruent and similar. However, specific parameters for the types of packaging that are used for the storage of organic foods are not included in these legislations. In the United States, there is a recommendation for biodegradable packaging.

2. General objective

The present study aimed to demonstrate the importance of consumer awareness and how it can act in a global paradigm shift demanding greater environmental responsibilities from those who produce the food, using meta-analysis qualitative for research. In addition, provide subsidies to regulatory bodies for the certification of organic products so that organic production is incorporated into the concept of a circular economy by adopting specific packaging requirements. This strategy is unheard of in Brazil.

2.1 Specific objectives

- Evaluate the issue of certification and the use of environmentally friendly packaging;

- To carry out a qualitative goal regarding trends in consumer behavior toward the packaging of food products in retail;

- List the types of biodegradable materials most commonly used in food packaging;

- Relate the circular economy with the production of organic foods.

3. Material and methods

In this study, an approach using the qualitative meta-analysis searching in the knowledge databases for research on the behavior of consumers of organic food in front of food packaging [5]. On the research bases, studies on ecologically correct packaging were also evaluated. Search base: Web of Science, ACM Digital Library, Cambridge Core, Journal Citation Reports, Scopus, Science Direct, *Periodicos Capes (Brazil)*.

Note: Research carried out by the author using keywords: packaging; organic product packaging, sustainable packaging+organic food; organic food waste; biological food waste, and biological packaging concerns the packaging of organic products [6].

4. Regulation in developed and developing countries

Rules and regulatory requirements for products in general are the effects of certain programs that are established by governments. The creation of these programs is due to a set of actions, within the public policies established by the governments, in compliance with certain sectors for the well-being of civil society. They are also the main instruments that the government uses to promote integration between sectors, standardize production criteria, establish means of supervision and control, etc.

Regulations arise from specific technical meetings of certain committees formed by specialized technicians. These committees meet for a certain time to discuss, evaluate, and establish what requirements and criteria should be followed by stakeholders. In the case of Brazil, the requirements and regulations for organic production were established within a National Organic Production Program, specifically in the Special Commission. This commission was initially formed to propose standards for the certification of organic products [7]. Technicians and specialists from the Ministry of Agriculture, Livestock and Supply participated together with the Federal Government. As a result, law 10.831/2003 was published [8], considered the legal framework of Brazilian organic agriculture. Law 10.831/2003 set up a framework for each theme in organic agriculture. This framework is made up of decree ordinances and specific normative instructions.

In the same manner, the United States forms its normative framework within public policy programs, also established by technicians and experts [9]. In the European Community, each member country had its own regulations. It was then necessary to align the regulations of each member country and standardize them so that the free trade in products between the member countries would comply with a single regulation. Currently, the European Council governs and is responsible for the management of the national organic production program [10].

The main regulatory requirements that are part of the regulations for organic products are described in **Table 1**. To illustrate the developing countries, Brazilian legislation was used as a reference because Brazil is a country with increasing growth in organic production. For developed countries, the United States is considered a country that has very strict legislation and is the largest market for organic products in the world, followed by the European Community, represented here as well.

Analyzing **Table 1**, basically, the described requirements are quite similar and congruent. Brazilian legislation is analogous to European legislation, while American legislation is more discerning and stricter.

5. Organic seal of certification

For food to be labeled as organic, it must undergo a certification procedure supervised by a notified agency. The primary function of the notifying body is to ensure that the entire production system conforms to the regulatory bodies' normative requirements. For example, the institutions in charge of regulating this type of agriculture are the USDA in the United States, the European Community Council in Europe, and the Ministry of Agriculture, Livestock [11], and Supply in Brazil. Once certified, the "organic" product must receive the certification seal of each country on its label, assuring the buyer of its legitimacy and dependability.

Countries/ requirements	Brazil	United States	European community
Definition	Organic: • Natural resources; • Respecting cultural and social aspects; • Biological, cultural and mechanical methods.	Organic or biological Biological, cultural and mechanical methods.	Organic or biological or ecological • Natural resources; • Respecting cultural and social aspects; • Biological, cultural and mechanical methods.
Seeds and seedlings	Only organic seeds		
Conversion time	12–18 months according to kind of culture	About 3 years	2–3 years
Composting	Use permitted provided approved by the certifying body	It follows strict and specific criteria for use	Use permitted provided approved by the certifying body
Seal of conformity	Mandatory		
Fertilization	Natural methods such as: crop rotation, and promotion of soil microbial activity. Use of external substances only with natural agricultural inputs permitted by the certifiers		
Pest control	Natural methods and natural agricultural inputs permitted by the certifiers		
Soil analysis to verify pesticide contamination	Only in case of suspicion by the regulator or certifiers	Performed periodically	Only in case of suspicion by the regulator or certifiers
Water	Conservation and chemical and biological control of irrigation water		
Packaging for marketing	There are no requirements for materials used	It follows specific regulations, is recommended the use of biodegradable materials	There are no requirements for materials used
GMO	Forbidden use		
Parallel production—organic and conventional in the same production unit	Allowed as long as the productive unit ensures full control so that it does not have cross-contamination. The certifier assesses the risk and approves the use of parallel production		
Organic management plan (field or manual notebook)	mandatory use where all production practices, controls, inputs etc. are recorded		
Product preservation	The production unit shall ensure the integrity and traceability of the product throughout the production process to the point of sale		
Bilateral equivalence between countries	Brazil and Chile	The United States and European Community	The European Community and the United States

Table 1.
Regulatory requirements for the organic foods for developed (references: USA and EU) and developing countries - adapt for [1].

These regulatory entities, which are country-specific, create the regulations and requirements that must be satisfied in order for the producer to ensure the conformity of his product. The certification procedure is a great technique for verifying and validating legal conformity. As each country has its own laws and certification processes, each country also has its own conformity seals.

Regrettably, unlike ISO standards (International Standard Organization), there is no single certification standard for all countries. As a result, each country has its own set of rules. The lack of regulatory equality makes free trade difficult, and those farmers who want to export their products, they have to pay the cost of certification in each country.

In the case of the United States and the European Union, there is already a bilateral agreement that allows certain products to be accepted. Except for Chile, the organic seal (**Figure 1**) for Brazil is not recognized in any other country.

In Brazil, all certified organic products sold in commercial establishments must have on their labels a seal that identifies compliance with current regulations (**Figure 1**). This requirement is included in the normative instruction of the Ministry of Agriculture, Livestock and Supply, IN 50/2009 [13].

In the United States, certification is based (**Figure 2**) on rules set by the USDA-NOP National Organic Program—Part. 205-311 [14]. In Brazil, products must also have the seal of conformity shown in **Figure 1**.

The member countries of the European Union also require labeling containing the seal of compliance, as decided by the European Council [10]- item 25 issued on 06/28/2007, and No. 889/2008 issued on 09/05/2008 and No. 834/2007 [15].

The European Council set out all the requirements for certification that are applicable to all member states. The Seal of Conformity, shown in **Figure 3**, is used in the whole European Union on products that are packaged for retail sale and on items that are imported from countries that are not part of the European Community. They must be certified in accordance with Community regulations.

Organic products are those that comply with the regulations attested by the certification bodies. Products processed with organic raw materials are recognized as organic if at least 95% of their content is composed of organic ingredients [16–18]

Figure 1.
Brazilian organic seal [12].

Figure 2.
USDA organic seal [14].

Figure 3.
European Community biological (organic) seal.

In the United States, the USDA-NOP National Organic Program - Part 205 classifies products as (a) "100% organic", (b) "Organic"—with at least 95% organic ingredients, and (c) "made with organic ingredients"—composed at least of 70% organic ingredients, and maximum 30% from conventional agriculture [14].

Similarly, in Brazilian regulation, a product having composition from 70% to 95% as organic ingredients shall be labeled as a "product with organic ingredients". In this case, it is not a normative requirement to report which ingredients are organic, but some producers prefer to explicitly put them on the label [18].

Although each country's regulations establish specific rules for the use of seals, they are still not representative. This is because many consumers have difficulties recognizing these seals and what they exactly mean. Aggravated, most consumers are unaware of the regulatory requirements that food must meet to bear the seal of compliance.

This lack of consumer awareness stems from a lack of information and visibility of certification seals. One way to leverage this market would be the more effective participation of public authorities. In countries where the government has drastically influenced the adoption of public policies in this sector, organic production has grown and consequently, consumption has grown, as is the case in Ethiopia [19] and Australia [20]. In Ethiopia, the government took strong steps and gave subsidies to encourage the change of conventionally farmed land into organically farmed land.

Visibility and awareness actions will make the seals play an important role in the consumer's choice of product at the time of purchase.

Another critical challenge is making consumers know how to associate the seal with all the difficulties that producers have to overcome to meet numerous regulatory requirements. It is not only about financial difficulties, but also that farmer have many difficulties in maintaining the integrity and seasonality of the crop, defeating diseases and pests, and using natural methods of control and combat. When a product arrives at retail with an organic seal on the label, it is an achievement for every producer, and this achievement, without a doubt, deserves to be recognized by consumers [21].

6. The importance of the consumer in the production chain

Conquering more and more consumers through the seal of compliance is a major challenge. In this context, Janssen and Hamm [21] demonstrated that there is often a lack of consumer knowledge about organic certification schemes. Most consumers are unaware that there are several types of seals and product labels, as well as that organic manufacturing, is subject to a control system.

Furthermore, customer evaluations of organic labeling systems are often subjective and, as a result, not supported by knowledge of what the seal of compliance actually means Janssen and Hamm [21]. Essoussi and Zahaf [22] show that only people who buy organic products regularly in North America know how to tell the difference between labels and what they mean.

The fact that natural products sold directly to the consumer, such as in street fairs, do not need to be labeled nor have the seal of conformity affixed to them, ends up contributing to their lack of knowledge [11, 23].

Along with the study by Janssen and Harm [21], the knowledge profile on labeling in the European Community was subpar and insufficient, and it also lacked objective consumer information. In the United States, the mark of conformity affixed to the food label distinguishes the product as organic and is easier for customers to recognize [22].

It is important to emphasize that different nations have different levels of confidence and trust in organic products. According to Sønderskov and Daugbjerg [24], label confidence is higher in areas where the state and government are actively involved and have a significant impact. For instance, there is a significant amount of trust placed on labels in Scandinavian nations [24–26]. Label trust is not as well-articulated in the UK and Brazil, where it is also fairly shaky [26].

The number of different types of labels and seals that exist on foods often confuses the consumer. There are organic or biological products, natural, denomination of controlled origin, fair trade, vegan, cruelty-free, HACCP—Hazard Analysis and Critical Control Point, Certified Humane, etc. [27]. Some are obtained voluntarily and others on a compulsory basis. Many of these seals are difficult for the consumer to understand. Customers in Brazil, for instance, still don't know how to tell the difference between organic and agroecological products [11].

Researchers should further investigate consumer behavior in relation to labels, seals, and certifications' precise meanings. Due to a lack of knowledge about seals, it is extremely difficult for consumers to determine what an organic product actually is. Further action is required in order to address the issue of the consumer's unfamiliarity with the relevance of the seal of conformity. Ensuring the consumer's perception of identifying a seal, and understanding what exactly it means, can be a useful tool for improving the offer [28].

Ensuring consumer perception by identifying a seal and understanding exactly what it means can be a useful tool to improve the offer of organic products. Thus, valuing this cognitive perception will make the consumer to choose more organic foods than conventional foods in the purchase decision process [2]. In fact, retail chains and commercial establishments can adopt strategies and actions to promote products of this nature. Even educational campaigns. In the United States, for example, the number of establishments dedicated exclusively to organic products is growing. In Brazil, this reality is still far from happening [29]. Unfortunately, the consumer's cognitive perception still associates the consumption of organic products with those of a more favored social class.

Another important point and rarely studied by the scientific community, given the lack of articles in the databases.

If the perception of the eventual consumer regarding the seal of conformity is already difficult, imagine his/her perception regarding the use of non-sustainable packaging. However, when addressing this issue to active consumers accustomed to consuming products of this nature, the issue of packaging needs to be discussed.

The benefits generated by the growing practice of consuming organic products are evident. Concern about the materials used to package these foods is still incipient, both among producers and consumers and among regulatory bodies. If organic production is so concerned with soil and water preservation, it is a counterpoint not to worry about disposing of plastics and Styrofoam in household waste.

In nearly every country, the demand for organic food has increased exponentially over the past several years. This is a result of people's growing concern about their health and that of their families. Specifically, the COVID-19 pandemic has spurred the search for healthier foods [30]. In Brazil, according to the Brazilian Association of Supermarkets—ABRAS the market grew by an average of 30% in 2020 and the global market, 11,5% [31, 32], reaching 106 billion euros in 2019 [33].

Therefore, for this reason, the discussion of packaging by regulatory bodies is essential and fundamental. Boosting the use of innovative technologies in the search for biodegradable materials is essential.

6.1 Global trends in consumer behavior

Consumers are increasingly looking for the information they need to make their choices at the time of purchase as much as their individual needs for change and their environmental concerns. However, there is still a significant lack of understanding about what can and cannot be recycled. Therefore, the study conducted by Mintel [34] published in 2022, shows that the majority of consumers hold companies accountable for providing information clearly and objectively, as well as recycling and the impact of packaging materials on the environment.

The following data were taken from the Mintel report [34]:

a. *54% of consumers in China, say labeling that measures environmental impact using understandable quantifiers would inspire them to buy;*

b. *65% of Canadian consumers approve their behavior can make a positive difference to the planet;*

c. *52% in Germany think there is not sufficient information on how to correctly dispose of compostable food packaging;*

d. *36% of Indians strongly agree that there is not good information available and how they could act to reduce household waste;*

e. *36% of USA consumers made their choice because the packaging was more environmentally friendly than a product from another brand;*

f. *55% of German consumers aged 35–44 do not trust organizations to be honest about their environmental impact;*

g. *33% in Brazil don't trust companies to be honest about their environmental impact;*

h. *47% of USA consumers believe companies are responsible for increasing the amount of packaging that is recycled;*

i. *42% of UK consumers believe companies can do more than governments to change the world;*

j. *76% of Japanese consumers agree that it's necessary for brands/companies to make big efforts to protect the environment.*

This means that brands need to take advantage of the situation and promote smart labels and packaging by providing a positive carbon footprint of the products. As caring about the environment turns into doing the right thing, consumers will want to be in charge by getting clear information about products that fully meet their environmental values and about how to get rid of packaging in a responsible way.

Education and incentives to consumers to combat household waste packaging through communication activities and on-pack marketing are critical and companies' responsibility. For many consumers, understanding the impact of their consumer choices is the first step in making positive changes. Consumers will require a clear and objective system printed on food packaging that demonstrates environmental responsibility for the product and provides comparable metrics to enable them to make clear and direct comparisons between products.

> *"Consumers, experiencing the environmental impact of their actions, are increasingly looking to make environmentally responsible choices that address concerns such as plastic pollution and climate change. Brands should aim to provide clear on-pack information that highlights understandable measures of the environmental impact of both the product and packaging." [34]*

7. The correct and incorrect points: the increase in consumption and discarded packaging

A consumer society symbolizes the twenty-first century. In recent years, the amount of domestic waste generated has increased dramatically and in an uncontrolled way.

Although the use of packaging protects food and prolongs its shelf life, it is also responsible for a substantial amount of household waste generated [35, 36]. Waste recycling is a matter of home education, and the less developed a country is, the less prevalent this habit is.

In addition, people often do not evaluate the type of packaging in which the food is stored at the time of purchase. This lack of perception and evaluation should be considered during the purchase process. In addition, it should also be the subject of discussions in the technical regulatory commissions. It is necessary to discuss the issue of the use of non-sustainable packaging in organic production. If specific rules are laid down for the use of biodegradable packaging, it is likely that farmers will initially need help. Educational guidance and subsidy policy actions helped producers in this transition.

In Brazilian regulations, certification criteria assess the management and production system. Ensuring traceability is a mandatory condition of control, describing the types of diseases and pests that attack the crop, as well as the measures to protect and combat them, is evaluated and authorized by the certifying body. It is also part of these requirements, the formalized labor and social issues, and within the laws. There are no requirements regarding the use of sustainable packaging, which is a counterpoint to the philosophy of environmental preservation [23]. Although the origins of organic food can be attributed in part to environmental concerns, there is no significant concern about the amount of household waste produced, particularly when it comes to packaging [37].

Commonly used materials such as non-biodegradable plastic packaging, Styrofoam, aluminum, and steel increase household trash on the globe and produce polluting, occasionally hazardous materials. That runs counter to one of the main goals of organic farming, which is to save and preserve the environment. In addition to the package itself, the packaging-product system should be taken into account [35].

Based on this, Santos' research [5] specifically set out to look into the variables influencing customers' decision to buy organic goods in environmentally friendly packaging. The subjects of this study are residents of Portugal. The Theory of Planned Behavior was the basis for Santos's research, and it was expanded to include things like how consumers think they know and how they feel about the environment [5]. A questionnaire was the research tool, and 311 responses from various consumer profiles were obtained. Because of this, even though perceived environmental concern and knowledge account for 26% of the variance in attitudes of organic consumers toward sustainable packaging, their effects on purchase intent were minimal [5].

Santos' study emphasizes the significance of expanding our understanding of the correlation between preferences for organic foods and the selection of sustainable packaging. In the end, it raises questions about how, even though organic customers express environmental concerns and believe they are aware of environmental issues, more needs to be done to improve production systems as a whole and create a cleaner supply chain, particularly in relation to packaging [5].

Even if the scenario is less than ideal, advances in the food market, shifts in customer preferences, and a balance between food quality, protection, and environmental impact have all resulted in new packaging requirements over time.

It can also be noted that, over the last few years, the use of ecological materials such as biodegradable plastics, for example, has been expanding more and more. But it is still insignificant and the cost is high when compared to regular plastic.

Slowly, the industry's use of non-biodegradable plastic is becoming less prevalent. This fact generates a degree of environmental optimism. Innovation in the food industry predicts more significant use of bioplastics, smart packaging, and "green" packaging in the coming years [38]. Active packages (AI) are those that, for instance, warn about the condition of the food and whether it is still safe for consumption. Sustainable or green packaging (SOGP) or Intelligent or smart packaging (IOSP) decreases carbon emissions [39] .

Geueke et al. say that packaging should be looked at as a sustainable way to promote a circular economy because of how it affects the environment. This would help get rid of the "take-make-throw-away economy" of packaging, which is hurting many ecosystems [40].

The circular economy is a system that incorporates a production and consumption model that involves sharing, leasing, reusing, repairing, refurbishing, and recycling old materials and products for as long as possible, as stated by the European Parliament [41].

Therefore, extending the products' life cycle as much as possible is the goal.

In practice, the circular economy means cutting down on waste all along the manufacturing chain, even after the product has been used. As according to van Herpen et al. [42], "*Wherever practicable, a product's materials are preserved inside the economy when it reaches the end of its life*". These can be effectively used again and time again, adding to their value.

For the sake of the environment and ecosystems, the economic production model of the past, which involved taking, manufacturing, and discarding without giving anything back, must end. For this purpose, the circular economy is a crucial instrument for preventing and minimizing environmental harm caused by human activity [40, 41]. In this regard, organic production, whose pillars promote the preservation of biomes, and producer awareness, must consider the significance of employing sustainable packaging within the circular economy, including the use of clean production techniques, recycled materials, and renewable energy sources.

As explained by Santos et al. [5], there has been a considerable increase in the number of scholars studying the relationship between food and packaging. In the study "Sustainable Packaging for Supply Chain Management (SPSC)," activists talked about how the packaging is handled, protected, and kept safe.

However, these studies are more directed to the use of new materials whose objective is food conservation and safety, without much focus on the degradation of these materials in nature.

There are not many policies governing the interaction between products and packaging in the food and beverage industry. What is evident is that there is more concern about how food deteriorates in storage than about the amount of waste that will be produced with the disposal of its packaging.

From the industry's point of view, as it is a sector where innovation is increasingly present, the difficulty in adopting ecological packaging would not be as felt as in the case of small rural producers. Anyway, the reduction of domestic waste is a crucial issue for environmental preservation, and consequently, everyone should prioritize their choices. Changes are initially slow and laborious, but once awareness-raising actions are started, they become habits [36].

According to studies by LOHAS-Lifestyles of Health and Sustainability [43], the organic food and beverage market is expanding three to four times faster than that of non-organic products, according to studies by LOHAS. As a result, it has already aroused the interest of large manufacturers, who are looking for ways to increasingly exploit this market, preserving the brand's hegemony by attracting a specific audience, the organic niche [36].

As Joseph Schumpeter's theory of creative destruction says, which he highlights in this situation, markets are essentially a process of industrial mutation that continually revolutionizes the economic structure from within. Creative destruction refers to the continuous process of replacing obsolete production units with new ones as a result of product and process innovation [44] .

With innovations in the industry, the use of smart packaging, the development of new ecological materials allied to changes in the population's habits, and being increasingly ecologically conscious will bring the change that the planet so much needs.

8. Eco-packaging or sustainable packaging

The packaging used in food provides benefits to society in many respects. Furthermore, by 2020, it is estimated that the number of packaged goods sold globally will reach 3.844 billion [45]. This number of available products tends to grow strongly in the coming decades, being an essential indicator in generating jobs, industrial growth, and demand for new technologies.

The use of packaging allows the marketing sector to leverage the market of a given brand, favoring the differentiation between products. It also has the role of educating and informing the consumer while protecting food during production, transportation, and acquisition. On the other hand, disposal can have negative consequences if made of unsustainable material [42]. SPC believes that new technologies will be more present and more affordable in the future, with the development of smart and eco-friendly packaging. This is why the packaging industry is considered a fast-growing sector that incorporates new packaging design techniques and innovation [45].

Now, in the process of being considered ecological or green, you must follow certain criteria. According to Sustainable Packaging Coalition [46] sustainable packaging needs to be:

"A. Is beneficial, safe & healthy for individuals and communities throughout its life cycle
B. Meets market criteria for performance and cost
C. Is sourced, manufactured, transported, and recycled using renewable energy
D. Optimizes the use of renewable or recycled source materials
E. Is manufactured using clean production technologies and best practices
F. Is made from materials healthy throughout the life cycle
G. Is physically designed to optimize materials and energy
H. Is effectively recovered and utilized in biological and/or industrial closed loop cycles."

To leverage the market for innovative and sustainable packaging, it is necessary to first study consumer behavior in order to identify the presence of ecological awareness at the time of purchase. This information is important for making an ecological map of how people treat the environment. According to Nguyen et al. [47], based on systematic reviews of 261 articles, the objective was to evaluate research on packaging ecodesign in the literature through three perspectives, positivism, interpretivism, and consumer transformative research. Nguyen apud Koenig-In Lewis et al. [47], an exploratory study of consumer behavioral assessment showed that purchase intention is significantly affected by consumer concerns about the environment. As people learn more about the environment, they end up wanting products that are better for them.

Nevertheless, it is logical that this sustainable behavior is quite vulnerable, that is, it is characterized by impotence and dependence. This is because it depends on particular situations, such as demographics, socioeconomics, and personal (age and gender). For consumers, these situations have a direct impact on their behavior [48]. Often, the lower the level of education or income power, the higher this vulnerability. The existing literature on research *on Transformative consumer research* (TCR) cited by Zeng et al. [48] is precisely the link between packaging and vulnerability. Some studies in the field of CRT have investigated the impacts of packaging innovation on vulnerable groups at risk, for example, older consumers or children, or even biological, psychological, and social factors. According to research conducted by Zeng et al. [48], there is a strong indication of a growing interest among researchers in topics related to sustainability and packaging innovation in the last 10 years. In relation to the CRT, the study points out three main topics addressed: consumer vulnerability, consumer health status, and sustainability.

With regard to the consumer of organic products, this vulnerability should be minimal considering that it is a niche market aimed at a consumer much more aware of the environment. There is little research on consumer perceptions of environmentally friendly packaging. In fact, environmentally friendly packaging has never been a clear concept in consumer behavior literature [48]. This point is corroborated by the fact that there are no metrics available in the databases consulted for this evaluation. Specific studies are needed to trace the behavior of consumers of organic products and their relationships in the face of the packaging used. These kinds of studies are very important for making producers aware of the right kinds of packaging to use.

8.1 Eco-friendly packaging—global trends

Some types of packaging, such as cardboard and glass, can even be considered sustainable as long as they are destined, after use, for recycling. However, they depend on several factors. Users are not always educated to recycle garbage. Another factor that contributes to inadequate disposal is the absence of selective collection in some establishments, as well as the absence of effective educational campaigns. Cardboard, for example, could be used as compost in organic production.

With the lack of an adequate recycling process and environmental concerns, new materials have emerged that are considered eco-friendly or green packaging. Although the availability of propagation is still a faraway reality and it is not possible to trace a promising gift, the development of these new materials has grown significantly in the packaging sector. The market is expected to account for US $29.7 billion by 2026 [49]. Still being a more costly technology, bio-based polymers are 20–100% more expensive than plastics derived from petroleum polymers, which still makes them a little less accessible to the food industry [34].

Moreover, unfortunately, it is observed that most of the time there is no concern, whether from industry or producers, about providing food in packaging made of sustainable raw materials of low degradation time in nature. Moreover, unfortunately, it is observed that most of the time there is no concern, whether from industry or producers, about providing food in packaging made of sustainable raw materials of low degradation time in nature [50]. Generally, the concern of this sector is to have packaging that ensures a longer shelf life of food, conserves its integrity and allows a greater shelf-life.

On the other hand, there is optimism, the result of environmental concern and climate change caused by the growing generation of household waste. Because of this, some industries that make things are looking for new materials that are better for the environment and keep the quality of food the same as plastics. Ordinary plastics can take 450–1000 years to decompose, causing harm to the ecosystem and affecting fauna and flora. Biodegradable plastics decompose on average 60% in 180 days [47, 48]. For this reason, packaging made of polymers considered "green" is the current major trend in the packaging industry. They can be used in various sectors, not only in food but also automobiles and toys, for example [49]. Green polymers are sustainable polymers of plant origin that during their synthesis or processing have fewer impacts on the environment. They are polymers whose raw materials are not of fossil origin but of renewable matter such as corn, sugarcane, or cellulose. Recycling a ton of green polymer bottles can save up to 1.5 tons of CO_2 emissions. They can be recycled or reused, and even if discarded, they do not have a negative impact on the environment because they decompose with carbon-neutral emissions. As a denomination, these polymers are added to the word "green" after the citation of their nomenclature: PVC (polychloride vinyl) green, PP (polypropylene) green, PE (polyethylene) green [50].

Although they are environmentally correct, their production scale is still very small. This is because its cost is still expensive when compared to petroleum-processed polymers. Currently, the cost of biodegradable plastics varies from US$ 2/kg to US$ 2/kg as compared with that of traditional plastics, which is approximately, US$ 1/kg to US$ 2/kg [49, 51].

It comprises this range of packaging of bioplastics or biopolymers or green polymers, those also made from algae, mushrooms, shrimp shells, organic waste, fibers, and vegetable resins etc. [52].

8.2 The importance of packaging being sustainable

Consequently, the consumer of organic products plays a crucial role in this production chain, and their awareness of this role can encourage the shift in favor of Schumpeter's theory of creative destruction. To create a circular economy, people need to use packaging that is good for the environment.

According to van Herpen et al., it has become standard practice in many supermarkets in Europe to sell products without plastic wrapping in the fruit and vegetable section. The removal of packaging at the time of purchase is a recent trend looked [42]. In countries like Brazil, this reality is still difficult to find.

van Herpen et al.'s research [42] looked at whether removing the primary packaging of fruits and vegetables would encourage more people to buy them. Plastic might seem to go against the idea of a sustainable environment, which is why this is a big deal for organic fruits and vegetables. Consistent with the study's findings, the typical consumer prefers organic fruits and vegetables that are not packaged in plastic. However, the packaging is frequently associated with wastefulness and the "throw-away society" in the eyes of consumers.

Kafel et al. [35] evaluated the types of materials most commonly used in the European Community to package organic products. **Table 2** shows the results obtained by Kafel et al. [35].

Even though the data indicates that plastic packaging is used a little more than cardboard packaging, 58.7% (**Table 2**) is quite significant. The results of the study by Kafel et al. [35] indicate the high use of plastic in the food industry.

In order to have a fully functioning circular economy, it is necessary to guarantee the preservation of soil and water throughout the production chain and the product's life cycle.

Additionally, the materials indicated in **Table 2**, Styrofoam is a common material widely used by Brazilian manufacturers (**Figure 4**). **Figure 4** shows how organic products on store shelves are sold to customers.

Plastic, as seen in **Table 2**, is widely used and one of the agents responsible for the large amount of waste that remains in nature in the long term. It pollutes the soil, rivers, lakes, and oceans [54]. In addition, several animals lose their lives after consuming plastics found in nature waste by men. For this reason, consumers need to be increasingly informed and cautious when selecting products at the time of purchase. It is necessary to raise awareness of the way food is presented, and the kind of waste it will produce after consumption.

The use of plastic in the food sector is common, even for reasons of price, accessibility, and availability. Changing Kafel's data is a crucial issue.

Plastic packaging	Paper/ paperboard	Glass	Multilayer laminate materials	Biodegradable materials	Metal
58.7%	47.8%	39.1%	23.9%	23.9%	8.6%

Table 2.
Materials used for packaging of organic products [35].

Figure 4.
Brazilian organic products at the supermarket—photo by National Institute of Technology—INT [53].

Finally, there is reason to believe that environmental preservation is gaining more supporters. More and more organic producers are realizing that the exponential rise in household trash is almost impossible to keep up with.

9. Organic production, consumer eco-friendly packaging, and circular economy

Undoubtedly, the packaging is essential to maintain the integrity of food, avoiding diseases and chemical contamination. Now, the food production system, whether organic management on farms or in the processing industry, needs to worry about going beyond maintaining food conservation. It is necessary to innovate, preventing these packages from becoming household waste with environmentally persistent materials, and in the long run, they will become part of the virtuous circle of the circular economy [55]. Among the potential solutions, the production of biodegradable polymers or bioplastics from agro-food or vegetable products will ensure that when discarded they return to the soil and, instead of polluting the land, they will function as nutrients. With regard to organic production, the use of packaging made of bioplastics would be ideal and perfect. It would ensure that the entire life cycle of the product would be in accordance with the circular economy, not generating pollutants, properly using natural resources, and prioritizing durable and renewable resources, as illustrated in **Figure 5**.

10. Conclusion

It is unacceptable, nowadays, to accept organic food being sold in non-biodegradable packaging, harmful to ecosystems. Maybe this situation would not change until the courts demand it and the government has to step in.

The fact is that there are consumers making a difference by being more demanding and defending environmental preservation. However, although they are gradually increasing and becoming representative groups, they are still a minority. They are not

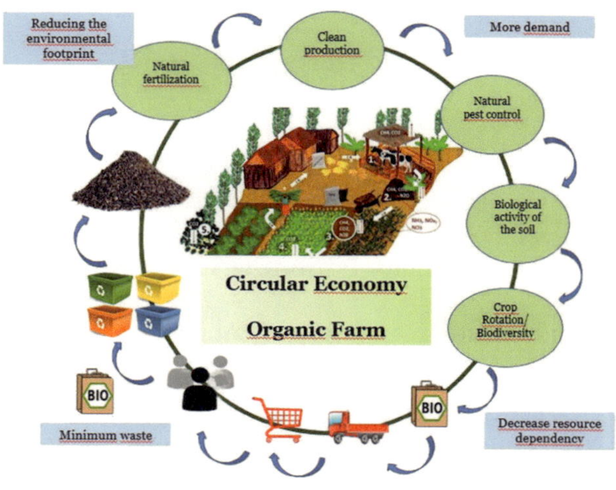

Figure 5.
Elaborada pelo autor—a figura dentro do círculo foi retirada do site Vectorstock [56].

enough to generate significant changes in the production chain. It is necessary that public, technological, and environmental institutions debate these issues more and more; they affect the food-packaging interaction with the environment.

Thus, it is important that the results of these debates generate informative and educational campaigns, in addition to pressuring the packaging industry to manufacture ecologically correct products, reducing the carbon footprint. Another important point in the present study is the fact that more in-depth research is needed on the role of the consumer in relation to conscious, behavioral, and ecological consumption. Lastly, to show that actions need to be taken right away, consumer society can no longer avoid taking responsibility.

On the other hand, while regulatory and standard-setting bodies are unaware of the importance of requiring organic food producers to use ecological packaging, the generation of domestic waste on the planet resulting from organic production, composed of aluminum, plastic, Styrofoam, and metal, has no end. Incorporating the mandatory use of biodegradable packaging into legislation and certification mechanisms is urgent and necessary. As organic production is based on the preservation of soil and water, it is not acceptable, therefore, its consumption generates polluting domestic waste. At the same time, the role of the consumer in this scenario is fundamental. Being more conscious at the time of purchase, adopting sustainable choice practices, recycling garbage, and demanding biodegradable packaging are necessary and fundamental changes in habits.

The qualitative meta-analysis carried out in this study showed that, although eco-consumers are gradually increasing and becoming a representative group, they are still a minority. For the time being, they are still not enough to generate significant changes in the production chain and to integrate organic production within the circular economy cycle. Therefore, it is necessary that public, technological, environmental institutions, associations, and cooperatives increasingly address these issues and begin to demand specific legislation for the proper use of packaging, especially for certified organic foods.

Author details

Rosana Carvalho Esteves
Department of Certification/National Institute of Technology, Rio de Janeiro, Brazil

*Address all correspondence to: rosana.esteves@int.gov.br

IntechOpen

© 2022 The Author(s). Licensee IntechOpen. This chapter is distributed under the terms of the Creative Commons Attribution License (http://creativecommons.org/licenses/by/3.0), which permits unrestricted use, distribution, and reproduction in any medium, provided the original work is properly cited.

References

[1] Esteves RCA. Vendramini AL, Accioly F. A qualitative meta-synthesis study of the convergence between organic crop regulations in the United States, Brazil, and Europe. Trends in Food Science & Technology. November 2020;**107**:343-357. DOI: 10.1016/j.tifs.2020.10.044

[2] Seufert V, Ramankutty N, Foley JA. Comparing the yields of organic and conventional agriculture. Nature. 2012;**485**:229-232. DOI: 10.1038/nature11069

[3] Mader P. Soil fertility and biodiversity in organic farming. Science (1979). 2002;**296**:1694-1697. DOI: 10.1126/science.1071148

[4] Läpple D, Van Rensburg T. Adoption of organic farming: are there differences between early and late adoption? Ecological Economics. 2011;**70**:1406-1414. DOI: 10.1016/j.ecolecon.2011.03.002

[5] Santos V, Gomes S, Nogueira M. Sustainable packaging: Does eating organic really make a difference on product-packaging interaction? Journal of Cleaner Production. 2021;**304**. DOI: 10.1016/j.jclepro.2021.127066

[6] Portal .periodicos. CAPES - Acervo n.d. Available from: https://www-periodicos-capes-gov-br.ezl.periodicos.capes.gov.br/index.php/buscador-primo.html. [Accessed: 12 July 12 2022]

[7] Revbea S, Paulo V. MARCO REGULATÓRIO DA AGRICULTURA ORGÂNICA: PROTEÇÃO, EFICÁCIA, DIFICULDADES E INTERAÇÕES COM A EDUCAÇÃO AMBIENTAL. vol. 4. 2021

[8] Brazil. BRAZIL. Republic Presidency. Civil House. Law No 10,831, of 23 December 2003. Rules on organic agriculture, and other matters. Official Gazette of the Federative Republic of Brazil. Brasília, DF, 24 December 2003. Section 1, p. 8. 2003

[9] eCFR :: 7 CFR Part 205 -- National Organic Program n.d. Available from: https://www.ecfr.gov/current/title-7/subtitle-B/chapter-I/subchapter-M/part-205?toc=1. [Accessed: 3 July 2022]

[10] European Council. European Council (23 and 24 October), Conclusions, EUCO 169/14, Brussels. European Commission 1999

[11] Candiotto LZP. Organic products policy in Brazil. Land Use Policy. 2018;**71**:422-430. DOI: 10.1016/j.landusepol.2017.12.014

[12] BRAZIL. Ministry of Agriculture, Livestock and Supply. Ministry of the environment. Joint normative instruction No. 17 of 18 June 2014. Changes the Normative n° 46, of 6 October 2011 (Establish the Technical Regulation for Organic production systems, as w 2014

[13] Instrução Normativa N° 50 de 05 de Novembro de 2009 (Selo Federal do SisOrg).pdf — Português (Brasil) n.d. Available from: https://www.gov.br/agricultura/pt-br/assuntos/sustentabilidade/organicos/arquivos-organicos/legislacao/portugues/instrucao-normativa-no-50-de-05-de-novembro-de-2009-selo-federal-do-sisorg.pdf/view. [Accessed: 9 June 2022]

[14] National Organic Program | Agricultural Marketing Service n.d. Available from: https://www.ams.usda.gov/about-ams/programs-offices/national-organic-program. [Accessed: 11 March 2021]

[15] EUR-Lex - 32007R0834 - EN - EUR-Lex n.d. Available from: https://eur-lex.europa.eu/legal-content/EN/TXT/?uri=celex%3A32007R0834. [Accessed: 12 July 2022]

[16] 7 CFR § 205.311 - USDA Seal. | CFR | US Law | LII / Legal Information Institute n.d. Available from: https://www.law.cornell.edu/cfr/text/7/205.311. [Accessed: 9 June 2022]

[17] BRAZIL. Ministry of agriculture, livestock, and food supply. Normative instruction No. 46 of 06 October 2011. Establishes the Technical Regulation for Organic systems of the plant, and Animal Production, as well as the lists of substances permitted for use in Org. 2011

[18] Portaria nº 52, de 15 de março de 2021—Português (Brasil). BRAZIL Ministry of Agriculture, Livestock, and Food Supply - Portaria No 52, de 15 de Março de 2021—Português (Brasil) (n.d) Retrieved August 24, 2022, from https://WwwGovBr/Agricultura/Pt-Br. 2021

[19] IFOAM. The IFOAM Norms Organic Production Processing. International Federation of Organic Agricultural Movement 2014; version 20:1-252. Available from: https://doi.org/3-934055-58-3

[20] Australian Organic Market Report 2017, Australian Organic Ltd. www.austorganic.com. Available from: https://monikasorganics.com.au/wp-content/uploads/2017/09/AO_Report_2017.pdf. ISSN 2207-5135 [Accessed: 2022-06-21]

[21] Janssen M, Hamm U. Product labelling in the market for organic food: Consumer preferences and willingness-to-pay for different organic certification logos. Food Quality and Preference. 2012;**25**:9-22. DOI: 10.1016/j.foodqual.2011.12.004

[22] Essoussi LH, Zahaf M. Decision making process of community organic food consumers: an exploratory study. Journal of Consumer Marketing. 2008;**25**:95-104. DOI: 10.1108/07363760810858837

[23] Brazil. BRAZIL. Republic Presidency. Civil House. Law No 10,831, of 23 December 2003. Rules on organic agriculture, and other matters. Official Gazette of the Federative Republic of Brazil. Brasília, DF, 24 December 2003. Section 1, p. 8. Presidência Da República Casa Civil Subchefia Para Assuntos Jurídicos, 2003:8

[24] Sønderskov KM, Daugbjerg C. The state and consumer confidence in eco-labeling: organic labeling in Denmark, Sweden, The United Kingdom and The United States. Agriculture and Human Values. 2011;**28**:507-517. DOI: 10.1007/S10460-010-9295-5

[25] Hjelmar U. Consumers' purchase of organic food products. A matter of convenience and reflexive practices. Appetite. 2011;**56**:336-344. DOI: 10.1016/J.APPET.2010.12.019

[26] EconPapers: The character of demand in mature organic food markets: Great Britain and Denmark compared n.d. Available from: https://econpapers.repec.org/article/eeejfpoli/v_3a33_3ay_3a2008_3ai_3a5_3ap_3a406-421.htm. [Accessed: 12 July 2022]

[27] Selos Distintivos de Qualidade e Origem - Portal Embrapa n.d. Available from: https://www.embrapa.br/alimentos-e-territorios/areas-de-atuacao/selos-distintivos-de-qualidade-e-origem. [Accessed: 12 July 2022]

[28] Margariti K. "White" space and organic claims on food packaging: communicating sustainability values and affecting young adults' attitudes and purchase intentions. Sustainability. 2021;**13**:11101. DOI: 10.3390/SU131911101

[29] Cs SI, Ergi EM, En NGTR. The World of organic agriculture Technology Innovation Platform of IFOAM – Organics International (TIPI) A Global Vision Organic Farming. n.d. ISBN: 9788583170181. ISBN-10: 8583170185

[30] Bhattacharjee S, Reddy KP, Sharma V. Covid stimulating organic food consumption: Exploring factors of consumer buying behaviour – Indian Journal of Ecology. Scopus n.d. 2021;**48**:59-63. Available from: https://pesquisa. bvsalud.org/global-literature-on-novelcoronavirus-2019-ncov/resource/pt/covidwho-1329359. [Accessed: 21 June 2022]

[31] Setor de orgânicos cresce 30% no Brasil em 2020 | Clipping | ABRAS n.d. Available from: https://www.abras.com.br/clipping/geral/72392/setor-de-organicos-cresce-30-no-brasil-em-2020. [Accessed: 12 July 2022]

[32] Reganold JP, Wachter JM. Organic agriculture in the twenty-first century. Nature Plants. 2016;**2**:15221. DOI: 10.1038/nplants.2015.221

[33] Cresce o consumo de alimentos orgânicos - Centro de Inteligência em Orgânicos. Centro de Inteligência em Orgânicos. (n.d). Available from: https://ciorganicos.com.br/noticia/cresce-o-consumo-de-alimentos-organicos/ [Accessed: 22 June 2022]

[34] Packaging Trends Mintels global packaging experts explore how the 2022 Global Consumer Trends will impact the CPG packaging industry in the years ahead. (n.d.). Mintel - The world's leading market intelligence agency. Available from: https://www.mintel.com/global-packaging-trends

[35] Kafel P, Nowicki P, Wojnarowska M. Packaging's of the organic farming products in the context of circular economy. 2021. Available from: https://www.shs-conferences.org/articles/shsconf/abs/2021/03/shsconf_glob20_05011/shsconf_glob20_05011.html. DOI:10.1051/shsconf/20219205011

[36] Gilding P. A grande ruptura - como a crise climática vai acabar com o consumo e criar um novo mundo. Editora Apicuri; 2014

[37] Brennan L, Langley S, Verghese K, Lockrey S, Ryder M, Francis C, et al. The role of packaging in fighting food waste: a systematised review of consumer perceptions of packaging. Journal of Cleaner Production. 2021;**281**:125276. DOI: 10.1016/J.JCLEPRO.2020.125276

[38] Biodegradable Packaging Material Market to 2032 - Fact.MR n.d. Available from: https://www.factmr.com/report/2063/biodegradable-packaging-material-market. [Accessed: 12 July 2022]

[39] Top 9 Sustainable Packaging Trends | Sphera n.d. Available from: https://sphera.com/spark/top-9-sustainable-packaging-trends/?utm_source=Google&utm_medium=Text&utm_campaign=Spark&keyword=sustainable%20packaging&matchtype=b&device=c&gclid=CjwKCAjw2rmWBhB4EiwAiJ0mtdgCLCQJGZ49i-ybE4jQnoeZ-Wn2S-JSot463ElXgY6tB-H-XlL4vFhoCZ4AQAvD_BwE. [Accessed: 12 July 2022]

[40] Geueke B, Groh K, Muncke J. Food packaging in the circular economy: overview of chemical safety aspects for commonly used materials. Journal of Cleaner Production. 2018;**193**:491-505. DOI: 10.1016/J.JCLEPRO.2018.05.005

[41] Circular economy: definition, importance and benefits | News | European Parliament n.d. Available from: https://www.europarl.europa.eu/news/en/headlines/economy/20151201STO05603/

circular-economy-definition-importance-and-benefits. [Accessed: 1 July 2022]

[42] van Herpen E, Immink V, van den Puttelaar J. Organics unpacked: the influence of packaging on the choice for organic fruits and vegetables. Food Quality and Preference. 2016;**53**:90-96. DOI: 10.1016/j.foodqual.2016.05.011

[43] Choi S, Feinberg RA. The lohas (lifestyle of health and sustainability) scale development and validation. Sustainability (Switzerland). 2021;**13**: 1-17. DOI: 10.3390/SU13041598

[44] Burlamaqui L, Nelson D, Block F, Wray R, Kregel J, Proença A, et al. Texto para discussão Schumpeter's Creative Destruction as a Radical Departure: A New Paradigm for Analyzing Capitalism CEBRI- Centro Brasileiro de Relações Internacionais Nucleo de Economia Política - Março 2022. Available from: https://cebri.org/media/documentos/arquivos/Leonardo_Burlamaqui_-_Schumpet622927e254f92.pdf. [Accessed: 16 June 2022]

[45] Home | PACKMEDIA - news and reports on trends, best practices, and new technologies for packaging, labeling, coding n.d. Available from: https://packmedia.net/.[Accessed: 28 July 2022]

[46] Definition of Sustainable Packaging I. A Vision for Sustainable Packaging The Sustainable Packaging Coalition ® (SPC), a project of GreenBlue ®. (n.d.). Available from: http://www.environmentalleader.com/2010/01/05/pike-research-finds-global-sustainable-packaging-market-to-double-by-2014/. [Accessed: 29 July 2022]

[47] Nguyen AT, Parker L, Brennan L, Lockrey S. A consumer definition of eco-friendly packaging. Journal of Cleaner Production. 2020;**252**:119792. DOI: 10.1016/J.JCLEPRO.2019.119792

[48] Zeng T, Deschênes J, Durif F. Eco-design packaging: an epistemological analysis and transformative research agenda. Journal of Cleaner Production. 2020;**276**:123361. DOI: 10.1016/J.JCLEPRO.2020.123361

[49] Bioplastics & Biopolymers Market Global Forecast to 2026 | Marketsand Markets n.d. Available from: https://www.marketsandmarkets.com/Market-Reports/biopolymers-bioplastics-market-88795240.html. [Accessed: 28 July 2022]

[50] Landim APM, Bernardo CO, Martins IBA, Francisco MR, Santos MB, de Melo NR. Sustentabilidade quanto às embalagens de alimentos no Brasil. Polímeros. 2016;**26**:82-92. DOI: 10.1590/0104-1428.1897

[51] Johnson A, Gabler B, Stevenson M, Trepegnier C, Tsuji A, Designer M, et al. Environmental Technical Briefs of Common Packaging Materials, prepared by the Sustainable Packaging Coalition®, summarize the environmental, human health and other relevant issues associated with the production, use and end-of-life of common packaging materials. Contents. 2009

[52] Syed Ali Ashter, 10 - New Developments, Editor(s): Syed Ali Ashter, In Plastics Design Library, Introduction to Bioplastics Engineering, William Andrew Publishing. 2016:251-274, ISBN 9780323393966, DOI: 10.1016/B978-0-323-39396-6.00010-5. Available from: https://www.sciencedirect.com/science/article/pii/B9780323393966000105. [Accessed: 29 July 2022]

[53] INT—Português (Brasil) n.d. Available from: https://www.gov.br/int/pt-br. [Accessed: 12 July 2022]

[54] Plastic pollution facts and information. (n.d.). Retrieved July 12, 2022, from National Geographic

Articles Website. https://www.nationalgeographic.com/environment/article/plastic-pollution. [Accessed: 12 July 2022]

[55] Rosentrater KA, Jiménez A, Guillard V, Gaucel S, Fornaciari C, Angellier-Coussy H, et al. The next generation of sustainable food packaging to preserve our environment in a circular economy context. 2018;**5**:121. DOI: 10.3389/fnut.2018.00121

[56] Agriculture and organic farm fresh line art vector image on VectorStock n.d. Available from: https://br.pinterest.com/pin/843018242632992O3/. [Accessed: 29 July 2022]

Chapter 9

The Perception of Intelligent Packaging Innovation: The Latest Process and Technological Progress

Erika Loučanová, Ján Parobek and Martina Nosáľová

Abstract

As a result of global change and progress in recent decades, the approach to utilizing product packaging materials has changed. Subsequently, innovative packaging is the result of creative thinking beyond the usual thinking framework. A complete understanding of the customer's needs is an indispensable requirement for the ability to develop packaging with optimized performance. The study deals with the perception of intelligent and active packaging by respondents in Slovakia. The approach of Kano model was applied for the study of customers' attitudes to the individual functions of active and intelligent packaging. Firstly, the requirements of the packaging functions among the monitored age categories were identified. Subsequently, the innovation status within the individual age categories was evaluated. Thereafter, a 3D simulation was used to figure out the resultant perception of intelligent and active packaging functions in Slovakia. Based on the research results, we can conclude that the awareness of customers in Slovakia about intelligent packaging innovations is at a low level and is oriented towards a weak green strategy.

Keywords: intelligent and active packaging, consumer perception, Kano model, Slovakia

1. Introduction

The global trends highlight the continual accelerating pace of innovation entrances, growth of global competitiveness, and technologically oriented innovation. As a result of global change and progress in recent decades, the approach to packaging materials and packaging techniques also changed. According to Magnusson et al. [1], innovation strategies are crucial for corporate success and should be a top priority for packaging companies, and new methods for innovative packaging development are needed. Innovative packaging is the result of creative, unconventional thinking beyond the usual thinking framework.

The study deals with the perception of intelligent and active packaging by respondents in Slovakia. The aim of the study describes the customers' attitudes to the packaging functions and the innovation status within the individual age categories according to the KANO model and identifies an innovation green strategy for intelligent packaging in Slovakia based on the research findings.

2. The packaging: importance, functions and environments

The packaging is one of the most important parts of forming the product. Its size, shape, design, selected color, and font significantly influence the consumer decision-making process and thereby affect the marketability of the product itself [2]. The packaging can be understood as the tool or the set of tools protecting the product from potential damage. It allows for better handling and facilitates the sales and consumption of products [3]. The importance of packaging is growing, and the reasons are increasing logistical costs, the progress of packaging technology and increased environmental consideration [4].

Packaging performs a series of different tasks: it protects its contents from contamination and spoilage, makes it easier to transport and store goods and provides uniform measuring of contents [5]. In conventional terms, the packaging is expecting as implies of assurance, conservation, dealing with, transport and capacity of goods. These days, the other capacities of packaging, such as getting the consideration of clients and brand communication, are getting to be more recognizable. As a result, the current intrigued of the company is in this manner to offer a client a packaging that will meet the showcasing prerequisites and market wants of dealing with and transport at the same time [6–9].

Right now, packaging is a fundamental component in the market from the products point of view and ensures to protect the quality of nourishment goods. It too plays a key part by ensuring stuffed goods against outside conditions, influencing the quality and well-being security of nourishment goods and making transportation, capacity and apportioning of goods easier [10].

Although the function of packaging varies depending on its type [11], three main purposes of packaging can be found in literature: protection, convenience and communication [12]. The traditional perception of packaging classifies the main functions of packaging into four basic categories: protection, communication, convenience and containment [13]; nevertheless, these functions are not totally exclusive – for example, the communication function of the package can also help to enhance food protection and convenience. These four functions are interconnected, and all of them should be assessed and considered simultaneously in the package development process [14].

The package is applied to:

- protect the good against the deteriorative impacts of the outside environment,
- communicate with the buyer as a promoting tool,
- give the buyer the ease of utilization and time-saving comfort,
- contain goods of different sizes and shapes [13].

This model of four traditional functions of packed serves as the basis for many studies. However, according to the literature review of Dopico-Parada et al. [15], it does not consider certain key aspects for consumers today such as the environmental impact of packaging or how packaging meets social needs. Nor does it consider the economic function of packages or the added value of intelligent and active packaging that can upgrade security and provide information. In addition, a package ought to not as it were meet all these capacities but ought to moreover meet market criteria for costs [15].

However, it is necessary to certify know-how on how features and properties of packages affect sustainable development in general. To satisfy the requirements of the society in connection to sustainable development, the packages ought to meet the economic and social and environmental dimensions [16–20]. Many authors define the different divisions of the packaging functions. Lindh et al. [16] proposed to establish uniform terminology of packaging functions for better understanding and communication leading to their development and decision-making processes. In the research, they divided the functions according to the environmental, social and economic dimensions. The authors Zeman [3] and Kačenák [17] divided the packaging function into six key functions: protection, guarantee and rationalization, economic, communication and ecological functions. Packaging must comply with the transport and storage and must be suitable for the store and prevent stealing; it includes promotional and informational functions. Dzurová [21] lists five functions of packaging based on Schulte (in Dzurová [21]), namely: protection, storage, transport, handling and information. For this particular research, we used the dividing by Calver [22] and Loučanová [9]: handling, protective, informative, economic, ecological and promotional and recently often mentioned social function.

According to Dopico-Parada et al. [15], packaging can be seen as an solution that protects goods from external impacts, ensuring their security and quality, protracting good life, encouraging transport, taking care of and capacity, working as a bolster for data and improving the experience. Furthermore, all these packaging characteristics should be ensured at the lowest possible cost for consumers and in a sustainable way, complying with social and environmental responsibility.

The packaging has to perform its functions in three different environments [12]: physical, human and ambient type of environment. A physical environment is one in which physical damage can be caused to the product (shocks from drops, falls, damage during transportation or storage, etc.). Also, an ambient environment surrounds the package, damage to the product can be caused by gases, water, light, temperature, as well as microorganism. The third type is the human environment, in which the package interacts with people. The package must contain the information required and communicate with people in the appropriate way. Failure in considering all these three types of the environment during the packaging development will result in poorly designed packages, increased costs, consumer complaints and even rejection of the product by consumers [14].

The packaging process generates cooperation between product and packaging with the aim of fulfilling the needs of the product end-user as well as the manufacturer and the distributor. In principle, the packaging is just a clever way of constructing a container out of a selected material or combination of materials, where a wide variety and choice are available. A range of parameters, varying from product characteristics to consumer requirements and trends, affect this selection. These parameters can be grouped into three categories [23]:

- Parameters in the micro or product environment.
- Parameters in the ambient or distribution environment.
- Parameters in the macro or market environment.

2.1 Packaging innovation

When developing innovations, it is essential to think almost the article at diverse levels while each level increments its value to the client [24]. With respect to the development, it is fundamental to screen packaging capacities and discover out in the event that it can be utilized moreover. In the case of innovation, it is important to target it to the specific customer segment, and therewith it is in the company's interest to properly identify the target groups of the packaging innovation. Companies should monitor changes in consumers' preferences and also focus attention on new technology of packaging when selecting and introducing the packaging to the market.

Joseph Schumpeter (1931, In Zaušková and Loučanová [24]), has the classic approach to the classification of innovation; at first, he considered them as any positive alter within the generation living being. Cogliandro [25] is one of the essential authors of the concept of shrewdly advancement, who in his research does not consider innovation as a strategy or rapidly obtained wealthy program for the buyer, but as an original issue solving, intuitive sense for the market and the pursuit of success. Innovation attempts to provide and explain information about the status and concurrently they are able to manage it.

Luo Zongwei [26] characterized intelligent innovation within the field of pharmaceuticals as cleverly computational strategies, which are these days elite and essential for the generation and optimization of products (objects, goods). Modern and innovative computing instruments are reliably created and connected to develop modern goods elements and their components. The term intelligent innovation is not so regular in Slovakia; however, innovation is display in many research studies. Loučanová et al. [26–28] based on the elementary definitions of innovation characterize intelligent innovation as 'any autonomic alter with a positive effect to the customer'. They grow the comfort of the buyers and concurrently represent a more conservative, compelling and safe solutions.

According to Robertson [14], there are several drivers for packaging innovation: one is the fast-changing social trends and increasing consumers' demand for convenience and safety. The other is growing environmental awareness, while profitability and differentiation are also important for food companies seeking to attract consumer attention. Because consumers want innovation and value novelty, the packaging industry must continue to innovate or risk stagnation.

The food packaging industry is largely driven by market drivers to satisfy the needs of society and the economy. The most needed packaging innovations are those that can lead to practical solutions to fulfil socio-economic needs. The majority of those socio-economic needs are, according to Yam and Lee [29], consumer lifestyle, value, profits, food safety, food packaging regulations and environmental concern. Consumers' lifestyle recently is influenced by the aging of the population, an increasing number of smaller families or single-person households, and as a result, consumers are increasingly demanding food products that are convenient, taste good, are safe and nutritious.

In principle, the trends affecting packaging development and use can be divided into four main areas – business dynamics, distribution trends, trends in consumption and legislation [23]. These four areas tend to be also the main drivers of packaging innovation. The development in business dynamics is particularly related to fundamental changes in the packaging functions and extended perception of functions. The distribution and retail chains are continuing to internationalize and are constantly developing new consumer product logistics, processes and trends. This not only has an effect on the functional packaging requirements, it also affects the way in which products are presented. Different markets mean different perceptions, different consumption habits and different cultural and social values. And last but not least, over the last 5–10 years, environmental legislation has become one of the major drivers for innovation in packaging [23].

2.2 Active and intelligent packaging

The result of creativity is innovation in packaging. It is an alternative approach outside the typical context of thinking [13]. The effect of this method is the construction of communicating features of packaging. Essentially, the market provides two different possibilities of innovative packaging systems: active and intelligent packaging. They emphasize to increase new and standard features to meet present buyer requests, enlarged requests of regulatory and security. Intelligent and active packaging can be described as [30]:

- Active packaging changes the condition of goods and draws out the life or increments security, whereas keeping up the quality of the packaged nourishment.

- Intelligent packaging screens the status of goods nourishment and gives data on quality amid transportation and storage.

Active packaging is characterized as change in the awareness to the functions. The key features can be considered as the security, which has been moved from passive to active. A passive fence between the goods and atmosphere is considered as the old-style understanding of the package protection function. Active packaging lets changes the condition of the goods [13]. It function is to spread shelf life and improve goods safety, while keeping the quality [31]. According to the above-mentioned definitions of active packaging, they could be grouped according to the approach in which they affect the quality of the goods like this: emitters – active systems according to the release of materials and absorbers according to absorption [32].

Intelligent packaging is an approach which has the ability to carry out features (for example, recording, communicating, tracing, detecting) to simplify decision-making to give information, extend shelf life, increase quality and safety and warn about possible complications [13]. Kačeňák [31] considers intelligent packaging as a system for observing circumstances around the goods and delivering data about the quality of goods during logistic operation such as storage and transport. He included among these features namely: indicators of oxygen and carbon dioxide, time–temperature indicators, pathogen and color temperature indicators and indicators after warranty [32].

Following the literature review, the significance of intelligent and active packaging mainly represents the development of the two main features: 'information' which provides all data monitoring from the conditions of goods and protective function – particularly in shift from passive to active protection of goods. The demand of buyers to the packaging innovation focuses on: the majority of costumers expect that the packaging will be environmentally friendly with sufficient available information on the packed and protection functions. Our results approve the practicality of an active and intelligent packaging system in terms of essential functions and consequently innovation in packaging technologies, thus responding to the present conditions on market.

2.3 Customers' perception

Innovation management and innovation penetration cannot exist without examining consumer preferences and attitudes, because new product acceptance is ultimately an important factor in success. This approach can help clarify, predict or influence adaptation or refusal of innovation as regards [33]:

- The inclination to agree with innovation of the products or goods features failed for the reason that these innovations were not sufficiently explained to customers.
- Discovering the psychological awareness of invention – the aim is identifying customers' upcoming welfares of goods innovation.
- Investigation of consumer attitudes to innovations.

Firms must analyze the development of consumers' tastes to choose the marketing and packaging strategy. From the other point of view, they also have to pay attention to the new technologies of the competition. Last century, design have changed about every 15 years; but nowadays, it is much often due to the market environment. There is significant relationship between firms and the environment due to packages strategy [34–39]. Therefore, it is not only important to create packaging, but also there is a responsibility for how it can be re-used, i.e. recycling.

Buyers' attitude towards the goods packaging has developed, whereas, according to previous studies, the packaging is defined as a crucial part of the goods. On the other side, contemporary studies point to that it is an attractive part of the goods. Though, the first impression of the package for buyers does not continue if the packaging is not sufficiently functional and user-friendly. It must be for customers simple from the utilization point of view and must offer appropriate information [40]. Many consumer behavior studies investigate the perception of packaging by consumers – product rating influences the purchasing intentions and attitudes of consumers [41–43]. Brennan and Crandison [44] described common utilized active packaging materials in the Australia, Japan and USA, but much less in European countries. On the other side, some technologies utilizing intelligent packaging materials are more common in Europe. The differences are due to difference cultural peceptions and lack of accepting of benefits. Similar results aimed on above mention topic were concluded by Zhang et al. [45]. They analyzed the topic focused on innovations in packaging design. A reference system for the design of a goods packaging framework that can be connected to e-operations was analyzed by Regattieri et al. [46]. They described the e-commerce solutions for innovative packaging mathematical model.

Limited knowledge about the importance of packaging in the supply chain and for the end consumer represents an important challenge for the packaging industry. For instance, few product manufacturers appear to understand the important communicative role of packaging [47]. As a result of this knowledge gap, packaging's great potential for value addition is often overlooked and neglected [1, 4, 11]. A complete understanding of the customer's need is a requirement for the ability to develop packaging with optimized performance [1].

3. Methodology

In the framework of similar research, due to its anonymity and the least forced form of feedback for determining the attitudes, satisfaction and perception of products by customers, questionnaire inquiry is one of the most used methods. However, in addition to these and other advantages, questionnaire inquiry also has a number of disadvantages [48]. In an attempt to solve the inherent shortcomings of the traditional inquiry method, this research proposes to use the Kano analytical model to investigate customer satisfaction. The Kano model is designed to include not only quantitative but also qualitative measures. The priority of the researched packaging innovations is justified by the use of the Kano model, which uses both customer satisfaction and producer capacity [49]. According to Goodpasture [50], the purpose of the Kano model is to identify buyers' attitude according to the desires of an analyzed item.

The approach contained of several phases, **Figure 1**.

Firstly, the questionnaire is created to recognize actual statements. The questionnaire contained a pairs of negatively and positively comprehended questions. The methodology identifies the answers (sentence). The scale is from 1 to 5 and represents strong agreement to strong disagreement. Subsequently, questionnaire measures were identified. In Slovakia, the sample of respondents was set at 1231. For each given age category, we selected questionnaires to keep a similar proportion of respondents, the questionnaires were distributed through the internet by electronic forms.

According to the cross rule valid in the Kano model methodology [51], our received answers are evaluated in the subsequent analyses. Based on age categories, the

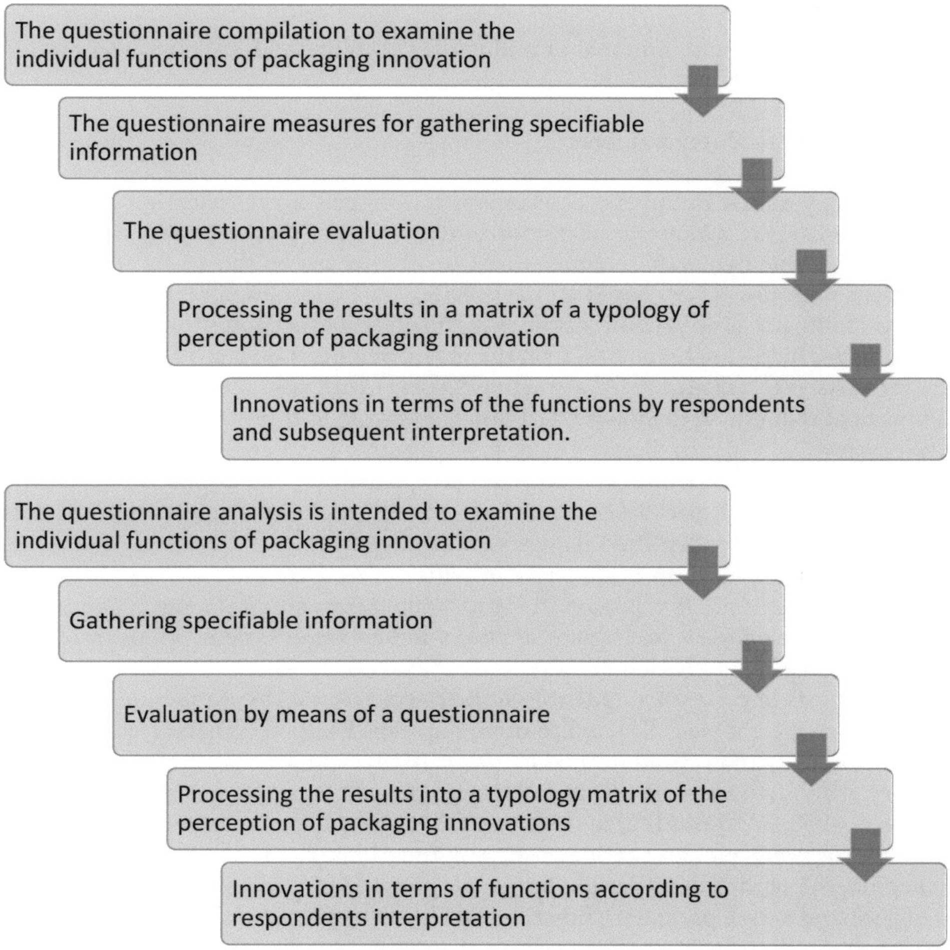

Figure 1.
The main methodological steps of the research.

answers are consequently analyzed by two-factor analysis. The Kano model described results in the following categories (how consumers perceived packaging) [51, 52]:

- M – are obligatory requests that buyers reflect as standard and are routinely expected. These requirements are described as basic or primary. They deal with buyers in the event of non-compliance. Recognizing them is a fundamental significance mostly for even nevertheless their fulfillment is reproduced in satisfaction, their debit is interpreted as buyer's dissatisfaction.

- O – are one-dimensional requests – are characterized by those goods features that lead to fulfillment and satisfaction in the event of nonconformity to buyer dissatisfaction.

- A – are attractive requests have a pure effect on buyer satisfaction for the reason that these requests buyers do not expect.

- R – according to theory, they are reverse or contrary requests.

- I – are requests with minimal or null impact on buyers. So-called irrelevant requests.

- S – are skeptical requirements.

A typology matrix of buyers (consumers) is designed as the following step. There is adapted matrix of awareness of innovations in terms of the packaging functions. The results define two main features, there are age groups on the x-axis and innovation status is on the y-axis. Based on the results of the questionnaire, the innovation status is identified as an amount of the recognized requests compulsory on the new packaging by their functions. Based on the given weights, Loučanová [8] 'M' obligatory requests with amount 3, 'A' attractive requests with amount 2, 'O' one-dimensional requests with amount 1, 'I' indifferent requests and without requests have amount 0, 'R' contradictory requests have amount − 1 and 'S' skeptical with amount − 2 (in different age groups). The recognized requests impact on new technologies and materials packaging according to their functions. They are defined as the average weight of the identified requests percentage.

Subsequently, the comparison analysis, which aims to identify and measure comparable data, was used. It was used to identify the differences between customers' perceptions of intelligent packaging functions and customers' perceptions of active packaging functions.

The analysis of comparison identified the buyer requests by Kano model. Consequently, each weight was allocated to these requests. Each recognized request characterizes amount 1, and it is multiplied by particular weight according to recognized category as follows: must be = 3, attractive = 2, one-dimensional = 1, indifferent = 0, reverse = −1 [8, 52]. Resulting in the sum of all values, the analysis compares buyer awareness of intelligent and active packaging functions. For intelligent and active packaging, the model identifies the target age group. Afterwards, the perception of intelligent and active packaging functions in Slovak conditions is designed by 3D simulation. Subsequently, model describes a portfolio matrix. The matrix takes into account the environmental-focused companies focusing on buyer perceptions and satisfaction with innovations. The designed matrix analyses influence of innovation on the buyer satisfaction coefficient. This coefficient is designed according to methodology of the Kano model, and the results describe the share of individual buyer requests as follows [53]:

$$\text{Satisfaction coefficient} = \frac{(A + O)}{(A + O + M + I)} \tag{1}$$

The coefficient describes the impact of buyer satisfaction requests ranging from 0 to 1 (the coefficient closer to 1, meaning the better satisfaction of buyers with the requests).

Consequently, the satisfaction coefficient of the each observed parameter is given according to the groups describing of the green buyer and the ability to separate the goods by the innovation. In each group, there is the arithmetic mean of the parameter and describes the final significance of the phenomenon. These values are applied to the portfolio matrix. Based on their location in the four quadrants, according to Bloom and Ginsberg [54], respondents adopt four environmental-oriented enterprise policy strategies: weak green, defensive green, extreme green and shadow green strategies.

4. The perception of intelligent and active packaging in Slovakia

In the research, we focused on intelligent and active forms of packaging, the occurrence of which we have mostly noticed in the Slovak market. The Kano model was used for the study of customers' attitudes to the individual functions of active and intelligent packaging. Firstly, we examined the perception of their availability and their functionality, as well as other customer requirements for these forms of packaging. The requirements of the packaging functions among the monitored age categories were identified.

Consequently, the analysis calculated the competitiveness and the impact of the innovation within the specific categories of age by classifying the innovation status. The innovation status is deliberated as the sum of the specific recognized categories of the examined factors. The influence of the status of innovation for each age category is stated as the weighted average of the specified buyer requests identified by the Kano model. The innovation perception typology matrix describes innovation status in (**Table 1**).

Obviously, as apparently from **Table 2**, active and intelligent packaging has dissimilar influences on buyers in different age. The positive influence (growing the competitiveness of goods through active and intelligent packaging is in the age category 18–30 and 31–40. The results highlight that in the case of goods designed for these consumers, the implementation of intelligent elements for innovation creation represents an increase in the competitiveness of these products. On the contrary, these innovations are differently perceived by the elderly respondents. The age categories 41 and more and especially elder consumers are particular by experiencing. They often experience innovation with negative satisfaction. The dissatisfaction is higher with increasing age (**Figure 2**). The results confirmed the theoretical knowledge that noted buyers' fears of innovation [55].

Subsequently, the identified innovation status and the impact of innovation on individual age categories are illustrated in the innovation perception typology matrix, see **Figure 2**.

The results confirmed that the respondents included in the age category of 41–50 years have the specific requirements. They are aware of what is important to them

Parameters/Age	**18–30**		**31–40**		**41–50**		**51–60**		**61 <**	
Conception of innovation packaging	**A**	**2**	**A**	**2**	**I**	**0**	**R**	**−1**	**R**	**−1**
Accessibility	I	0	I	0	I	0	I	0	I	0
Perception	R	−1	I	0	R	−1	R	−1	R	−1
Functionalities	I	0	I	0	I	0	I	0	I	0
Voice control	II	0	II	0	II	0	II	0	II	0
The attractiveness of the packaging	I	0	I	0	I	0	I	0	I	0
Advertisement	II	0	II	0	II	0	II	0	II	0
Freshness indicators	Q	0	Q	0	Q	0	Q	0	Q	0
The price of packaging	I	0	R	−1	I	0	I	0	R	−1
Innovation status		1		1		−1		−2		−3
Age/Parameters	**18–30**		**31–40**		**41–50**		**51–60**		**61 <**	
Conception of innovation packaging	40.72	2	38.5	2	37.63	0	32.43	−1	36.97	−1
Accessibility	55.20	0	46.52	0	51.55	0	41.65	0	56.1	0
Awareness	44.34	−1	42.25	0	46.91	−1	36.19	−1	46.67	−1
Functionalities	57.46	0	60.97	0	59.79	0	52.91	0	61.82	0
Voice control	51.14	0	58.29	0	55.15	0	35.38	−1	57.58	0
The attractiveness of the packaging	47.06	0	52.4	−1	52.06	0	55.09	0	49.7	0
Advertisement	47.51	0	56.15	0	45.88	0	35.77	0	49.09	0
Freshness indicators	32.58	0	31.55	0	43.30	0	41.23	0	35.15	0
The price of packaging	52.94	−1	43.32	−1	44.33	0	34.31	0	53.33	−1
Factor size		4.12		3.54		5.21		7.62		15.21

Table 1.
Basic data for compiling a typology matrix focusing on intelligent and active packaging.

when choosing products, and they also recognize what packaging function is an indispensable part of the product. As the most important factor, they consider the protective and informative function of a package. The handling and environmental functions are attractive. The above-mentioned age group is the target group for new packaging innovations, which follows from our intended the highest innovation status with the influence size of 54.

Very similar attitudes can be observed in the category of respondents in the age 51–60 years, who are also very well aware of their specific requirements and what they consider unnecessary. In this age category, consumers consider as important the handling, protection, informative, economic and environmental functions of packaging. These functions are one-dimensional requirements for them. For one-dimensional requirements applying that the higher rate of these requirements fulfillment, the consumers are more satisfied. However, in comparison to the mandatory requirements, customers do not expect one-dimensional requirements automatically. With regard to this age category of consumers, companies should focus on packaging innovations in the context of general functions, but nevertheless the specific packaging innovations, what is obvious from the high innovation status.

Packaging functions	Ages														
	18–30			31–40			41–50			51–60			61+		
	Requirement	Active packaging function	Intelligent packaging function	Requirement	Active packaging function	Intelligent packaging function	Requirement	Active packaging function	Intelligent packaging function	Requirement	Active packaging function	Intelligent packaging function	Requirement	Active packaging function	Intelligent packaging function
Containment	I	0	—	R	−1	—	A	2	—	O	1	—	A	2	—
Protection	I	0	—	O	1	—	O	1	—	O	1	—	I	0	—
Communication	I	—	0	I	—	0	O	—	1	O	—	1	I	—	0
Convenience	O	—	1	O	—	1	A	—	2	O	—	1	I	—	0
Total	—	0	1	—	0	1	—	3	3	—	2	2	—	2	0

Table 2.
The comparison analysis of perceptions of intelligent and active packaging functions in Slovakia.

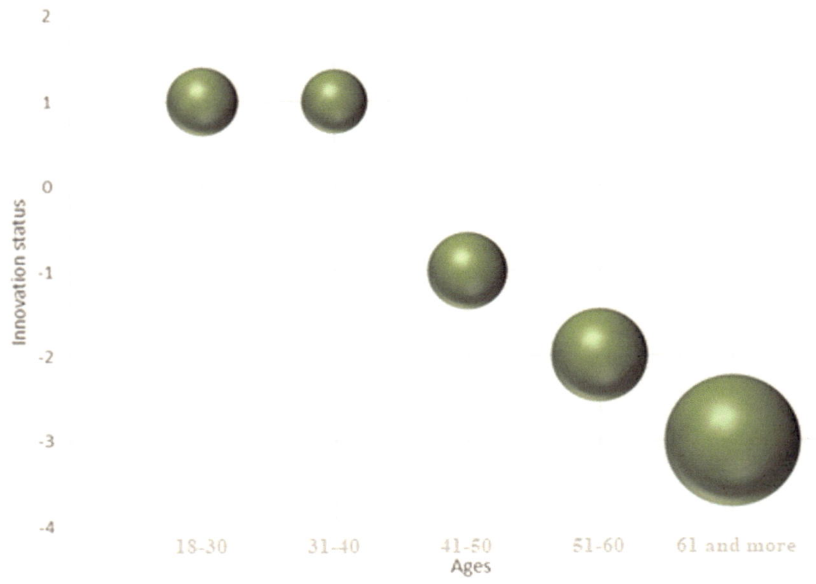

Figure 2.
A matrix of intelligent innovation intelligence typology, focusing on intelligent and active packaging.

And finally in the last examined age category (61 and more), the consumers identified only the handling function as important. That is the reason why the firms have to focus on that issue how to simplify goods handling. Though, the innovation status of this age category is low. Everything stated is clearly connected to the most common purchase problems related to the age of customers. Also, for example, according to Lesakova [56], the increasing age is associated with increasing mobility problems of older people and those are becoming dependent on assistance when they come to the purchase.

Overall, the global aging of the populations calls for more age-friendly approaches to be implemented in packaging. The population aging may be perceived as a challenge to prepare for these developments in such a way that older people can benefit from age-friendly strategies in packaging innovation.

To conclude the survey, results indicate different attitudes to intelligent and active packaging according to the age, but the majority of respondents agreed that the packaging should be ecological and should meet the informative and protective functions. These three functions can be attributed to innovations that are most preferred among all respondents.

From the point of view of the perception of the functions of intelligent and active packaging, the research results indicate different attitudes according to age categories (**Table 2**).

Active packaging can be considered valued above all for the buyers in the age of 41 and older. It characterizes active packaging features that lead to fulfillment and satisfaction. The younger consumers (< 40 years) are not influenced to such an extent by active packaging and their functions. The functions are indifferent for them. It involves the attributes that are not critical for customers, and their pass or fail does not affect their satisfaction or dissatisfaction, **Figure 3**.

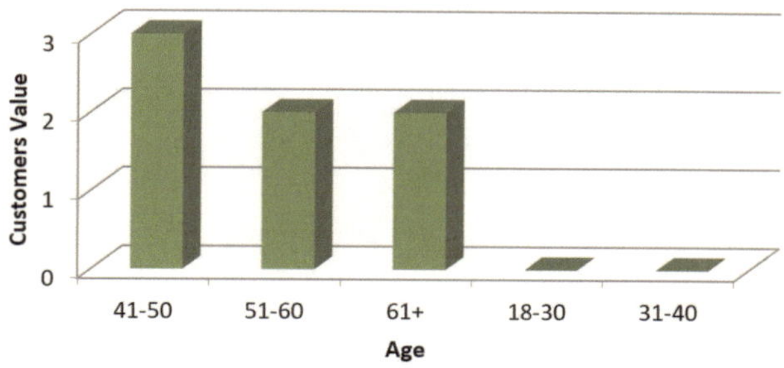

Figure 3.
The customer value of active packaging functions.

Regarding intelligent packaging, the most affected group is age category 41–50, followed by the category 51–60. Principally the innovations of intelligent packaging affect the buyers in the age 18–60 very similarly. These consumers have one-dimensional requests. The consumers from age category 61 and older are indifferent and do not react to innovation in packaging, **Figure 4**.

The evaluation of the value of intelligent and active packaging functions indicates the dissimilarities in customers' values, **Figure 5**. The younger consumers are more focus on the intelligent packaging. The generation of middle-aged is interested in both forms of packaging innovation equally. And finally the older age categories rather prefer only active packaging.

Active packaging functions are most valuable for the older buyers in the age of 41 and older. They represent a group that intelligent and active packaging is attractive and have one-dimensional requests. There are features that lead to satisfaction in the

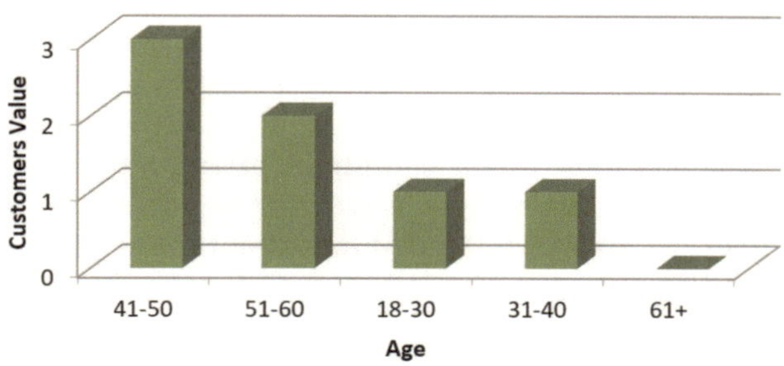

Figure 4.
The intelligent packaging functions from the consumers point of view.

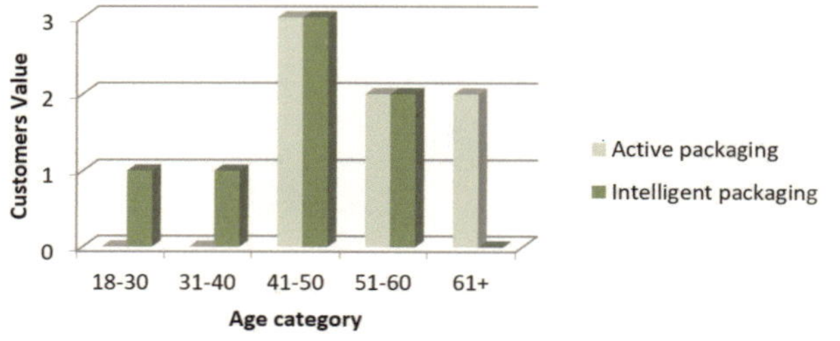

Figure 5.
The comparison of the value of active and intelligent packaging functions for the customers.

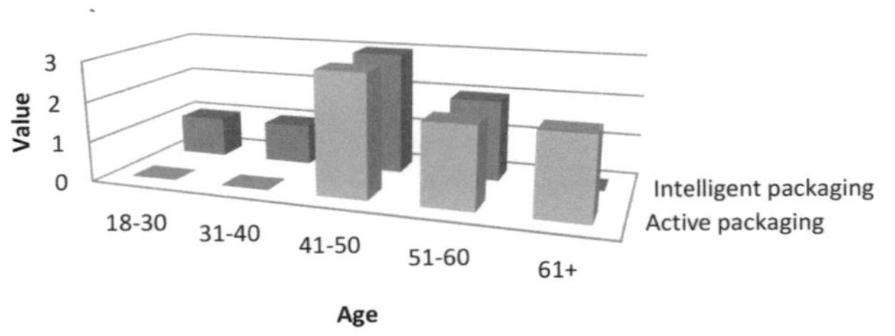

Figure 6.
Age and perceptions of active and intelligent packaging functions.

case of non-compliance to buyers' dissatisfaction. On the other hand, the young consumers in the age under 40 years are more fascinated in intelligent functions. On the contrary, these consumers are not interested in active packaging, and they have indifferent awareness, **Figure 6**.

The evaluation of intelligent and active packaging features for the buyers shows the discrepancy in different consumers groups. To summarize it, the younger consumers are more focused on the intelligent packaging, and the middle-aged groups are fascinated in all innovations in packaging and the older age categories prefer only innovations focused on active packaging. Finally, two individual functions of packaging – convenience and containment are important for consumers, as well as the communication function, **Figure 7**.

The monitored parameters in the study were divided into groups characterizing the nature of the green consumer and the consumer's ability to distinguish product packaging from competing products, i.e. to distinguish intelligent packaging innovations from traditional packaging. **Table 3** presents a summary of the examined parameters and their satisfaction coefficients.

The values of the satisfaction coefficient point to the fact that consumers are satisfied with the concept of intelligent packaging innovations. Other investigated parameters have low satisfaction coefficient values. We have identified a green innovation strategy of intelligent packaging by graphically representing the determined values of the nature of the green customers and ability to distinguish the packaging. Slovak consumers perceive intelligent packaging innovations as weakly green [57], see **Figure 8**.

Based on the identified green strategy of intelligent packaging innovations, we can conclude that Slovak consumers perceive intelligent packaging innovations as

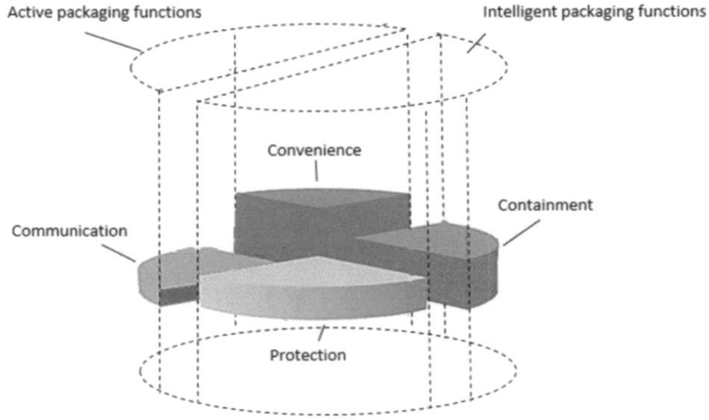

Figure 7.
Perceptions of intelligent innovation packaging by respondents in Slovakia – 3 D simulation.

Monitored parameter	Identified requirement	Percentage	Coefficient of satisfaction
Concept of innovation packaging	A	31.03	0.457
Accessibility	I	52.41	0.086
Perception	R	44.46	0.103
Functionalities	I	59.84	0.153
Nature of the green consumers		0.200	
Voice control	I	55.41	0.101
The attractiveness of the packaging	I	50.20	0.246
Advertisement	I	49.54	0.180
Freshness indicators	Q	35.59	0.412
The price of packaging	I	45.11	0.079
Ability to distinguish the packagings		0.203	

Table 3.
The Kano model values and environmental focus groups.

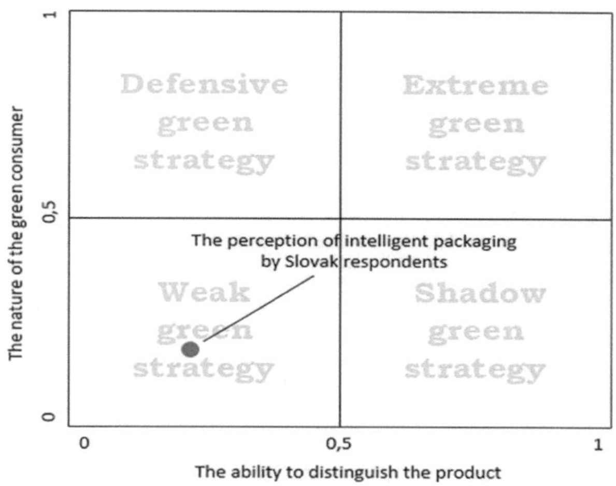

Figure 8.
Portfolio matrix of green strategies – perception of intelligent packaging innovations in Slovakia.

follows: Companies try to be environmentally responsible, but they do not promote these environmental initiatives and do not use green marketing to a sufficient extent. Businesses in Slovakia are more oriented towards reducing costs and increasing efficiency in order to increase competitive advantage by means of low costs. They use innovations precisely to make their processes more efficient. They do not consider intelligent packaging innovations to be a high-profit segment. With this strategy, companies fear that their products will not be able to differentiate themselves from competing products, and the introduction of smart packaging innovations would increase the price of their products. From the customers' point of view, customers are trying to be environmentally responsible, but the limits are in their awareness and the price. Price is one of the main factors in their purchasing decision.

5. Conclusion

Currently, society focuses on ecological strategies and reducing the society's impact on the environment. That is why packaging is approached more innovatively and creatively. The research results of the presented study point to the fact that the main target group for innovative packaging solutions (intelligent and active packaging) is consumers in the age category of 41–50 years. From the perspective of innovation status and age category, younger consumers are more focused on the smart functions of the packaging. The younger age group appreciates the packaging as an intelligent communicator. In this age category, they have the highest requirements for packaging innovations. As consumers get older, they are more oriented towards the active functions of packaging.

The study points to the fact that the concept of packaging innovations (intelligent and active packaging) is attractive to Slovak consumers, but in several aspects and especially the awareness of Slovak customers about these packaging innovations is at a low level – indifferent requirements, with a weak green strategy.

In general, intelligent and active packaging is easy to use and provides benefits for consumers, companies throughout the food chain and society.

These packaging innovations generally increase product safety and reduce food waste. In addition, with the help of data carriers of intelligent packaging, better management of the entire supply chain is possible. For some types of smart and active packaging, the main negative is the price, because the costs of development and production are still very high – they can represent 50–100% of the total cost of the product. And as several studies have identified, the limit for packaging costs in companies is 10% of the value of the products. From the point of view of retailers, smart packaging has identified potential negative changes in consumer purchasing behavior: Customers are more likely to prefer products with packaging indicating a more honest product. Even the difference in the color of the indicator can lead to a decrease in trust in the given brand. This change in consumer behavior could lead to an increase in the amount of unsold food. But at the same time, with the help of these indicators, it is possible to optimize the classic principle of displaying goods, when the retailer first sells products with a shorter shelf life and then with a longer shelf life, which reduces food waste [58].

Based on the theoretical analysis of the literature and research results, smart packaging presents the following advantages and disadvantages (**Table 4**).

Considering the results of the research and the mentioned facts, it is necessary:

- Increasing awareness of packaging innovations (intelligent and active, because as stated by Rogers [55], consumers accept innovations if they are sufficiently informed and do not feel a threat. Because as stated by Odecka and Bråthen [59], consumers often have a negative attitude towards innovations in case of lack of information and sufficient explanation.

- It is necessary to increase the attractiveness of smart packaging innovations, because according to Helus [60] and Kopaničová and Klepochová [61], customers also make decisions based on the attractiveness of the packaging. The packaging represents the customer's first contact with the product and therefore represents a mental accelerator, triggers cognitive processes and influences purchasing decisions.

Advantages	Disadvantages
Generally	
provides the user with relabeling and correct information on the conditions of the food, the environment and the packaging integrity	extra cost
enables the detection of calamities and possible abuse through the entire supply chain	possible migration issues of complex packaging materials into product
reducing food loss and waste	lack of recyclability of disposable packages
prevent unnecessary transport and logistics from an early stage	possible mistrust/confusion of technology
enhancing food safety and biosecurity	
enhancing food quality assurance	
From the perspective of customer perception in Slovakia	
concept of intelligent and active packaging	Price
	Awareness

Table 4.
The advantages and disadvantages of the intelligent packaging.

- Compatibility must be ensured so that both producers and consumers can identify intelligent and active packaging for a given product and that it is useful and represents an advantage. Therefore, it is necessary to identify the product for which these packages are advantageous and represent an increase in their sales or reduces the impact.

- It is necessary to define and clarify the recycling of intelligent and active packaging – from the point of view of their production, installation and use. This will include packaging innovations (intelligent and active packaging) in the green innovation strategy.

- It is necessary to carry out further research for the further development of packaging leading to the reduction of costs and the expansion of their benefits and wider use, improvement of product quality and safety, etc.

In the future, intelligent and active packaging has the potential to represent a competitive advantage for products to meet customers' needs and to increase their satisfaction. Definitely, innovation diffusing and management cannot exist without customer research because acceptance of innovation is ultimately an important factor of innovation success not only in the domestic but also global market.

In addition to the above and based on these analyses, the paper provides both theoretical and practical benefits in the form of recommendations for innovators in regard to active and intelligent packaging. This can then be reflected in the performance of companies and their investment decisions as stated by Ipate et al. [62] and Borlea et al. [63]. The benefit is the possible application of acquired theoretical knowledge in their implementation in practice, leading to an increase in the performance of companies in their investment decisions in the context of innovation processes. According to intelligent innovation, the company must understand the customer's needs and attitudes and then subsequently find the right marketing communication tools with customers. Definitely innovation diffusing and management cannot exist without customer research since innovation acceptance is ultimately an important factor of innovation success.

The results of the analysis can be beneficial in designing future packaging innovations as well as in the selection of marketing communication tools with regard to the specifics of the perception of individual groups of customers.

Acknowledgements

The authors are grateful for the support of the Scientific Grant Agency of the Ministry of Education, Science, Research and Sport of the Slovak Republic. Grant No. 1/0494/22 Comparative advantages of the wood-based sector under the growing influence of the green economy principles.

Conflict of interest

"The authors declare no conflict of interest."

Author details

Erika Loučanová, Ján Parobek* and Martina Nosáľová
Technical University in Zvolen, Zvolen, Slovakia

*Address all correspondence to: parobek@tuzvo.sk

IntechOpen

© 2022 The Author(s). Licensee IntechOpen. This chapter is distributed under the terms of the Creative Commons Attribution License (http://creativecommons.org/licenses/by/3.0), which permits unrestricted use, distribution, and reproduction in any medium, provided the original work is properly cited.

References

[1] Magnusson A, Olander Roese M, Olsson A. Finding Methods for Innovative Packaging Development: The Card Approach. Packaging Logistics. Lund: Lund University; 2013. p. 51

[2] Kotler P. Marketing Management. Praha: Grada Publishing; 2001. p. 720

[3] Zeman S. Balenie a obalová technika. Nitra: Slovenská poľnohospodárska univerzita; 2005. p. 177

[4] Azzi A, Battini D, Persona A, Sgarbossa F. Packaging design: General framework and research agenda. Packaging Technology and Science. 2012;**25**(8):435-456. DOI: 10.1002/pts.993

[5] Hine T. The Total Package: The Evolution and Secret Meanings of Boxes, Bottles, Cans, and Tubes. 7th ed. Boston: Little Brown & Co; 1995. p. 286

[6] Loucanova E, Kalamarova M, Parobek J. The innovative approaches to packaging – comparison analysis of intelligent and active packaging perceptions in Slovakia. Studia Universitatis "Vasile Goldis" Arad – Economics Series. 2017;**27**(2):33-44

[7] Loučanová E, Kalamárová M, Parobek J. Inteligentné inovácie a ich riešenia ako nástroj konkurencieschopnosti. Zvolen: Technická univerzita vo Zvolene; 2017. p. 139

[8] Loučanová E. Inovačné analýzy a stratégie. Zvolen: Technická univerzita vo Zvolene; 2016. p. 149

[9] Loučanová E, Parobek J, Kalamárová M. The perception of respondents of packaging innovations in Slovakia. Studia Universitatis "Vasile Goldiş" Arad - Economics Series. 2016; **26**:33-43

[10] Wyrwa J, Barska A. Innovations in the food packaging market: Active packaging. European Food Research and Technology. 2017;**243**:1681-1692. DOI: 10.1007/s00217-017-2878-2

[11] Hellström D, Nilsson F. Logistics-driven packaging innovation: A case study at IKEA. International Journal of Retail & Distribution Management. 2011;**39**(9):638-657

[12] Lockhart H. A paradigm for packaging. Packaging Technology and Science. 1997;**10**(5):237-252

[13] Yam KL, Takhistov PT, Miltz J. Intelligent packaging: Concepts and applications. Journal of Food Science. 2005;**70**(1):1-10

[14] Robertson GL. Food Packaging: Principles and Practice. Boca Raton: CRS Press; 2013. p. 733

[15] Dopico-Parada A, López-Miguens MJ, Álvarez-González P. Building value with packaging: Development and validation of a measurement scale. Journal of Retailing and Consumer Services. 2021;**63**:102685. DOI: 10.1016/j.jretconser.2021.102685

[16] Lindh H, Olsson A, Williams H. Consumer perceptions of food packaging: Contributing to or counteracting environmentally sustainable development? Packaging, Technology and Science. 2016;**29**:3-23

[17] Loučanová E, Nosáľová M. Eco-innovation performance in Slovakia: Assessment based on ABC analysis of eco-innovation indicators. BioResources. 2020;**15**(3):5355-5365

[18] Loučanová E, Paluš H, Báliková K, Dzian M, Slašťanová N, Šálka J. Stakeholders' perceptions of the innovation trends in the Slovak forestry and forest-based sectors. Journal of Business Economics and Management. 2020;**21**(6):1610-1627

[19] Kalamárová M, Parobek J, Loučanová E. Consumer perception of active and intelligent packaging in Slovakia. In: Management and Economics in Manufacturing: Global Scientific Conference; 5th - 6th October 2017; Zvolen. Zvolen: Technical University in Zvolen; 2017. pp. 25-30

[20] Parobek J, Loučanová E, Nosáľová M, Dovčíková A. Perception of wood-based packaging innovation applying smart solutions. In: Digitalisation and Circular Economy: Forestry and Forestry Based Industry Implications: Proceedings of Scientific Papers; 11 – 13 September 2019; Varna. Zagreb: WoodEMA, i.a.: University of Zagreb, Faculty of Forestry; 2019. pp. 207-212

[21] Dzurová M. Obal a balenie ako súčasť logistiky. Bratislava: Eurounion; 1997. p. 144

[22] Calver G. What is Packaging Design? Hove: RotoVisio; 2004. p. 256

[23] Sonneveld K. What drives (food) packaging innovation? Packaging technology and science. Packaging Technology and Science. 2000;**13**:29-35

[24] Zaušková A, Loučanová E. Inovačný manažment. Zvolen: Vydavateľstvo Technickej university vo Zvolene; 2008. p. 91

[25] Cogliandro JA. Intelligent Innovation: Four Steps to Achieving a Competitive Edge. Fort Lauderdale: Ross Publishing; 2007. p. 232

[26] Zongwei L. Smart Manufacturing Innovation and Transformation: Interconnection and Intelligence. Business Science Reference, Hershey: IGI Global; 2014. p. 407

[27] Loučanová E, Kalamárová M, Olšiaková M. The need of introduction of eco-innovation aspects in packaging from the customers' perspective. In: The path forward for wood products: A global perspective: Proceedings of scientific papers; Louisiana, 5 – 8 October 2016. Zagreb: WoodEMA, i.a. – International Association for Economics and Management in Wood Processing and Furniture Manufacturing. 2016. pp. 57-60

[28] Loučanová E, Parobek J, Kalamárová M, Paluš H, Lenoch J. Eco-innovation performance of Slovakia. Procedia Economics and Finance. 2015; **26**:920-924

[29] Yam KL, Sun Lee D. Emerging Food Packaging Technologies: Principles and Practice. Cambridge: Woodhead Publishing; 2012, 2012, 512

[30] Ahvenainen R. Novel food packaging techniques. Cambridge: Woodhead Publishing Limited; 2000. p. 590

[31] Kačenák I. Balenie tovaru. Bratislava: Ekonóm; 1996. p. 320

[32] Sosnovcová J. Aktivní a inteligentní obalové systémy pro balení potravin. Brno: Státní zdravotní ústav; 2008. p. 15

[33] Trommsdorf V, Steinhoff F. Marketing inovací. Praha: C. H. Beck; 2009. p. 291

[34] Pajtinková BG, Gubíniová K. Udržateľný marketingový manažment. Bratislava: 1AM press; 2012. p. 241

[35] Loučanová E, Šupín M, Čorejová T, Repková Štofková K, Šupínová M, Štofková Z, et al. Sustainability and branding: An integrated perspective of eco-innovation and brand. Sustainability. 2021;**13**(2):10

[36] Šupín M, Loučanová E, Olšiaková M. Forests, wood products and bioenergy in climate change adaptation and mitigation. In: Sustainability of Forest-based Industries in the Global Economy: Proceedings of Scientific Papers. Zagreb: WoodEMA, 2020. Zagreb: WoodEMA, i. a.: University of Zagreb, Faculty of Forestry; 2020. pp. 19-23

[37] Kaputa V, Olšiaková M, Maťová H, Drličková E. Do preferences for wood-framed houses' attributes change over time. In: Digitalisation and Circular Economy: Forestry and Forestry Based Industry Implications—Proceedings of Scientific Papers; 11 – 13 September 2019; Varna. Zagreb: WoodEMA, i.a.: University of Zagreb, Faculty of Forestry; 2019. pp. 69-74

[38] Šupín M, Loučanová E, Olšiaková M. Sustainable bioenergy policy for the period after 2020. In: Digitalisation and Circular Economy: Forestry and Forestry Based Industry Implications—Proceedings of Scientific Papers; 11 – 13 September 2019; Varna. Zagreb: WoodEMA, i.a.: University of Zagreb, Faculty of Forestry; 2019. pp. 315-320

[39] Paluš H. Survey of attitudes towards forest and chain of custody certification in the Slovak Republic. Drewno: Prace Naukowe, Doniesienia, Komunikaty. 2009;**52**:65-81

[40] Löfgren M, Witell L. Kano's theory of attractive quality and packaging. The Quality Management Journal. 2005; **12**(3):7-20

[41] Kauppinen-Räisänen H, Luomala HT. Exploring consumers' product-specific colour meanings. Qualitative Market Research: An International Journal. 2010;**13**:287-308

[42] Loučanová E, Kalamárová M, Parobek J. Inteligentné inovácie a ich riešenia ako nástroj konkurencieschopnosti. Zvolen: Technická univerzita vo Zvolene; 2017. p. 139

[43] Prakash G, Pathak P. Intention to buy eco-friendly packaged products among young consumers of India: A study on developing nation. Journal of Cleaner Production. 2017;**141**:385-393

[44] Brennan JC, Crandison AS. Food Processing Handbook. Weinheim: Wiley; 2011. p. 826

[45] Zhang ZG, Song RH, Wu P, Yao MQ, Chen L. Innovation packaging design of Mogan Huangya tea. In: 12th IEEE International Conference on Computer-Aided Industrial Design and Conceptual Design. 27 – 29 November 2011; Chongqing. New York: IEEE; 2011. pp. 149-153

[46] Regattieri A, Santarelli S. The important role of packaging in operations management. Operations Management. 2013;**3**:10-5772

[47] Wells LE, Farley H, Armstrong GA. The importance of packaging design for own-label food brands. International Journal of Retail & Distribution Management. 2007;**35**(9):677-690

[48] Madáč J, Mateides A, Pohančaník P. Spokojnosť zákazníka 2. Banská Bystrica: EF UMB; 2007. p. 65

[49] Xu Q, Jiao RJ, Yang X, Helander M, Khalid HM, Opperud A. An analytical

Kano model for customer need analysis. Design Studies. 2009;**30**(1):87-110

[50] Goodpasture J. Quantitative Methods in Project management. USA: J. Ross Publishing; 2003. p. 288

[51] Grapentine, T. Why the Kano model wears no clothes. Quirks Marketing Research Media. [Internet]. 2015 Available from: http://www.quirks.com/articles/2015/20150407.aspx [Accessed: 2000-MM-DD]

[52] Ducár S, Naščáková J, Malák M. Návrh systému merania spokojnosti zákazníkov Kano modelom. Transfer Inovácií. 2006;**9**:137-139

[53] Krnáčová P, Lesníková P. Kano model spokojnosti zákazníka v neziskovom sektore. In: Vedecké State Obchodnej Fakulty 2012. Bratislava: Ekonóm; 2012. pp. 368-382

[54] Bloom PN, Ginsberg JM. Choosing the right green marketing strategy. MIT Sloan Management Review. 2004;**46**(1):79-84

[55] Rogers EM. Lessons for guidelines from the diffusion of innovations. The Joint Commission Journal on Quality Improvement. 1995;**21**(7):324-328. DOI: 10.1016/s1070-3241(16)30155-9

[56] Lesakova, D. Determinanty úrovne, štruktúry a tendencií v osobnej spotrebe a spotrebiteľskom správaní seniorov v kontexte cenovej a dôchodkovej politiky v SR. In: Vedecké state Obchodnej fakulty 2012. Bratislava: Ekonóm. 2012. pp. 422-429

[57] Loučanová E, Parobek J, Nosľová M, Dopico A. The perception of respondents of intelligent packaging in Slovakia as ecological innovations. Studia Universitatis Vasile Goldiș Arad, Seria Științe Economice. 2019;**29**(1):58-67

[58] Müller P, Schmid M. Intelligent packaging in the food sector: A brief overview. Food. 2019;**8**(1):16

[59] Odeck J, Bråthen S. On public attitudes toward implementation of toll roads-the case of Oslo toll ring. Transport Policy. 1997;**4**:73-83

[60] Helus Z. Sociálni psychologie pro pedagogy. Praha: Grada; 2015. p. 400

[61] Kopaničová J, Klepochová D. Consumers in new millennium: Attitudes towards adoption of new technologies in purchasing process. Studia Commercialia Bratislavensia. 2016;**9**(33):65-74

[62] Ipate N, David KG, Ipate I, Bogdan A. The bioeconomy model in future sustainable development. Studia Universitatis "Vasile Goldiş" Arad Economics Series. 2015;**25**(2):115-125

[63] Borlea SN, Mare C, Achim MV, Puscas A. Direction of causality between financial development and economic growth. Evidence for developing countries. Studia Universitatis "Vasile Goldiş" Arad Economics Series. 2016;**26**(2):1-22

Chapter 10

Overview of Food Antimicrobial Packaging

Małgorzata Mizielińska and Artur Bartkowiak

Abstract

Acutely due to awareness that food products are highly vulnerable to microbial contamination, the food industry constantly tries to uncover new methods for the preservation of their products in order to guarantee their goods and processes continue to offer the highest quality and uphold safety standards throughout the production, storage, and distribution chain. Antimicrobial food packaging can play an important role in food shelf-life extension through the inhibition of microorganism growth present on the surface of food products. Antimicrobial packaging materials containing active substances incorporated into a polymer matrix or as surface coatings have begun to receive more attention for their use as antimicrobial control agents in food packaging systems. The most commonly used packaging materials are paper and plastics. However, from the ecological point of view, biopolymer-based materials have recently garnered more attention in the development of antimicrobial packaging as an alternative, due to their nontoxic biodegradability. In addition, the ongoing global spread of the pandemic caused by the SARS-CoV-2 has led to a preference for fresh food packaged in single-use food coverings. In order to address customer concerns and safeguard their health, the packaging industry could implement additional health safety measures, such as active packaging with antiviral properties.

Keywords: antimicrobial and antiviral packaging, active packaging, active coatings, active films food packaging

1. Introduction

Packaging, as an essential component of both the food manufacturing process and the entire food supply chain, plays an essential role in the safeguarding of food products [1, 2]. A fundamental and vital role of food packaging is to preserve food quality and safety, reduce food waste and foodborne diseases, and limit the negative impact of uneaten food on the environment and the global economy [2]. It should be noted that the packaging itself is a coordinated system for product safety, as well as being an efficient and practical method for shipment, marketing, storage, retailing, and consumption to increase sales and profits for producers and retailers and maintain food quality for consumers. Packaging must be able to meet all quality requirements while being cost-effective and efficient. It is important to permit brands to form and standardize packaging and also to create effective advertising and make a large-scale distribution

IntechOpen

and global retailing possible [1]. However, it should not be forgotten that food packaging systems must primarily protect food from environmental contamination, shock, outside odor, dust, physical damage, and mechanical force, as well as temperature, moisture, gas release, light, microorganism contamination, water and water vapor, in addition to other external environmental factors during processing, transport/distribution, storage, and marketing. It should maintain the basic attributes of food, such as texture, color, taste, and quality of the food products, as well as microbial purity resulting from the increase in food shelf-life, and subsequently the reduction of food waste. The main causes of food deterioration, such as microbial spoilage or oxidation, may be avoided by the application of the appropriate packaging systems [3–5]. The oxidation of food products may result in decreased nutritional value, energy content, flavor, and color, thus decreasing the quality of food. On the other hand, microbial spoilage might not only result in a decrease in food quality [3, 4]. Foodborne diseases through the consumption of food products containing pathogenic bacteria or fungi can contribute to serious health issues or even the death of consumers [1, 3]. Fresh food and animal-originated food products are mostly contaminated by bacteria [3, 6]. Most frequently health issues are caused by pathogenic bacteria, such as *Salmonella* spp., *Escherichia coli, Listeria monocytogenes, Staphylococcus aureus, Clostridium* spp., and *Campylobacter jejuni* [3, 7]. Filamentous fungi are also considered a severe pathogen in food products, due to their ability to penetrate and break down food using extracellular enzymes [1]. Vegetables, fruits, and grain crops are vulnerable to fungal strain contamination, mostly from *Penicillium* spp., *Aspergillus* spp., *Botrytis* spp., *Sclerotinia* spp., *Alternaria* spp., *Fusarium* spp., *Rhizoctonia* spp., *Geotrichum* spp., *Phytophthora* spp., and *Pythium* spp. [1, 8]. To avoid food oxidation or pathogen entry and any growth inside food products with extended shelf life during their storage, various active packaging systems can be used [1, 3, 4].

1.1 Packaging materials covered with active coatings

The modification of various polymers, including biopolymers, to obtain antimicrobial packaging that mainly includes polymer blending with bioactive components. Unfortunately, the negative effects of thermal processing on active agents during their incorporation into the polymer matrix can be seen. One solution to this problem is the coating of the polymer packaging surface. Fabrication of one-layer, bi-layer, tri-layer, and/or multilayer films through the use of a surface coating/covering technology is a promising strategy to improve the mechanical, barrier, or antimicrobial properties of traditional packaging materials for food packaging. Moreover, the surface coating method is more conducive to the migration of active substances and thus effectively protects the product [9, 10]. In the case of antimicrobial packaging, two important components, namely a polymer-based coating carrier, and an antimicrobial agent are very important. Antimicrobial coatings use many biobased carriers, such as carbohydrates, proteins, and lipids. Carboxymethyl cellulose (CMC) is a water-soluble derivative of cellulose that has the ability to form coatings and films in order to obtain an antimicrobial coating which may lead to increased food shelf-life. CMC is a linear polymer, it is long-chain and high molecular weight, make it a suitable polymer as a coating [11]. Starch as an inexpensive biodegradable polysaccharide is also used as a coating carrier, due to its renewable, nontoxic, multifunctional, biodegradable, and film-forming properties [12]. Among coating carriers, polysaccharides, such as pullulan, cellulose, or chitosan and its derivatives, are very attractive to provide barrier properties. Additionally, chitosan has been found to be stable and effective against a wide spectrum of microorganisms, where its antimicrobial activity depends on the concentration, molecular weight, and degree of deacetylation. Chitosan solutions in various organic acids may be prepared so that on drying, they form clear, flexible, and tough films/coatings [9, 13]. Mao et al. [9] noted that polymeric coatings based on caseinate, chitosan (CS), and polydopamine were used as carriers to obtain antibacterial packaging. As active agents, natural plant polyphenols and essential oil were used, and the effectiveness of these packaging materials was observed. Poly(vinyl alcohol) (PVA) and blends of PVA with starch have also been used as coating carriers due to their good film-forming and other functional properties, such as mechanical strength, water-solubility, oxygen barrier properties, transparency, and degradability [9]. In contrast to biopolymer and polymer films, paper-based packaging materials are mainly composed of fibrous cellulose pulp. This is a hydrophilic and porous material with a low barrier against water vapor and oxygen in comparison to plastic-based packaging materials. To improve these properties, covering a surface with coatings is necessary. Hydrophobic coating materials, such as paraffin wax and poly(butylene terephthalate) (PBAT), are considered promising candidates as biodegradable coating materials, due to their full biodegradability, thermoplasticity, low water vapor permeability, high flexibility, and good processability. The incorporation of active agents, such as antimicrobials and antioxidants, can make the coatings more valuable. Antimicrobial, active coating materials are usually prepared by blending or mixing antimicrobial substances with biopolymeric base carriers [14]. Recently, nanocellulose-based coatings are also used in food packaging due to their unique properties, such as biodegradability, mechanical properties, transparency, and antimicrobial activity against foodborne pathogens, including *E. coli*, *S. aureus*, *S.* Typhimurium, and *L. monocytogenes* [15–17]. The covering of polymer films or paper with functional bio-based, active coatings is a promising approach to improve packaging characteristics (e.g., antimicrobial, water vapor, and gas barrier properties) without compromising the biodegradable and/or recyclable features of packaging materials. The active coatings may demonstrate their effectiveness as antimicrobial packaging [11, 15–17]. In order to prepare antimicrobial coatings using the selected coating carriers, several categories of antimicrobial compounds, (described below in the next section) can be used [11].

1.2 Antimicrobial agents as additives to active packaging materials

Natural antimicrobial agents/compounds refer to a class of substances extracted from plants and animals or produced by microorganisms. These active agents may perform antagonistic actions against bacteria, viruses, yeast, or molds [18]. They also show anti-insect and antioxidant activity [19]. According to their biological origin, they can be divided into three categories: plant-derived antimicrobial agents, animal-derived antimicrobial agents, and microbial-derived natural antimicrobial agents. The food industry has used typically mainly chemical preservatives, such as hydrogen peroxide, sorbate, sorbic acid, benzoate, benzoic acid, and nitrite, to inhibit the growth of microorganisms responsible for food spoilage. Commercial preservatives may extend the shelf-life of food products; however, they may have unfavorable effects on the sensory properties of food [18, 20]. In order to extend the shelf-life of food and reduce health hazards, natural antimicrobial compounds, such as essential oils, propolis, lactoferrin, glucose oxidase enzyme bacteriocins, and probiotics extracted from animals, plants or produced by microorganisms could replace typical chemical preservatives. They could be used by the food packaging industry due to their nontoxic character [18]. Essential oils (EOs) play an important role in the design of antimicrobial packaging materials, as they exhibit a high and specific antimicrobial efficacy against a broad range of foodborne pathogens [3]. EOs are extracted from leaves, bark, flowers, and seeds of aromatic plants and are "generally recognized as safe," (GRAS) [19]. Eos, such as thyme, clove, cinnamon and tea tree, peppermint, oregano, lemongrass, and citronella, are plant-derived compounds that exhibit promising inhibitory effects [3, 18, 21]. EOs from the *Myrtaceae* family such as cloves, eucalyptus, galbanum, thyme, and tea tree contain eugenol and terpinen-4-ol as major bioactive components. It has been reported that these active agents offer antifungal effects. The results indicate that they inhibit glycolysis, which in turn influences cell energy metabolism, therefore disturbing the normal physiological activity of fungal pathogens [21]. According to several studies, EOs from *Lauraceae* and *Lamiaceae* (e.g., thyme, rosemary, oregano, cinnamon, etc.), exhibit essential antibacterial activities against foodborne pathogens and are generally rich in phenolic compounds, such as carvacrol, thymol, cinnamaldehyde, or eugenol [20]. Typically, the antibacterial mechanism is based on chemical interaction with the bacterial membrane and mitochondria leading to altering of their permeability, destroying their structural order, resulting in a massive loss of cell contents, important ions and molecules, and eventually leading to the death of the cell [18, 20, 21]. Moreover, many studies have demonstrated that the minor components of EOs play an outstanding role in the antimicrobial activity of essential oil, probably by a synergistic effect with the other EO components. One known example is the synergistic interaction between p-cymene and carvacrol, while p-cymene barely inhibits bacterial cell growth. On the other hand, carvacrol itself has a proven antibacterial effect against a wide range of microorganisms. It has been shown that the growth of microorganisms was significantly inhibited by a mixture of p-cymene and carvacrol. Interestingly, this activity was significantly lower when each terpene acts separately on bacteria growth [20, 22]. To summarize, at present, according to their high antimicrobial activity, they are mainly essential plant oils that act as antimicrobial and antioxidant compounds and they are widely used in smart or bioactive packaging materials to prevent the surface growth of microorganisms in foods [18, 23]. Natural antimicrobial agents can be extracted from plants, such as black currant, apple, pomegranate, grape, and quince, as well as chokeberry, bilberry, raspberry, mulberry, blueberry, yerba mate, green tea, sour cherry, walnut, rosemary, thyme, cinnamon, oregano, cumin, and many others. Plant extracts contain a wide range of bioactive components that include, polyphenols, iridoids, amides, saponins, alkaloids, and glycosides, as well as tannins, terpenoids, and quinones, which all have been reported to have a broad spectrum of antimicrobial properties. [18, 24–27]. The generic and quantitative contents of active compounds in plants vary widely. There are many factors that influence the composition and concentration of the active substances in plants and plant extracts, such as the organ, cultivar, and many various growth conditions, including weather. These factors all have a significant influence on the antimicrobial effect of the plant extracts [26, 27]. There are three main antimicrobial action mechanisms by plant extracts: (a) the inhibition of cytoplasmic membrane function (the destruction of the cell membrane and membrane proteins and causing damage to the cell wall of the microorganisms);

(b) inhibition of nucleic acid synthesis (extracts may suppress DNA synthesis by inhibiting DNA gyrase activity); (c) inhibition of energy metabolism (extracts can inhibit ATP synthesis). Finally, plant extracts may influence biofilm formation by influencing the quorum sensing mechanism, pigment production, and bacteria swarming motility, as well as altering the structure of the biofilm itself. Although active compounds from extracts can have an inhibitory effect at higher concentrations, in the case of lower concentrations they may have a stimulatory effect, indicating a bacterial defense mechanism [18, 26, 27]. Due to strong antimicrobial and antioxidant properties, plant extracts rich in active compounds, such as polyphenols, can be used within the food industry as natural preservatives and limit the nowadays use of chemical preservatives. They can also be used within the food packaging industry to extend the shelf life of food products by inhibiting microorganism growth and spoilage processes [18, 26, 27]. Antimicrobial peptides (AMPs) can be synthesized artificially in the laboratory or produced by bacteria [28]. Antimicrobial peptides are mostly composed of 12 ~ 60 amino acids that offer antimicrobial activity and participate in the host defense system [18, 28]. AMPs may be synthesized via three main methods: enzymatic synthesis, chemical synthesis, and biosynthesis, using a DNA recombinant technique. Among these methods, chemical synthesis is very common and has attracted increasing attention in the food packaging industry. AMPs obtained from bacteria are known as bacteriocins [28]. These AMPs can offer broad activity to directly inhibit the growth of yeasts, molds, bacteria, viruses, or even cancer cells [18, 28]. Typically, so-called antibacterial peptides (ABPs) have been mainly found to be active against bacteria. They have a clear influence on the bacterial cell membrane and create pores on their surface, resulting in the leakage of the intracellular matrix. They may also penetrate the cell membrane and interact with intracellular structures and disturb many activities including DNA/RNA or protein synthesis, resulting in bacterial cell death. ABPs are mainly positively charged molecules and they exhibit a high ratio of hydrophobic amino acids, allowing them to selectively bind to negatively charged bacterial membranes. The action mechanism of ABPs leads to the perforation of the cell membranes and their death [18, 29]. This is quite different from the bactericidal mechanism of antibiotics and does not lead to the microorganism becoming resistant. Furthermore, AMPs do not easily bind to mammalian cell membranes, which finally could be very harmful. It should be emphasized that antimicrobial peptides generally do not have any toxic side effects, though some have antioxidant functions and have been seen to scavenge free radicals. Lysostaphin and nisin are good examples of bactericidal peptides that suppress the growth of gram-positive bacteria. Antifungal peptides (AFPs) also function as antimicrobial proteins. They interact with the cell membrane, resulting in the disruption and finally death of the cell. Good examples of AFPs are echinocandin, defensin, and heliomycin [28]. As emphasized in the work of Ramos [30], innovative solutions have been proposed to improve the structural and functional properties of biopolymer-based packaging materials, including the incorporation of low amounts of specific nanoparticles (NPs) devoid of any relevant alteration of their migration to ensure their suitability for their application in food packaging. The authors stressed that the use of metallic-based nanoparticles (NPs), rather than antimicrobial organic agents, offered some advantages, such as high antimicrobial activity, lack of negative influence on food sensory properties, and compatibility with harsh polymer processing conditions, making NPs highly suitable for food spoilage control. Many nanomaterials have been used for food packaging, such as silver NPs, copper NPs, zinc oxide NPs, and titanium dioxide

NPs, as well as silicon dioxide NPs or mixtures of antimicrobial agents containing NPs. Several studies have confirmed that Ag-NPs, ZnO-NPs, and TiO2-NPs are often used in the packaging industry as antimicrobial agents [4, 30–34]. Ag-NPs were found to exhibit antioxidant activity and offer antibacterial effects against gram-positive bacteria, such as *S. aureus,* including methicillin-resistant *S. aureus* (MRSA), gram-negative *E. coli,* and *Pseudomonas aeruginosa* [35]. It has been shown that nano-biocomposite films based on PLA with modified cellulose nanocrystals (s-CNC) and Ag-NPs demonstrated high antimicrobial activity against *E. coli* and *S. aureus.* These materials showing homogeneous Ag dispersion in the polymer matrix, while also not affecting the PLA transparency, showed a significant improvement in barrier properties and antimicrobial activity [36, 37]. One should stress that consideration regarding human safety, as well as the environmental effects of packaging materials containing Ag-NPs in direct contact with food, have led several studies to report that these types of substances could be used for food preservation, due to their quality and safety [30, 36, 37]. The authors of several studies demonstrated that Ag nanoparticles migration levels were significantly below the legislative migration limits in Europe set by EU Regulation No. 10/2011 (for plastic materials intended to come into contact with food), such as in PLA nanocomposites and in poly(vinyl chloride) (PVC) [30, 38–40]. One of the most attractive nanoparticles that could be used in the packaging industry is nanosized ZnO. ZnO-NPs with unique morphologies, such as nanohelix and nanorings, may be easily synthesized and are cost-effective [32]. These nanoparticles have been explored as antimicrobial substances, used in antimicrobial food packaging as one of five various zinc compounds, which are regarded as being safe (GRAS) by the United States food and drug administration (USFDA, 21CFR182.8991) [41–43]. It was reported [32] that ZnO-NPs were highly reactive and induced reactive oxygen species (ROS), which causes single-stranded DNA breaks at a relatively low concentration of 10 mg/ml. Due to their high activity, ZnO nanoparticles offered a broad bactericidal effect on gram-positive and gram-negative bacteria and bacterial spores that typically are resistant to high pressure and high temperature [44–46] as well as yeasts and molds (Noshirvani et al. 2017). An additional advantage of ZnO-NPs is their UV-blocking properties [32]. The application of nanoparticles can improve the UV-shielding of all respective packaging film materials [47–49]. TiO2 nanoparticles have been reported by several authors [33, 50] to exhibit antimicrobial activity when exposed to UV light by generating reactive oxygen and hydroxyl radical (OH•) species (ROS) on its surface, resulting in the oxidation of the polyunsaturated phospholipids of microorganism cell membranes. As a consequence, the microorganism was inactivated. TiO2 nanoparticles have been used to inactivate a wide spectrum of microorganisms, such as *E. coli, P. aeruginosa, Enterococcus faecalis, Cyanobacteria, Lactobacillus helveticus, Legionella pneumophila, Clostridium perfringens, Salmonella enterica Choleraesuis, Vibrio parahaemolyticus* and *L. monocytogenes* [50]. The process of microorganism species inhibition by CuO-NPs was influenced by the concentration and size of nanoparticles. The CuO-NPs were confirmed to be active against gram-negative and gram-positive bacteria by transiting the microorganism cell membrane and then destroying their enzymes. The antifungal action of CoO-NPs was also observed. It should be mentioned that copper is very important in the case of active packaging because of its high activity against a wide spectrum of microorganisms. When copper ions are attached to a microbial cell they immediately donate and accept electrons, as a result, they show increased redox ability and the capability to inactivate the cell components and kill them [51].

1.3 Active packaging

Food packaging materials are very important in the food industry. The packaging material should isolate and preserve food from the external environment. It must be a nontoxic and impermeable physical barrier [4]. An important fact is that food packaging material should have very good mechanical, physical, chemical, water, and light barrier properties, and should be thermally stable and specifically processable during whole production and food packaging process. Additionally, it should be effective against microorganisms responsible for food spoilage [4, 51]. Various materials, such as glass, paper, paperboard, and metal have been used as packaging materials for many years because they are cheap, lightweight, and ecologically friendly [4]. Over the last decades synthetic, plastic materials have become an effective and dominant packaging material due to their light weight, high transparency, cost-effectiveness, and versatility. Moreover, these synthetic polymers have good mechanical, thermal, and barrier properties [4, 52–54]. Conventional food packaging is a passive, inert barrier that should protect food from the external environment. To extend the shelf-life of food products, food packaging has begun to evolve from simply passive to innovative interactive strategies, including intelligent, active, and smart packaging. This preservation is aided by active packaging that can even improve the quality of food through interactions between food products, surroundings, and the covering materials. Active packaging (AP) is defined as packaging that interacts with the packed food and environment dynamically to maintain both food product quality and extend shelf-life [1]. AP materials include additional ingredients in the packaging material or the package headspace to enhance system performance. **Figure 1** shows the forms of active packaging which may be used for food product preservation [55].

Form A (**Figure 1**) shows active packaging materials covered with coatings containing active compounds that are heat-sensitive or incompatible with the polymer matrix. One of the most common types of food AP systems is form B. The uniform distribution of the active agents makes the incorporation of active compounds into the polymer matrix

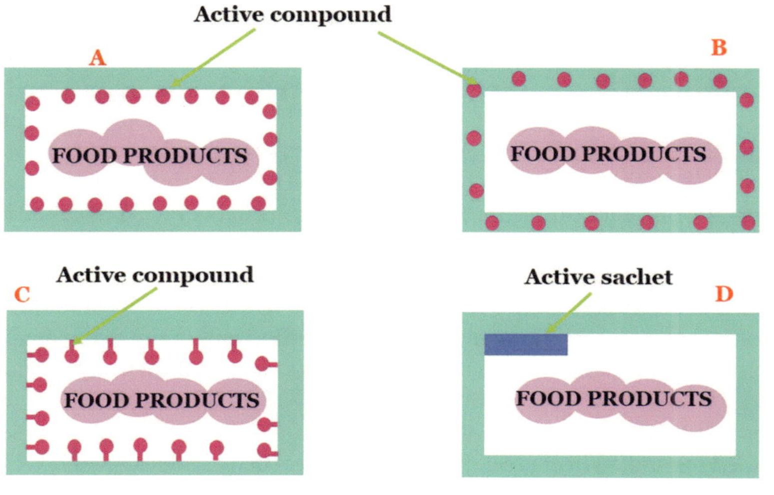

Figure 1.
Typical forms of active food packaging systems according to Almasi et al. [55] (A) polymer covered with a coating containing active agents, (B) active agents incorporated into the polymer matrix, (C) active agents immobilized on the polymer surface, and (D) active sachets inside the food packaging.

possible. It should be noted that active substances must have good compatibility and high resistance to polymer processing conditions and no adverse effects on polymer properties. It is also very important that the matrix should release these compounds gradually in control way to the food. The immobilization of active substances on the polymer surface by ion or covalent linkages requires the presence of functional groups on both the active compound and the polymer. This form (C) makes the strong bonding of active agents onto the polymers possible. It also allows the slow release of these substances into the food. The third form of active packaging is the introduction of pads or sachets containing active substances into the package [55]. Pads or sachets may be put in the packaging headspace in a free form, that alters the interior environment of the packaging by various mechanisms like an absorption and/or evaporation process, that in turn inhibit the growth of microorganisms or other food and hinder the deterioration processes. This form of active packaging, as well as the forms described above, are classed as "releasing systems" or "absorbing systems" based on their mode of action [1, 54]. "Absorbing system" means that the active components absorb unwanted substances from the food surface or the packaging during storage. Ethylene and oxygen scavengers were found to be very important examples of absorption systems. In the releasing system, antioxidants or antimicrobial agents migrate to the surface of food products, preventing food spoilage. They may be released by direct contact between the food and packaging material (leaching systems) or through gas-phase diffusion from the packaging layer to the food surface (volatile systems) [1, 55]. The type of food product, the gas composition, the particular packing machinery, type of package, headspace, additives (including antimicrobial additives), and storage temperature are significant factors that influence the shelf-life and the quality of packaged fresh or processed food. Considering both the type of packaging material and the type of food, the product can be stored under aerobic conditions in a modified atmosphere, or in a vacuum [1, 54, 55]. The selection of the proper storage conditions depends on the barrier against gasses and water vapor, hardness, stability, heat resistance capabilities, market requirements, and most important the cost of the packaging materials [1, 55]. The modified atmosphere packaging (MAP) preservation method has been used to protect food products against deterioration to extend their shelf-life [56]. MAP refers to the use of high-barrier packaging materials to package products in which gases, such as CO_2, N_2, and O_2, are mixed in selected ratios and introduced into the packaging material to inhibit the growth of microorganisms to reduce the enzymatic reaction and decrease the rate of lipid oxidation [56–59]. Nitrogen, oxygen, and carbon dioxide are the most commonly used gases in MAP. The ideal CO_2 and N_2 concentration in MAP depends on the food products and the different gas atmospheres that have been used for specific food products. The shelf life of many food products is limited by microbiological growth in the presence of O_2 [56, 57]. Generally, the effect of MAP is conditioned by the concentrations of CO_2 available in the packaging. Carbon dioxide inhibits the growth of the microorganisms responsible for food spoilage growing in normal air conditions, such as *P.* sp. and *Shewanella putrefaciens* [56–59]. The reduction in the amount of oxygen in the package may be a possible solution to preserve food and maintain its quality [57, 59]. Another option for extending the shelf life of food is to use vacuum packaging. As an example, a) the mixture of gases containing 40% CO_2, 30% N_2, and 30% O_2 was recommended for low-fat fish [57], b) the atmosphere of 40% CO_2, 10% N_2, and 50% O_2 was recommended for chilled beef steak [60], c) mixtures such as 80% O_2, 20% CO_2; 50%O_2, 20% CO_2, and 30% N_2, or 20% O_2, 20% CO_2, and 60% N_2 was recommended for dark-cutting beef stored under chilled and superchilled conditions [61], d) vacuum packaging and the mixture of gases for MAP containing 40% CO_2, 20% N_2, and 40% O_2 was recommended for pork steaks [62]. Packaging functionality depends on the maintenance of the modified atmosphere or vacuum inside the package for as long as possible, hindering gas diffusion through the material and damage from external forces. This happens through a layered composition of selected distinct polymer films [63]. A multilayered composition should provide improved food protection during storage. MAP packaging and vacuum typically comprises of low-density polyethylene (PE-LD), linear low-density polyethylene (PE-LLD), polyethylene terephthalate (PET), polypropylene (PP), polyamide (PA), and ethylene vinyl alcohol (EVOH) as an oxygen barrier [63, 64]. PE films as flexible materials are often used to maintain a moisture barrier. PET is used as a semi-effective gas and moisture barrier that imparts rigidity and strength to the packaging [64]. There are two common methods for producing multilayer packaging materials: layer-by-layer deposition and coextrusion methods. In the layer-by-layer technique, the layers may be formed one by one. It is important that each layer is chemically or physically bonded to each other to form the multilayer film. In the coextrusion method, polymers or biopolymers with different properties may be combined. The good adhesion of layers and the appropriate controlling of the thickness are the advantages of coextrusion techniques. In this technology, the packaging materials are coextruded together through separate screws. Therefore, these polymers/biopolymers would not be mixed and the final coextrudes would have a layered structure [55]. It should be noted that multilayer packaging can consist of even up to 11 ultra-thin layers offering different barrier properties [53]. In addition, the combination of different, blended, or multilayered polymer films which are used as food packaging, pose a particular challenge for recycling because there are made up of a large group of polymers [65]. Multilayer packaging cannot be recycled using traditional recycling technologies, because these technologies exist to recycle single-component plastics. Multilayer plastics have become a waste stream, additionally contaminated with food and other impurities. Another disadvantage is that they are in most cases nonbiodegradable. The accumulation of a packaging waste stream may lead to an environmental health crisis characterized by the accumulation of plastic waste in oceans [64].

1.4 Biodegradable, active packaging materials

Biodegradable packaging materials have emerged as an alternative to replace multilayer plastics left as waste. Biodegradable films are usually obtained from biopolymers of high molecular weight and classified according to the nature of their components. These biopolymers may be applied in the form of films, thermoformed, or as a cover/coating (thin layer), through immersion or spraying of a film-forming solution [52]. Starch was found to be one of the most promising biopolymers to replace nondegradable traditional plastics. Poly(butylene adipate-co-terephthalate) (PBAT), as an aliphatic-aromatic co-polyester, is also a promising biopolymer due to its ease of processing and good mechanical properties [66]. Biopolymers, such as PLA (polylactide), PHA (polyhydroxyalkanoates), or PBS (polybutylene succinate), also belong to the bio-based and biodegradable plastic family [53]. To obtain biopolymer packaging materials with antimicrobial activity, the active substances should be introduced directly into the polymer matrix. A good example is cassava starch chitosan films containing oregano essential oil incorporated in a polymer matrix, produced though an extrusion process. These packaging materials were found to be active against gram-positive bacteria (*B. cereus* and *S. aureus*) and gram-negative bacteria (*Salmonella enteritidis* and *E. coli*) [66]. Another example is starch/PBAT films incorporated with ε-PL by Gao et al. [67]. ε-PL is a homo-polyamide produced by *Streptomycetaceae* and *Ergot* fungi. As an antimicrobial agent, ε-PL exhibits a broad-spectrum of antimicrobial activity gram-positive and against gram-negative bacteria, molds, and yeasts. The authors indicated that starch/PBAT films with ε-PL inhibited the growth of tested microorganisms, such as *E. coli*, *S. aureus*, *and B. subtilis*. Biodegradable polymers, polyvinyl alcohol, and starch were used to prepare blends with natural additives, such as propolis extract and anthocyanin incorporated into the blend matrix. Boric acid was used as a cross-linker. Five different concentrations of propolis extract ranging from 0.5, 2, 5, 10, to 20% were used to develop active composites. It was demonstrated that an active film based on PE containing 20% of propolis was active against *E. coli* and methicillin-resistant *S. aureus*, respectively [68]. These novel materials having at least one dimension of just a few nanometers belong to polymer bio-nanocomposites. Bio-nanocomposites are novel, high-performance, lightweight, and eco-friendly materials that can replace traditional nonbiodegradable packaging materials obtained from synthetic materials [51]. Bio-nanocomposites are mostly constructed on biopolymer matrixes reinforced by nanofillers. These materials were found to have improved mechanical, barrier, thermal, and even antimicrobial properties attributed to the presence of the nanomaterials in the polymer matrix. The bond between the nanoparticle and biopolymer results in improved mechanical and thermal properties of the packaging materials [1, 69]. Several of the aforementioned nanomaterials were used in food packaging materials. There were nanoparticles of silver, copper, zinc oxide, titanium dioxide [1, 30–33], silicon dioxide [34], nanocellulose, nanoclays, and chitosan [30–33, 70, 71]. Thymol and silver nanoparticles (Ag-NPs) were used by Ramos and coauthors [30] to develop poly(lactic acid) (PLA)-based films with antibacterial activity. Various amounts of thymol (6 and 8 wt%) and 1 wt% Ag-NPs were introduced into the PLA matrix to produce active nano-biocomposites. The authors indicated that PLA-based nano-biocomposites showed dose-dependent slight antibacterial activity against *E. coli*. In addition, these films inhibited the growth of *S. aureus* 8325-A. Tarabiah and coauthors [72] used polyethylene oxide (PEO) as semicrystalline polymer and carboxymethyl cellulose (CMC) to develop the biodegradable, nontoxic PEO/CMC blend matrix (70/30 wt.%) as a host blend. ZnO nanorods as a filler were introduced into the polymeric matrix with various concentrations (upto 0.6 wt.%). The authors demonstrated that these PEO/CMC/ZnO nano-biocomposites can be used as a UV-mask. They also showed that these films were active against *S. aureus* and *E. coli*. De Souza and coauthors [73] used ZnO and Ag-ZnO nanoparticles as fillers and introduced them into the PBAT, which is an aliphatic-aromatic completely biodegradable flexible polyester, synthesized from 1,4-butanediol, adipic acid, and terephthalic acid. The authors established that such nano-biocomposite was active against *E. coli*. They also confirmed the synergistic effect between ZnO and Ag nanoparticles. The active packaging from chitosan and chitosan containing titanium dioxide nanoparticles was developed by Kaewklin and his team [33] to extend the shelf-life of climacteric cherry tomatoes. They indicated that tomatoes packaged in a chitosan package with nanoparticles showed lower quality changes than those in a chitosan film and control. The results suggested that the chitosan films containing titanium dioxide nanoparticles as active compounds exhibited ethylene photodegradation activity when exposed to UV light and consequently delayed the ripening process and any changes in the quality of the tomatoes. Hu and coauthors [71] developed chitosan/ZnO bio-nanocomposite and then introduced several concentrations of this bio-nanocomposite (0–5wt %) into the matrix of starch to produce antimicrobial starch-based composite films. The authors confirmed that such bio-nanocomposite films offered antimicrobial activity. They were active against *S. aureus* and *E. coli*. In addition, the antimicrobial activity of the films positively correlated with the amount of ZnO-chitosan nanoparticles introduced into the matrix. However, high concentration of ZnO-chitosan nanoparticles as fillers affected the barrier and mechanical properties negatively. If we consider all of the properties, like mechanical properties, barrier performance, and the antimicrobial effect of the composite films, 3.0 wt% of ZnO-chitosan nanoparticles was proven to be the optimal concentration of the active compound in the biopolymer matrix. Ascorbic acid blended with various biopolymers, such as whey protein, gelatin, and alginate or CuO nanoparticles, were used to prepare nanocomposite films in food packaging applications. CuO-NPs were loaded with ascorbic acid (CuO@Ascor) and incorporated into two different biopolymers, such as chitosan and xanthan gum to form chitosan-CuO@Ascor (A CA) and xanthan gum CuO@Ascor (X CA) to establish and show the antimicrobial activity of wrapping packaging material [74]. The incorporation of the active compounds into the polymer matrix may be performed using the solution casting method or via the melt blending method (to avoid solvents). The first step of melt blending method is a regranulation, but before this process, active compounds have to be mixed with the selected additives, such as plasticizers or compatibilizers. If nanoparticles are used as active agents, selected additives should also be added to prevent agglomeration of the nanoparticles and to provide their uniform distribution into the polymer/biopolymer matrix. Then, polymer/biopolymer pellets must be mixed with the prepared mixture of active compounds with additives. As next step, material has to be extruded with a twin-screw extruder for incorporating an active agent into the polymer/biopolymer matrix. The thermal profile of the process depends on the polymer matrix and on active compounds. To obtain active packaging material, the final, modified, and active pellets must be extruded through a flat die using chill-roll cast film extrusion line or using a film-blowing machine [50, 67]. **Table 1** shows additional examples of bio-nanocomposite films with an antimicrobial activity which may be used in food packaging.

1.5 Control release of active agent from antimicrobial package

Direct incorporation of active/antimicrobial compounds into the polymer matrix or into the biopolymer coating is the most common type of food active packaging system. In this system, the active substance and a biopolymer/polymer are combined to

NPs	Host matrix	Methods used for producing antimicrobial materials	Inhibition of microorganisms	Ref.
ZnO-NPs	Chitosan/ Cellulose/ Acetate /Phthalate	Active films with varying ratios of nano ZnO reinforcement were prepared by solvent casting method. They were stirred overnight on a magnetic stirrer prior to casting, then poured onto transparent sheets and dried for 9 days at room temperature.	*E. coli / S. aureus*	[75]
ZnO-NPs	Polyvinyl alcohol (PVA) / Starch	Powdered starch was dissolved in distilled water at a temperature of 95°C for 30 min. Then PVA was dissolved in distilled water at 95°C for 30 min under magnetic stirring. The solutions were mixed using magnetic stirring at 95°C for 30 min. Then ZnO-NP was added to obtain solution and stirred for another 20 min. The resulting solution was casted on a sterile glass petri plate and dried at 60°C for 24 h.	*S. Typhimurium*	[76]

NPs	Host matrix	Methods used for producing antimicrobial materials	Inhibition of microorganisms	Ref.
ZnO-NPs	Chitosan/ Carboxymethyl-cellulose	Chitosan was mixed with acetic acid solution at 60 °C. Separately, CMC was dissolved in distilled water at 100°C. The biopolymer solutions were mixed. Then, the proper amount of ZnO-NPs was added to the CH/ CMC solution and sonicated for 2 h at 25°C. The homogeneous suspensions were poured into a transparent glass petri dish and left at room temperature for 72 h.	*P. aeruginosa, E. coli, Candidia albicans*	[77]
ZnO-NPs	Chitosan/ Polyethylene (PE)	Chitosan powder was dissolved in acetic acid. Solution was kept for 24 h until powder was totally dissolved. 0.1% solution of commercial ZnO nanoparticles in ethanol was added to the prepared chitosan solution to obtain ZnO-chitosan nanocomposite. Then PE films were cleaned with 70% ethanol solution and dried at room temperature. The dried PE films were treated using a plasma instrument. After plasma treatment, chitosan-ZnO nanocomposite solution was sprayed on the PE surface and allowed to dry at room temperature.	*E. coli, S. aureus, Salmonella enterica*	[78]
ZnO-NPs	Polyhydroxy-alkanotes	The nanocomposites were prepared via the ultrasonication method followed by solution casting. The ZnO-NPs were dispersed in chloroform by ultrasonication at 100 W for 30 min. Subsequently, the PHBHHx was dissolved at 50°C in the dispersion of nano-ZnO. Then, the mixture was sonicated for 30 min. The obtained blend was poured into a glass petri dish and finally dried under vacuum for 48 h	*S. aureus, E. coli*	[79]
Ag-NPs	(PVA)-Montmorillo-nite clay ginger extract	PVA was dissolved in mL distilled water under stirring at 80°C, then the ginger extract was added to the PVA solution and mixed for 2 h at room temperature. Then montmorillonite clay solution was added into it under stirring for 2 h at room temperature. After stirring, the AgNO3 solution was introduced into the solution. As the next step, it was kept under sunlight for 15 min, to accelerate the green synthesis of Ag-NPs by ginger extract. The solution was then mixed in a magnetic stirrer for 30 min and then casted onto petri plates and dried.	*S. aureus, S. Typhimurium*	[80]
Ag-NPs	Poly-3-hydroxybuty-rate-*co*-3-hydroxyvalerate (PHBV3)	Active PHBV nanocomposite films were prepared by melting compound of the masterbatch of the Ag-NPs and PHBV3. The PHBV18/Ag-NPs masterbatch was melt mixed with the required amount of virgin PHBV3 pellets to obtain blends with 8% of the PHBV18/Ag-NPs. The PHBV blends were prepared in an internal mixer during 5 min at 60 rpm and 180°C. Then, the batches were subjected to a rapid cooling-down and they were subsequently compression molded into films using a hot-plate hydraulic press.	*S. enterica, Listeria monocytogenes*	[81]

NPs	Host matrix	Methods used for producing antimicrobial materials	Inhibition of microorganisms	Ref.
Ag-NPs	Polyvinyl chloride	PVC was dissolved under agitation using a magnetic stirrer in 20 mL tetrahydrofuran containing industrial epoxidized soybean oil as a plasticizer and kept at 65°C for 120 s. The homogeneous solutions were mixed with various concentration of quercetin and silver nitrate (Ag-NPs). Next, the solutions were poured onto flat-surface glass plates and kept at 25°C until solvent evaporation.	*Bacillus subtilis, Aspergillus niger, Fusarium solani*	[82]
Ag-NPs	Carboxymethyl-cellulose sodium alginate	CMC and sodium alginate and AgNO3 were dissolved in water, followed by the addition of aniline to the above reaction mixture. The solution was heated at a constant rate of 5 °C per min to the boiling point of aniline (184°C). The reaction mixture was then cooled and dried.	*Klebsiella pneumoniae, Streptococcus pyogenes*	[83]
TiO2-NPs	Polyvinyl alcohol (PVA)-chitosan	PVA was dissolved in distilled water at 80°C. Chitosan solution was prepared by dissolving chitosan in acetic acid solution, and stirred overnight using a magnetic stirrer. The two solutions were mixed at a 1:1 ratio and at room temperature. TiO2 nanoparticles were incorporated into the PVA-CHI film-forming solutions using a homogenizer to avoid aggregation of the nanoparticles. The blend solution with TiO2-NPs was ultrasonically degassed, casted onto the glass plate, and dried at room temperature.	*S. aureus, E. coli*	[84]
Nano-cellulose	Chitosan / Polylactic acid / Rosin	Solutions of chitosan in acetic acid and Z PLA in ethanol were prepared by mixing for 15 min at 90°C. Then rosin was added into the solution and mixed for 10 min at room temperature. Glycerol was added and stirred for 10 min for complete homogenization. As the next step, mixtures were poured into petri dishes and dried at 40°C for 24 h	*E. coli, B. subitilis*	[69]

Table 1.
Examples of bio-nanocomposite films in food packaging applications.

form a composite matrix. There are three types of release mechanisms of active additives from a matrix [55, 85] 1. A diffusion-induced release in which the active agents diffuse through the macro-porous or micro-porous structure of the matrix from the film/coating surface into the food. 2. A swelling-induced release in which the incorporated active compounds are unable to diffuse within the polymer matrix. Due to its diffusion coefficient being too low. In this case, if the polymer/biopolymer matrix/coating is placed in a compatible liquid medium, the polymer/biopolymer starts swelling because of the penetration of the fluids into the matrix/coating. The swelling causes an increase in the diffusion coefficient of the active substance, then diffusion of the active agents may begin. In addition, this type of release mostly occurs in moisture-sensitive packaging materials, such as polysaccharide-based or protein films/coatings. 3. A disintegration-induced release, which is caused by the cleavage, degradation, or deformation of a polymer/biopolymer. This type of release occurs in reactive nonbiodegradable polymers or biodegradable types, such as poly(lactide) (PLA), polyanhydrides, and poly(lactide-co-glycolide) [55]. Several approaches have been proposed to achieve a more controlled diffusion of active compounds from the polymer/biopolymer matrix. Each of them focuses on a factor that influences the release rate. There are techniques that may be used to improve active packaging material characteristics. Multi-discipline techniques, such as nano-reinforcements, micro- or nano-encapsulation, which alter the properties of the active agents and decrease their volatility, diffusivity, or a chemical modification of a polymer/biopolymer, such as irradiation, cross-linking with selected agents and the lamination of films can lead to a more controlled release of active/antimicrobial substances [55]. Physical techniques, such as corona discharge, ultraviolet (UV) radiation generating carboxylic acid groups, gamma-ray, electron beam, and plasma forms that break the covalent bonds at the surface, leading to hydrogen abstraction and the formation of surface radicals and laser treatments can change the chemical structure of polymers [86]. Wet chemical methods using strong acids, such as chromic, permanganate or nitric acid, and bases, such as potassium base, have also been found to be an effective surface modification that generated various groups, such as hydroxyl, carbonyl, and carboxylic acid groups [87]. A good example may be a modification performed by Mulla et al. [88] who modified a LLDPE film surface by chromic acid treatment and coating it with clove essential oil. The chromic acid made the surface of the film more porous, allowing it to be coated with clove essential oil as an antimicrobial agent. This packaging material was confirmed to be effective against *S. Typhimurium* and *L. monocytogenes*. Another example of active layer modification is described by Fajardo et al. [89] who fabricated gliadin films cross-linked with cinnamaldehyde, as a carrier of lysozyme. The gliadin film cross-linked with cinnamaldehyde preserved its integrity in water and led to a prolonged release of antimicrobial compounds. The authors confirmed that the cross-linking process led to a slower release of lysozyme exhibiting greater antimicrobial activity compared to an unmodified film.

1.6 UV-aging of antimicrobial packaging materials

In general, an active coated packaging material or packaging material containing antimicrobial agents incorporated into the polymer matrix should function during storage to inhibit microorganism growth to extend the shelf-life of the food product and maintain its quality. This means that the coatings and active films should offer sufficient resistance against ultraviolet (UV) radiation or be shielded against UV [90–92]. UV radiation is a part of the nonionizing region of the electromagnetic spectrum that comprises approximately 8–9% of total solar radiation. It can lead to a degradation/deterioration in the optical, physicomechanical, and antimicrobial properties of packaging materials. Introducing an active, antimicrobial agent sensitive to UV in a coating carrier or into a polymer matrix can lead to an inactivation of the coating or active film after UV-aging. Introducing an active compound that is resistant to UV in a coating carrier/polymer matrix, or adding a substance with shielding properties, can prevent the inactivation of the coating/active film after UV-aging [90, 92]. As a result of nanotechnology development, ZnO nanoparticles have been incorporated into the matrix of many polymers or many coating carriers to obtain active layers to enhance the properties of such films/coatings without significant influence on their transparency. Additionally, these nanoparticles have attracted great interest and the development of coating/film applications as agents to improve anticorrosion properties has increased, particularly as UV absorbers [90–92]. Many studies demonstrated that ZnO nanoparticles exhibited superior chemical stability under UV radiation compared to other UV absorbers [47–49, 93]. The aim of the work of Mizielińska et al. [92], was to examine the effect of accelerated UV-aging on the antimicrobial activity of PLA films containing ZnO nanoparticles incorporated into a polymer matrix against selected microorganisms. The active foil samples were irradiated with UV-A and Q-SUN. The authors were able to demonstrate that PLA films with incorporated zinc oxide nanoparticles did not inhibit the growth of *S. aureus, B. cereus, E. coli, Bacillus atrophaeus*, and *Candida albicans* cells, but decreased the cell number. Accelerated UV-A aging had no negative influence on the chemical composition and the antimicrobial activity of the active film against gram-positive bacteria but influenced the antimicrobial effectiveness of gram-negative and *C. albicans* cells. Q-SUN irradiation decreased the antimicrobial activity of films with incorporated nano-ZnO against *B. cereus*. The greater effectiveness of antimicrobial packaging materials containing ZnO nanoparticles was observed when the active agent was introduced into the methyl hydroxypropyl celluloses (MHPC) carrier which was used to coat PE films [90]. These active coatings completely inhibited the growth of *S. aureus, B. cereus, E. coli, P. aeruginosa* and reduced the number of *C. albicans* cells. The influence of accelerated UV-A and Q-SUN irradiation on the antimicrobial properties of these active layers was also analyzed. Accelerated Q-SUN and UV-A irradiation had no influence on the effectiveness of the covered foil samples against *S. aureus, B. cereus,* and *E. coli*. Q-SUN irradiation decreased the activity of the MHPC coating containing nanoparticles only against *P. aeruginosa* and *C. albicans* cells. The negative influence of UV-aging on the chemical composition of the coatings due to the presence of ZnO nanoparticles was not observed. Another example can be PLA films which were covered with an MHPC/cocoa butter carrier, containing *Eucomis comosa* extract as an active, antimicrobial agent [91]. The covered, active films were treated with accelerated UV-A and Q-SUN irradiation. Later, the authors analyzed the effect of UV-aging on the antimicrobial properties and chemical composition of the coatings. It was noted that MHPC/cocoa butter coating containing *E. comosa* extract reduced the number of *S. aureus, B. cereus*, and *B. atrophaeus*. The accelerated UV-A and Q-SUN irradiations altered the chemical composition of the coating, but they had no influence on the antimicrobial properties of *E. comosa* extract coatings against both *S. aureus* and *B. cereus*. It was confirmed that only Q-SUN irradiation decreased to a small degree the effectiveness of the coating against *B. atrophaeus*.

1.7 The antiviral properties of active food packaging

The ongoing global spread of a pandemic caused by the coronavirus, known as SARS-CoV-2, which caused severe acute respiratory syndrome (SARS), currently poses high risks to human health and the world economy. This virus belongs to a family of enveloped viruses with +ssRNA and crown-like spikes on their spherical surfaces. CoV virion is classified as a very pathogenic virus [94]. SARS-CoV-2 is the third virus in the coronavirus family that has globally caused serious ailments in humans [95]. The virus particles are transmitted through human-to-human contact or contact with infected individuals mediated through the eyes, mouth, nose, or through the inhalation of exhaled virus in respiratory droplets [94]. To prevent the transmission of the virus, the use of gloves and medical masks has become essential. For instance, the demand for disposable polymers/biopolymers is expected to increase by 40% in packaging. Safety concerns related to shopping in supermarkets and small markets

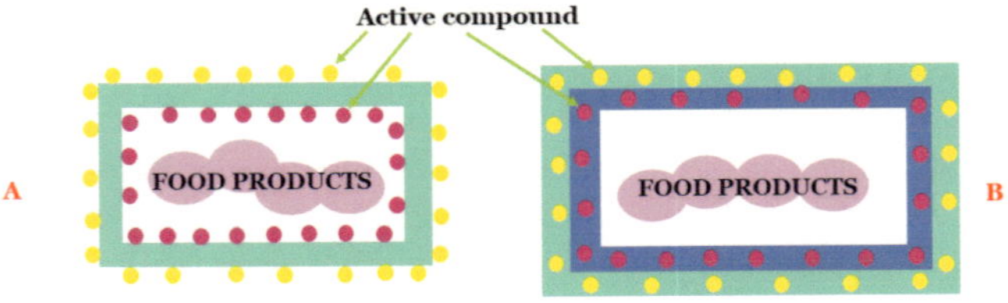

Figure 2.
Common forms of active food packaging systems with antiviral properties (A) polymer covered with the internal coating containing compounds active against microorganisms responsible for food spoilage and with the external antiviral coating and (B) two-layer film with an internal layer containing substances incorporated into the polymer matrix, active against microorganisms responsible for food spoilage and with an external layer containing substances incorporated into the polymer matrix, active against SARS CoV-2.

during the COVID-19 pandemic have led to the use of fresh-food products offered in polymer containers by suppliers and consumers, as well as the use of single-use food packaging materials and polymer bags to carry groceries [94, 96–98]. Multilayered active packaging systems are being developed to improve packaging properties, such as barrier properties, mechanical properties, antimicrobial effectiveness against bacteria, and yeast and molds responsible for food spoilage. However, adding antiviral materials as a layer containing antiviral agents to a coating carrier or incorporating them into a matrix of active films is a strategic route to develop antiviral packaging systems. Multilayered packaging systems are being developed through coextrusion, lamination, or covering with coatings [95]. This safe packaging should have an internal coating or extruded film layer to protect food products and an external coating or extruded film layer with antiviral compounds to protect customers (**Figure 2**) [94–98]. Additionally, this packaging-coated material or material containing antiviral substance should be active during storage, meaning it should offer sufficient resistance against UV aging or be shielded against ultraviolet light through the shielding properties of additives [94, 96–98].

In order to prepare antiviral coatings or films, many compounds, which are effective against viruses, such as SARS-CoV-2, such as ZnO nanoparticles, Ag nanoparticles, essential oils, and plant extracts, may be used [94–100]. As an example, Mizielińska et al. [94] developed an active coating based on nanoparticles of ZnO, geraniol, and carvacrol. Then PE films were covered with the active coating (coating carrier containing antimicrobial compounds mixture) using unicoater at a temp. of 25°C with a 40 μm diameter roller. The coatings were dried for 10 min at a temp. of 50°C. The authors analyzed the antibacterial and antiviral activity of these coatings. Additionally, the synergistic effect of the layer obtained was analyzed. Testing antiviral activity with human pathogen viruses, such as SARS-CoV-2, requires immense safety measures. Due to these concerns, the authors used phi 6 phage from the *Cystoviridae* family as a surrogate for the study of SARS-CoV-2. This phage was found to be similar (morphology, envelope, capsid size, and genome material) to the known pathogenic virus responsible for the COVID-19 pandemic [101]. The results of the study demonstrated that coatings containing an increased amount of geraniol or carvacrol and a very small amount of nano-ZnO were confirmed to be effective against gram-positive and gram-negative bacteria. It should be mentioned that a synergistic effect between these active agents was noted. To summarize, PE films covered with coatings containing geraniol or carvacrol and a very small amount of ZnO-NPs (as internal layer) may be used as packaging materials to extend the quality and freshness of food products. The same coatings can also be used as the external

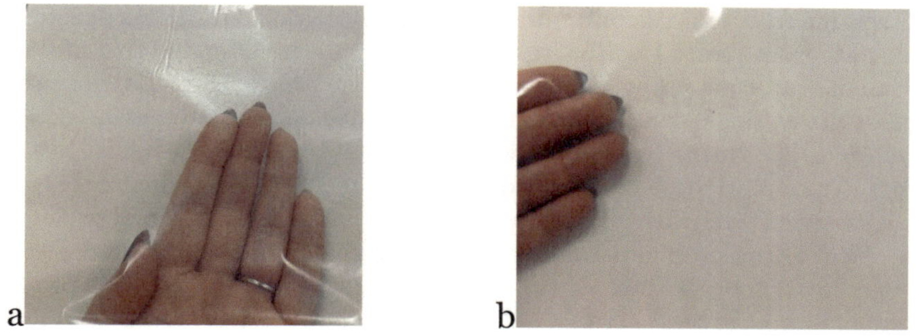

Figure 3.
a) an active coating based on nanoparticles of ZnO, geraniol, and carvacrol; b) an active coating based on a mixture of supercritical CO2 extracts of raspberry seeds, pomegranate seeds, and rosemary.

layers offering antiviral properties, as they showed moderate activity against the phi 6 phage. It was assumed by the authors that these coatings would also be active against SARS-CoV-2 particles. Similar results were obtained by Ordon et al. [96] who indicated that active coatings containing a mixture of supercritical CO2 extracts of raspberry seeds, pomegranate seeds, and rosemary showed bacteriolytic activity against *S. aureus* and *P. syringae* cells and bacteriostatic activity against *E. coli* and *B. subtilis* strains. The authors have also confirmed a synergistic effect in the active additives/compounds in the coatings. These coatings may also be used as internal coatings for packaging-covered films to extend the shelf life of food and to maintain their quality. All active coatings developed by the authors may also be used as external layers with antiviral effectiveness, as these coatings demonstrated significant effects against the phi 6 phage. Summarizing, the coatings obtained by the authors [94, 96] had antibacterial and antiviral properties. Their additional advantage was that they were transparent (**Figure 3**).

The lower activity of the mixture of CO2 extracts of raspberry seeds, pomegranate seeds, and rosemary when incorporated into a PE matrix [98]. The results of these tests demonstrated that the LDPE film containing a mixture of these extracts in a matrix inhibited the growth of *S. aureus* and this was also confirmed to be active against *B. subtilis.* This active film did not inhibit the growth of *E. coli* and *P. syringae* strains, however, the number of their cells decreased significantly. The LDPE active film was also confirmed to be active against phi 6 phage particles, meaning that the active foil sample had antiviral properties. A PE foil covered with active layers containing mixtures of *Styloscolex baicalensis* and *Glycyrrhiza* L. extracts as antibacterial compounds were found to have bacteriostatic activity against *S. aureus* cells and bacteriolytic effect on *B. subtilis* [97]. The layers were confirmed to be inactive against *E. coli* and *P. syringae* cells. This means that the coatings could be used as internal layers covering packaging film to preserve food products against gram-positive bacteria. The authors have also indicated that the coatings were highly effective against phage phi 6 phage, used as a SARS-CoV-2 surrogate. This means that the coatings could also be used as external layers to limit the spread of SARS-CoV-2 via human physical contact.

1.8 An application of antimicrobial packaging

Foodborne diseases caused by the consumption of food products contaminated with pathogenic microorganisms contribute to serious health issues in approximately 30% of the world population [102]. Additionally, the food industry has been facing huge losses due to microbial contamination for many years [3, 103]. As per the available data from World Health Organization [3], it was estimated that 600 million people fall ill after consuming contaminated food and around 420,000 die every year across the world. To avoid microorganisms' entry and growth inside food with extended shelf-life during their preservation, various packaging systems have been used. As a response to these needs, food packaging technology is constantly evolving from passive to innovative solutions, including active antimicrobial packaging [1]. The use of antimicrobial packaging is a practicable option to inhibit the growth of pathogenic microorganisms responsible for food spoilage and toxins in products throughout the postharvest period [1, 3]. The antimicrobial packaging described above may offer a potential solution for extending the shelf-life of packaged products without altering the food or the processes involved. The food could be healthier and free of preservatives, while still retaining all of the desirable qualities and food product safety requirements. The influence of antimicrobial packaging on the increased shelf-life of food products was proved by many researchers. Emamifar et al. [104] reported that LDPE nanocomposite packaging materials containing ZnO and Ag-NPs were conducive to prolong the shelf-life of fresh orange juice stored at 4°C. Several other researchers successfully developed an antimicrobial packaging material to preserve food products [105–107]. Li et al. [105] used coated materials, containing nano-ZnO particles to improve the shelf-life of freshly cut apples. Li et al. [107] used PLA-nanocomposite films to preserve cottage cheese. Zinoviadou et al. [108] used whey protein to isolate films containing antimicrobials to extend the shelf-life of fresh beef. **Table 2** shows more examples of the applications of antimicrobial packaging in food preservation.

1.9 Conclusions

Around 100 million tons of food products are wasted annually in the EU, which influences negatively on the environment. It was estimated that food waste would rise up to 200 million tons by 2050. Even if the relationship between shelf-life and food waste is not obvious, a huge part of food waste is related to the short shelf-life of fresh food products [2]. Antimicrobial packaging may be a solution to this problem. The fundamental role of antimicrobial packaging is to extend the shelf-life of food products by inhibiting the growth of microorganisms causing food spoilage. Longer shelf-life of food may lead to limit food waste and foodborne diseases and decreasing the negative impact of uneaten food on the environment and economy [3]. However, the long use of petroleum-based, antimicrobial packaging materials has also a negative influence on the environment because, after a single use of food packaging, 40% of these materials end up in landfill that corresponds to 9 million tons of plastic packaging waste accumulated in soils [3, 53]. Therefore, in 2015, the European Commission adopted a "circular economy action plan" with the goal to set the European Union on the course of the transition toward a more sustainable model for economic development [53]. Matthews et al. [53] mentioned that the main purpose of the EU's action plan is to maximize the usefulness of materials and resources and keep them in the economy for as long as possible to limit waste. The authors underlined that a circular economy could grow Europe's resource productivity by up to 3% annually by 2030. Two of the five important sectors identified in this action plan are food waste and plastics. The use of renewable/biobased sources for packaging materials has become a huge challenge. As a result of awareness in recent years, according to the latest data, the market for bio-based packaging material is predicted to increase from USD 81.70

Coating carrier/ polymer matrix	Active compound	Methods used for producing antimicrobial materials	Food application	Main results	Ref.
TPS/PBAT nanocomposite	ZnO-NPs	The cassava starch powder was dried in a hot oven at 50°C overnight before compounding. Starch, glycerol, and zinc oxide nanopowder were mixed for 10 min in a dough mixer at several ratios. The mixed materials were compounded in a twin-screw extruder by manual feeding. The heating profile and screw speed were set at 85°C to 150°C and 180rpm. The TPS-ZnO compound was cut into 2.5 cm pellets using a pelletizer, then the pellets were manually blended with PBAT pellets using a twin-screw extruder in the temperature ranging from 80–145°C with a 180 rpm screw speed to form 2.5 cm pellets. The TPS/PBAT/ZnO-NPs pellets were then blown using a temperature profile of 150–165°C with a screw speed and nip roll speed of 25–27 rpm and 2.7–3.2 rpm using a single-screw blown-film extruder	Pork	Microbial growth inhibition, shelf life extension	[109]
Chitosan nanocomposite	TiO2-NPs	Chitosan was dissolved in acetic acid solution with glycerol as plasticizer and the chitosan film-forming solution was shaken in a controlled-temperature water bath shaker at 90°C for 6 h. Then TiO2 nanopowders were added to the chitosan solution. As next step the solution was homogenized and subsequently degassed using a sonicator. The film-forming solution was casted and dried at 30°C.	Tomato-fruits	Extension of the shelf-life	[33]
Chitosan-potato protein-linseed oil	ZnO-NPs	Oil-in-water emulsion was prepared by adding linseed oil to potato protein solution. Then the mixture was stirred for 20 min at 10000 rpm to obtain emulsion. Then, chitosan solution (in acetic acid solution) was added to the emulsion. Subsequently, 1.5% glycerol was added as a plasticizer and homogenized by magnetic stirring. ZnO-NPs were added, and the final blend solution was sonicated for 20 min. Bubbles in the resultant solution was removed by vacuum processing. The solution was then poured into petri plates and dried at 40°C for 24 h.	Raw meat	Reduction of the number of bacteria	[110]
zein film	chitosan NPs encapsulated with pomegranate peel extract	Chitosan NPS were prepared by ionic gelation technique (chitosan was introduced into the acetic acid solution and mixed using magnetic stirrer followed by ultrasonication for 10 min. Then, sodium triphosphate pentabasic was added by drops into chitosan solution and mixed at 28°C for 2 h. Chitosan NPs were collected as a pellet after centrifugation. Pellets were washed with water and dehydrated using lyophilizer. For preparing PE doped chitosan NPs, pomegranate peel extract was added with the chitosan blend prior to the addition of sodium triphosphate pentabasic.). As next step, zein powder was dissolved in ethanol (96%) at 70°C. Then, glicerol plasticizer was added and mixed for 10 min. The temperature was reduced to 40°C and PE encapsulated chitosan NPs was included and mixed together for 30 min to obtain active nanocomposite film. Finally, the nanocomposite film solution was casted on sterile petri plates, and then dried at 50°C.	Fresh pork fish	Antimicrobial activity	[111]

Coating carrier/ polymer matrix	Active compound	Methods used for producing antimicrobial materials	Food application	Main results	Ref.
starch-based nanocomposite	Thyme EO*, montmorillonite nanoclay	Starch was dissolved in water, moderately stirred at room temperature, and then heated to 80°C for 30 min. After gelatinization, glycerol was added as a plasticizer. Montmorillinite powder was separately dispersed into distilled water and stirred at 500 rpm for 48 h. Then, the dispersion was added to the starch-glycerol suspension solution and then the mixture was mixed at 5000 rpm for 10 min. As next step, the thyme essential oil, with the tween 80 as an additive, was incorporated into the film forming solution at several concentrations. Samples were then homogenized at 20,000 rpm for 5 min. The active films were obtained via casting process in which the dispersion solutions were spread over a Teflon plate and then dried for 24 h at room temperature	Baby spinach leaves	Antibacterial activity	[112]
MC**/PC***/ alginate film	A) organic acids /rosemary extract/asian spice EO* B) organic acids /rosemary extract/asian spice EO*	Two types of films matrices based on 1) MC and 2) a blend of PC/alginate were used to obtain active films. Two antimicrobial formulations A: organic acids mixture + rosemary extract + Asian spice essential oil and B organic acids mixture + rosemary extract + Italian spice were introduced separately in each type of films during casting process.	Fresh broccoli	Antibacterial activity	[113]

Eos: essential oils.
**MC: methylcellulose.*
***PC: polycaprolactone.*

Table 2.
Some applications of antimicrobial packaging.

billion in the year 2020 to USD 118.85 billion by 2026 [3]. In fact, biodegradable and biobased/renewable polymers production capacity is estimated to grow from around 2.11 million tons in 2018 to 2.62 million tons in 2023 [114]. Furthermore, special research attention has to be paid to biodegradable and bio-based packaging materials with antimicrobial activity, where active compounds are incorporated into polymers/biopolymers to increase the quality and shelf-life of the packaged foods. To avoid food contamination and spoilage, active agents should exhibit a potent antimicrobial activity to meet the expectations of ideal antimicrobial packaging [3, 114, 115]. Extensive research on the development of innovative, antimicrobial, and eco-friendly packaging has been undertaken to control the growth of bacteria, yeast, and fungi in food products. Additionally, antimicrobial packaging materials should preserve not only food but also protect consumers' health. During a COVID-19 pandemic, not only SARS-CoV-2 virus particles may be present on the surface of the package. Frequent hand disinfection leads to the appearance of bacterial cells that are resistant to disinfectants and these bacteria may also be present on the package and consistently transmitted by hands. This review highlights that an external coating or external, extruded film layer of the packaging with antiviral and antibacterial properties may be a novel solution to protect customers against bacterial cells and viruses transmitted through packaging-to-human [94, 96–98].

Summarizing, it should be underlined that the global food packaging market reached a value of US$ 345.3 billion in 2021 [116], including only $9.6 billion (2.78%), which was reached by antimicrobial food packaging. However, it is projected that the antimicrobial food packaging market will hit around USD 19.7 billion by 2030 [117]. Many end-user sectors, including the beverage and food industry, are already interested in the various kinds of antimicrobial packaging, such as cartons and pouches, bags, trays, cans, cups, blister packs, and many more. There are companies that possess antimicrobial packaging materials in their offer as the answer to these demands. Dunmore a global manufacturer of laminated and coated films/foils developed and launched a new coated, polyester film (PET and BOPP) which is scratch-resistant and antimicrobial. This packaging material contains silver ions as active agents and it is effective against a wide spectrum of microorganisms [118, 119]. Klöckner Pentaplast Group is a leader in high-barrier protective packaging which extends the shelf-life of many food products. This company expanded its offer with new low-cost and effective antimicrobial performance films [118, 120]. Avient Corporation developed GLS TPEs (thermoplastic elastomer TPE) with antimicrobial additives, available as Versaflex™ and OnFlex™ grades. GLS TPEs with antimicrobial agents are commercially available in the United States and Asia. The company is focused on the development and additional antimicrobial formulations, which may be used as active additives to packaging materials [121, 122]. Cartro, a packaging business with headquarters in Mexico, and Mondi Ltd., one of the most important packaging manufacturers, partnered successfully in 2020. These companies decided to develop antimicrobial packaging for fresh and regional goods. Moreover, in the same year, Parkside Flexibles introduced new packaging materials covered with an antibacterial coating in partnership with Touch guard company [117]. Many companies are now focused on developing and launching antimicrobial packaging or active additives for packaging materials. Many global manufacturers offer high-barrier packaging that allows extending the shelf-life of food products by packing them in a MAP system or in a vacuum. The most important companies that have such packaging materials or active agents/additives in their portfolio are: PREXELENT®, Aptar CSP Technologies, BASF Group, Dow, BioCote, and MICROBAN [117, 118, 123–129].

Author details

Małgorzata Mizielińska* and Artur Bartkowiak
Faculty of Food Sciences and Fisheries, Center of Bioimmobilisation and Innovative Packaging Materials, West Pomeranian University of Technology Szczecin, Szczecin, Poland

*Address all correspondence to: malgorzata.mizielinska@zut.edu.pl

IntechOpen

© 2022 The Author(s). Licensee IntechOpen. This chapter is distributed under the terms of the Creative Commons Attribution License (http://creativecommons.org/licenses/by/3.0), which permits unrestricted use, distribution, and reproduction in any medium, provided the original work is properly cited.

References

[1] Akhila PP, Sunooj KV, Navaf M, Aaliya B, Sudheesh C, Sasidharan A, et al. Application of innovative packaging technologies to manage fungi and mycotoxin contamination in agricultural products: Current status, challenges, and perspectives. Toxicon. 2022;**214**:18-29

[2] Guillard V, Gaucel S, Fornaciari C, Angellier-Coussy H, Buche P, Gontard N. The next generation of sustainable food packaging to preserve our environment in a circular economy context. Frontiers in Nutrition. 2018;**5**:1-13

[3] Pandey S, Sharma K, Gundabala V. Antimicrobial bio-inspired active packaging materials for shelf life and safety development: A review. Food Bioscience. 2022;**48**:101730

[4] Perera KY, Jaiswal S, Jaiswal AK. A review on nanomaterials and nanohybrids based bio-nanocomposites for food packaging. Food Chemistry. 2022;**376**:131912

[5] Velásquez E, Vidal PC, Rojas A, Guarda A, Galotto MJ. Natural antimicrobials and antioxidants added to polylactic acid packaging films. Part I: Polymer processing techniques. Comprehensive Review in Food Science and Food Safety. 2021;**20**(4):3388-3403

[6] Luna Guevara JJ, Arenas-Hernandez MMP, Martĩnez Peña C, Silva JL, Luna-Guevara ML. The role of pathogenic *E. coli* in fresh vegetables: Behavior, contamination factors, and preventive measures. Hindawi International Journal of Microbiology. 2019

[7] Galié S, García-Gutiérrez C, Miguélez EM, Villar CJ, Lombó F. Biofilms in the food industry: Health aspects and control methods. Frontiers in Microbiology. 2018;**9**:898

[8] Frisvad JC, Hubka V, Ezekiel CN, Hong SB, Novakova A, Chen AJ, et al. Taxonomy of *Aspergillus* section Flavi and their production of aflatoxins, ochratoxins and other mycotoxins. Studies in Mycology. 2019;**93**:1-63

[9] Mao L, Bai Z, Yao J, Liu Y. Development of novel poly(lactic acid) active multilayer composite films by incorporating catechol-functionalized layered clay into chitosan/poly(vinylalcohol) coatings. Progress in Organic Coating. 2022;**170**:107000

[10] Zhu JY, Tang CH, Yin SW, Yang XQ. Development and characterization of novel antimicrobial bilayer films based on Polylactic acid (PLA)/Pickering emulsions. Carbohydrate Polymers. 2018;**181**:727-735

[11] Azhdari S, Moradi M. Application of antimicrobial coating based on carboxymethyl cellulose and natamycin in active packaging of cheese. International Journal of Biological Macromolecules. 2022;**209**(B):2042-2049

[12] Zhou X, Yang R, Wang B, Chen K. Development and characterization of bilayer films based on pea starch/polylactic acid and use in the cherry tomatoes packaging. Carbohydrate Polymers. 2019;**222**:114912

[13] Lan W, Du J, Wang M, Xie J. Effects of chitosan coating on quality and protein characteristics of large yellow croaker (*Pseudosciaena crocea*) during ice storage. Journal of Food Processing & Preservation. 2022;**46**:16649

[14] Shankar S, Rhim JW. Effects of poly(butylene adipate-co-terephthalate) coating on the water resistant, mechanical, and antibacterial properties

of Kraft paper. Progress in Organic Coating. 2018;**123**:153-159

[15] Nazrin A, Sapuan SM, Zuhri MYM, Ilyas RA, Syafiq R, Sherwani SFK. Nanocellulose reinforced thermoplastic starch (TPS), polylactic acid (PLA), and polybutylene succinate (PBS) for food packaging applications. Frontiers in Chemistry. 2020;**8**(4):213

[16] Ahankari SS, Subhedar AR, Bhadauria SS, Dufresne A. Nanocellulose in food packaging: A review. Carbohydrate Polymers. 2021;**255**(8):117479

[17] Kathuria A, Zhang S. Sustainable and repulpable barrier coatings for fiber-based materials for food packaging: A review. Frontier in Materials. 2022;**9**:929501

[18] Zhen N, Wang H, Li X, Xue J, Zhao Y, Wu M, et al. Protein-based natural antibacterial materials and their applications in food preservation. Microbial Biotechnology. 2021;**15**: 1324-1338

[19] Oun AA et al. Comparative study of oregano essential oil encapsulated in halloysite nanotubes and diatomaceous earth as antimicrobial and antioxidant composites. Applied Clay Science. 2022;**224**:106522

[20] Falleh H, Jemaa BM, Saada M, Ksouri R. Essential oils: A promising eco-friendly food preservative. Food Chemistry. 2020;**330**:127268

[21] Perumal AB, Huang L, Nambiar RB, He Y, Li X, Sellamuthu PS. Application of essential oils in packaging films for the preservation of fruits and vegetables: A review. Food Chemistry. 2022;**375**:131810

[22] Rattanachaikunsopon P, Phumkhachorn P. Assessment of factors influencing antimicrobial activity of carvacrol and cymene against *Vibrio cholerae* in food. Journal of Bioscience and Bioengineering. 2010;**110**:614-619

[23] Tajkarimi MM, Ibrahim SA, Cliver DO. Antimicrobial herb and spice compounds in food. Food Control. 2010;**21**:1199-1218

[24] Shiekh AK, Benjakul S. Melanosis and quality changes during refrigerated storage of Pacific white shrimp treated with Chamuang (Garcinia cowa Roxb.) leaf extract with the aid of pulsed electric field. Food Chemistry. 2020;**309**:125516

[25] Wang HN, Xiang JZ, Qi Z, Du M. Plant extracts in prevention of obesity. Critical Reviews in Food Science and Nutrition. 2020;**60**:1-14

[26] Efenberger-Szmechtyk M, Nowak A, Czyżowska A, Śniadowska M, Otlewska A, Żyżelewicz D. Antibacterial mechanisms of Aronia melanocarpa (Michx.), *Chaenomeles superba* Lindl. and Cornus mas L. leaf extracts. Food Chemistry. 2021;**350**:129218

[27] Efenberger-Szmechtyk M, Nowak A, Czyżowska A. Plant extracts rich in polyphenols: Antibacterial agents and natural preservatives for meat and meat products. Critical Reviews in Food Science and Nutrition. 2021;**61**:149-178

[28] Liu Y, Sameen DE, Ahmed S, Dai J, Qin W. Antimicrobial peptides and their application in food packaging. Trends in Food Science and Technology. 2021;**112**:471-483

[29] Zhang LJ, Gallo RL. Antimicrobial peptides. Current Biology. 2021;**26**: R14-R19

[30] Ramos M, Beltran A, Fortunati E, Peltzer M, Cristofaro F, Visai L, et al. Controlled release of thymol from poly(lactic acid)-based silver

nanocomposite films with antibacterial and antioxidant activity. Antioxidants. 2020;**9**(5):395

[31] Wu F, Harper BJ, Crandon LE, Harper SL. Assessment of Cu and CuO nanoparticle ecological responses using laboratory small-scale microcosms. Environmental Science: Nano. 2020;7(1):105-115

[32] Sruthi S, Ashtami J, Mohanan PV. Biomedical application and hidden toxicity of Zinc oxide nanoparticles. Materials Today Chemistry. 2018;**10**:175-186

[33] Kaewklin P, Siripatrawan U, Suwanagul A, Lee YS. Active packaging from chitosan-titanium dioxide nanocomposite film for prolonging storage life of tomato fruit. International Journal of Biological Macromolecules. 2018;**112**:523-529

[34] Guo ZY, Martucci NJ, Liu YZ, Yoo E, Tako E, Mahler GJ. Silicon dioxide nanoparticle exposure affects small intestine function in an in vitro model. Nanotoxicology. 2018;**12**(5):485-508

[35] Tanase C, Berta L, Coman NA, et al. Investigation of in vitro antioxidant and antibacterial potential of silver nanoparticles obtained by biosynthesis using beech bark extract. Antioxidants. 2019;**8**:459

[36] Dos Santos CA, Ingle AP, Rai M. The emerging role of metallic nanoparticles in food. Applied Microbiology and Biotechnology. 2020;**104**:2373-2383

[37] Fortunati E, Rinaldi S, Peltzer M, Bloise N, Visai L, Armentano I, et al. Nano-biocomposite films with modified cellulose nanocrystals and synthesized silver nanoparticles. Carbohydrate Polymers. 2014;**101**:1122-1133

[38] EU. Commission Regulation (EU) No 10/2011 of 14 January 2011 on Plastic Materials and Articles Intended to Come Into Contact with Food. Brussels, Belgium: EU; 2011

[39] Cushen M, Kerry J, Morris M, Cruz-Romero M, Cummins E. Migration and exposure assessment of silver from a PVC nanocomposite. Food Chemistry. 2013;**139**:389-397

[40] Chi H, Song S, Luo M, Zhang C, Li W, Li L, et al. Effect of PLA nanocomposite films containing bergamot essential oil, TiO2 nanoparticles, and ag nanoparticles on shelf life of mangoes. Scientia Horticulturae. 2019;**249**:192-198

[41] Akbar A, Anal AK. Zinc oxide nanoparticles loaded active packaging, a challenge studyagainst *Salmonella typhimurium* and *Staphylococcus aureus* in readyto-eat poultry meat. Food Control. 2014;**38**:88-95

[42] Jafarzadeh S, Alias AK, Ariffin F, Mahmud S, Najafi A, Ahmad M. Fabrication and characterization of novel semolina-based antimicrobial films derived from the combination of ZnO nanorods and nanokaolin. Journal of Food Science and Technology. 2017;**54**(1):105-113

[43] Wang Y, Ma J, Xu Q, Zhang J. Fabrication of antibacterial casein-based ZnO nanocomposite for flexible coatings. Materials and Design. 2017;**113**:240-245

[44] Dizaj SM, Lotfipour F, Barzegar-Jalali M, Zarrintan MH, Adibkia K. Antimicrobial activity of the metals and metal oxide nanoparticles. Materials Science and Engineering. 2014;**44**:278-284

[45] Oprea AE, Pandel LM, Dumitrescu AM, Andronescu E, Grumezescu V, Chifiriuc MC, et al. Bioactive ZnO coatings deposited by

MAPLE—An appropriate strategy to produce efficient anti-biofilm surfaces. Molecules. 2016;**21**:220-231

[46] Azizi S, Ahmad MB, Hussein MZ, Ibrahim NA. Synthesis, antibacterial and thermal studies of cellulose nanocrystal stabilized ZnO-Ag heterostructure nanoparticles. Molecules. 2013;**18**: 6269-6280

[47] Nguyena TV, Dao PH, et al. Effect of R-TiO2 and ZnO nanoparticles on the UV-shielding efficiency of water-borne acrylic coating. Progress in Organic Coating. 2017;**110**:114-121

[48] El-Feky OM, Hassan EA, Fadel SM, Hassan ML. Use of ZnO nanoparticles for protecting oil paintings on paper support againstdirt, fungal attack, and UV aging. Journal of Cultural Heritage. 2014;**15**:165-172

[49] Kairyte K, Kadys A, Luksiene Z. Antibacterial and antifungal activity of photoactivated ZnO nanoparticles in suspension. J. Photochem. Photobiol. Biol. 2013;**128**:78-84

[50] Bodaghi H, Mostofi Y, Oromiehie A, Zamani Z, Ghanbarzadeh B, Costa C, et al. Evaluation of the photocatalytic antimicrobial effects of a TiO2 nanocomposite food packaging film by in vitro and in vivo tests. LWT - Food Science and Technology. 2012;**50**(2):702-706

[51] Al-Tayyar NA, Youssef AM, Al-hindi R. Antimicrobial food packaging based on sustainable Bio-based materials for reducing foodborne pathogens: A review. Food Chemistry. 2020;**310**:125915

[52] Horst C, Pagno CH, Flores SH, Costa TMH. Hybrid starch/silica films with improved mechanical properties. Journal of Sol-Gel Science and Technology. 2020;**95**(1):52-65

[53] Matthews C, Moran F, Jaiswal AK. A review on European Union's strategy for plastics in a circular economy and its impact on food safety. Journal of Cleaner Production. 2021;**283**:125263

[54] Majid I, Ahmad Nayik G, Mohammad Dar S, Nanda V. Novel food packaging technologies: Innovations and future prospective. Journal of the Saudi Society of Agricultural Sciences. 2018;**17**(4):454-462

[55] Almasi H, Oskouie MJ, Saleh A. A review on techniques utilized for the design of controlled release food active packaging. Critical Reviews in Food Science and Nutrition. 2020;**2020**:1-21

[56] Zhang X, Pan C, Chen S, Xue Y, Wang Y, Wu Y. Effects of modified atmosphere packaging with different gas ratios on the quality changes of golden Pompano (Trachinotus ovatus) fillets during Superchilling Storage. Food. 2022;**11**:1943

[57] Garrido MD, Hernández MD, Espinosa MC, López MB. Enhanced quality characteristics of refrigerated Seabream (*Sparus aurata*) fillets packed under different systems (Modified Atmosphere vs. Vacuum). Journal of Aquatic Food Product Technology. 2016;**25**:156-168

[58] Yin C, Wang J, Qian J, Xiong K, Zhang M. Quality changes of rainbow trout stored under different packaging conditions and mathematical modeling for predicting the shelf life. Food Packaging and Shelf Life. 2022;**32**:100824

[59] Sivertsvik M, Rosnes JT, Kleiberg GH. Effect of modified atmosphere packaging and superchilled storage on the microbial and sensory quality of Atlantic Salmon (*Salmo salar*) fillets. Journal of Food Science. 2003;**68**:1467-1472

[60] Yang J, Yang X, Lin H, Liang R, Niu L, Zhu L, et al. Investigation of the relationship between microbiota dynamics and volatile changes in chilled beef steaks held under high-oxygen packaging enriched in carbon dioxide. Meat Science. 2022;**191**:108861

[61] Yang X, Zhang Y, Luo X, Zhang Y, Zhu L, Xu B, et al. Influence of oxygen concentration on the fresh and internal cooked color of modified atmosphere packaged dark-cutting beef stored under chilled and superchilled conditions. Meat Science. 2022;**188**:108773

[62] Li S, Guo X, Shen Y, Pan J, Dong X. Effects of oxygen concentrations in modified atmosphere packaging on pork quality and protein oxidation. Meat Science. 2022;**189**:108826

[63] Seier M, Archodoulaki VM, Koch T, Duscher B, Gahleitner M. Polyethylene terephthalate based multilayer food packaging: Deterioration effects during mechanical recycling. Food Packaging and Shelf Life. 2022;**33**:100890

[64] Walker TW, Frelka N, Shen Z, Chew AK, Banick J, Grey S, et al. Recycling of multilayer plastic packaging materials by solvent-targeted recovery and precipitation. Science Advances. 2020;**6**:eaba7599

[65] Dilkes-Hoffman L, Lane JL, Grant T, Pratt S, Lant PA, Laycock B. Environmental impact of biodegradable food packaging when considering food waste. Journal of Cleaner Production. 2018;**180**:325-334

[66] Pelissari FM, Grossmann MVE, Yamashita F, Pineda EAG. Antimicrobial, mechanical, and barrier properties of cassava starch-chitosan films incorporated with oregano essential oil. Journal of Agricultural and Food Chemistry. 2009;**57**:7499-7504

[67] Gao S, Zhai X, Wang W, Zhang R, Hou H, Lim LT. Material properties and antimicrobial activities of starch/PBAT composite films incorporated with ε-polylysine hydrochloride prepared by extrusion blowing. Food Packaging and Shelf Life. 2022;**32**:100832

[68] Mustafa P, Niazi MBK, Jahan Z, et al. PVA/starch/propolis/anthocyanins rosemary extractcomposite films as active and intelligent food packagingmaterials. Journal of Food Safety. 2020;**40**:12725

[69] Escamilla-García M, Calderon-Domínguez G, Chanona-Perez JJ, et al. Physical, structural, barrier, and antifungal characterization of chitosan-zein edible films with added essential oils. International Journal of Molecular Science. 2017;**18**:2370-2383

[70] Niu X, Liu Y, Song Y, Han J, Pan H. Rosin modified cellulose nanofiber as a reinforcing and co-antimicrobial agents in polylactic acid/chitosan composite film for food packaging. Carbohydrate Polymers. 2018;**183**:102-109

[71] Hu X, Jia X, Zhi C, Jin Z, Miao M. Improving the properties of starchbased antimicrobial composite films using ZnO-chitosan nanoparticles. Carbohydrate Polymers. 2019;**210**:204-209

[72] Tarabiah AE, Alhadlaq HA, Alaizeri ZAM, Ahmed AAA, Asnag GM, Ahamed M. Enhanced structural, optical, electrical properties and antibacterial activity of PEO/CMC doped ZnO nanorods for energy storage and food packaging applications. Journal of Polymer Research. 2022;**29**:167

[73] de Souza AG, Komatsu LGH, Barbosa RFS, Parra DF, Rosa DS. The effect of ZnO nanoparticles as Ag-carrier in PBAT for antimicrobial films. Polymer Bulletin. 2022;**79**:4031-4048

[74] Panneerselvam N, Sundaramurthy D, Maruthapillai A. Antibacterial/antioxidant activity of CuO impacted xanthan gum/chitosan @ascorbic acid nanocomposite films. Journal of Polymers and the Environment. 2022;**30**:3239-3249

[75] Indumathi MP, Sarojini KS, Rajarajeswari GR. Antimicrobial and biodegradable chitosan/cellulose acetate phthalate/ZnO nano composite films with optimal oxygen permeability and hydrophobicity for extending the shelf life of black grape fruits. International Journal of Biological Macromolecules. 2019;**132**:1112-1120

[76] Jayakumar A, Heera KV, Sumi TS, Joseph M, Mathew S, Praveen G. Starch-PVA composite films with zinc-oxide nanoparticles and phytochemicals as intelligent pH sensing wraps for food package application. International Journal of Biological Macromolecules. 2019;**1**(136):395-403

[77] Youssef AM, El-Sayed SM, El-Sayed HS, Salama HH, Dufresne A. Enhancement of Egyptian soft white cheese shelf life using a novel chitosan/carboxymethyl cellulose/zinc oxide bionanocomposite film. Carbohydrate Polymers. 2016;**151**:9-19

[78] Al-Naamani L, Dobretsov S, Dutta J. Chitosan-zinc oxide nanoparticle composite coating for active food packaging applications. Innovation in Food Science Emerging. 2016;**38**:231-237

[79] Diez-Pascual AM. Poly (3-hydroxybutyrate-co-3-hydroxyhexanoate) with zinc oxide nanoparticles for food packaging. Journal of Food Process Engineering. 2022;**45**:3814

[80] Mathew S, Snigdha S, Mathew J, Radhakrishnan EK. Biodegradable and active nanocomposite pouches reinforced with silver nanoparticles for improved packaging of chicken sausages. Food Packaging and Shelf Life. 2019;**19**:155-166

[81] Castro-Mayorga JL, Fabra MJ, Lagaron JM. Stabilized nanosilver based antimicrobial poly (3-hydroxybutyrate-co-3-hydroxyvalerate) nanocomposites of interest in active food packaging. Innovative Innov. Food Sci. Emerg. 2016;**33**:524-533

[82] Braga LR, Rangel ET, Suarez PAZ, Machado F. Simple synthesis of active films based on PVC incorporated with silver nanoparticles: Evaluation of the thermal, structural and antimicrobial properties. Food Packaging and Shelf Life. 2018;**15**:122-129

[83] Su CH, Velusamy P, et al. Studies of antibacterial efficacy of different bio-based polymer protected silver nanoparticles synthesized under reflux condition. Journal of Molecular Structure. 2017;**1128**:718-723

[84] Lian Z, Zhang Y, Zhao Y. Nano-TiO2 particles and high hydrostatic pressure treatment for improving functionality of polyvinyl alcohol and chitosan composite films and nano-TiO2 migration from film matrix in food simulants. Innovation in Food Science Emerging. 2016;**33**:145-153

[85] Mastromatteo M, Mastromatteo M, Conte A, Del Nobile MA. Advances in controlled release devices for food packaging applications. Trends in Food Science and Technology. 2010;**21**(12):591-598

[86] Goudarzi V, Shahabi-Ghahfarrokhi I. Development of photo-modified starch/kefiran/TiO2 bio-nanocomposite as an environmentally-friendly food packaging material. International

Journal of Biological Macromolecules. 2018;**116**:1082-1088

[87] Lee KT. Quality and safety aspects of meat products as affected by various physical manipulations of packaging materials. Meat Science. 2010;**86**(1):138-150

[88] Mulla MJ, Ahmed H, Al-Attar E, Castro-Aguirre YA, Arfat Y, Auras R. Antimicrobial efficacy of clove essential oil infused into chemically modified LLDPE film for chicken meat packaging. Food Control. 2017;**73**:663-671

[89] Fajardo P, Balaguer MP, Gomez-Estaca J, Gavara R, Hernandez-Munoz P. Chemically modified gliadins as sustained release systems for lysozyme. Food Hydrocolloids. 2014;**41**:53-59

[90] Mizielińska M, Łopusiewicz Ł, Mężyńska M, Bartkowiak A. The influence of accelerated UV-A and Q-SUN irradiation on the antimicrobial properties of coatings containing ZnO nanoparticles. Molecules. 2017;**22**:1556

[91] Mizielińska M, Kowalska U, Salachna P, Łopusiewicz Ł, Jarosz M. The influence of accelerated UV-A and Q-SUN irradiation on the antibacterial properties of hydrophobic coatings containing *Eucomis comosa* extract. Polymers. 2018;**10**:421

[92] Mizielińska M, Kowalska U, Jarosz M, Sumińska P, Landercy N, Duquesne E. The effect of UV aging on antimicrobial and mechanical properties of PLA films with incorporated zinc oxide nanoparticles. International Journal of Environmental Research and Public Health. 2018;**15**:794

[93] Marvizadeh MM, Oladzadab basabadi N, Nafchi AM, Jokar M. Preparation and characterization of bionanocomposite film based on tapioca starch/bovine gelatin/nanorod zinc oxide. International Journal of Biological Macromolecules. 2017;**99**:1-7

[94] Mizielińska M, Nawrotek P, Stachurska X, Ordon M, Bartkowiak A. Packaging covered with antiviral and antibacterial coatings based on ZnO nanoparticles supplemented with geraniol and carvacrol. International Journal of Molecular Sciences. 2021;**22**:1717

[95] Priyadarshi R, Purohit SD, Roy S, Ghosh T, Rhim JW, Han SS. Antiviral biodegradable food packaging and edible coating materials in the COVID-19 era: A mini-review. Coatings. 2022;**12**:577

[96] Ordon M, Nawrotek P, Stachurska X, Mizielińska M. Polyethylene films coated with antibacterial and antiviral layers based on CO2 extracts of raspberry seeds, of pomegranate seeds and of rosemary. Coatings. 2021;**11**:1179

[97] Ordon M, Nawrotek P, Stachurska X, Schmidt A, Mizielińska M. Mixtures of Scutellaria baicalensis and Glycyrrhiza L. Extracts as antibacterial and antiviral agents in active coatings. Coatings. 2021;**11**:1438

[98] Ordon M, Zdanowicz M, Nawrotek P, Stachurska X, Mizielińska M. Polyethylene films containing plant extracts in the polymer matrix as antibacterial and antiviral materials. International Journal of Molecular Sciences. 2021;**22**:13438

[99] Marquez R, Zwilling J, Zambrano F, Tolosa L, Marquez ME, Venditti R, et al. Nanoparticles and essential oils with antiviral activityon packaging and surfaces: An overview of their selection and application. Journal of Surfactants and Detergents. 2022;**2022**:1-35

[100] Wen CC, Kuo YH, Jan JT, Liang PH, Wang SY, Liu HG. Specific

plant terpenoids and lignoids possess potent antiviral activities against severe acute respiratory syndrome coronavirus. Journal of Medicinal Chemistry. 2007;**50**(17):4087-4095

[101] Prussin AJ II, Schwake DO, Lin K, Gallagher DL, Buttling L, Marr LC. Survival of the enveloped virus Phi6 in droplets as a function of relative humidity, absolute humidity and temperature. Applied and Environmental Microbiology. 2018;**84**:e00551-e00518

[102] Loo YY, Rukayadi Y, Nor-Khaizura MAR, Kuan CH, Chieng BW, Nishibuchi M, et al. In vitro antimicrobial activity of green synthesized silver nanoparticles against selected gram-negative foodborne pathogens. Frontiers in Microbiology. 2018;**9**:1555

[103] Verma M, Plaisier C, van Wagenberg CP, Achterbosch T. A systems approach to food loss and solutions: Understanding practices, causes, and indicators. Sustainability. 2019;**11**:579

[104] Emamifar A, Kadivar M, Shahedi M, Soleimanianzad S. Evaluation of nanocomposite packaging containing Ag and ZnO on shelf life of fresh orange juice. Innovative Food Science and Emerging Technologies. 2010;**11**:742-748

[105] Li X, Li W, Jiang Y, Ding Y, Yun J, Yao T, et al. Effect of nano-ZnO-coated active packaging on quality of fresh-cut 'Fuji' apple. International Journal of Food Science and Technology. 2011;**46**:1947-1955

[106] Li W, Zhang C, Chi H, Li L, Lan T, Han P, et al. Development of antimicrobial packaging film made from poly(lactic acid) incorporating titanium dioxide and silver nanoparticles. Molecules. 2017;**22**:1170

[107] Li W, Li L, Zhang H, Yuan M, Qin Y. Evaluation of PLA nanocomposite films on physicochemical and microbiological properties of refrigerated cottage cheese. Journal of Food Processing & Preservation. 2018;**42**:13362-13370

[108] Zinoviadou KG, Koutsoumanis KP, Biliaderis CG. Physical and thermo-mechanical properties of whey protein isolate films containing antimicrobials, and their effect against spoilage flora of fresh beef. Food Hydrocolloids. 2010;**24**:49-59

[109] Phothisarattana D, Wongphan P, Promhuad K, Promsorn J, Harnkarnsujarit N. Blown film extrusion of PBAT/TPS/ZnO nanocomposites for shelf-life extension of meat packaging. Colloids and Surfaces B: Biointerfaces. 2022;**214**:112472

[110] Wang C, Chang T, Dong S, Zhang D, Ma C, Chen S, et al. Biopolymer films based on chitosan/potato protein/linseed oil/ZnO NPs to maintain the storage quality of raw meat. Food Chemistry. 2020;**332**:127375

[111] Cui H, Surendhiran D, Li C, Lin Lin L. Biodegradable zein active film containing chitosan nanoparticle encapsulated with pomegranate peel extract for food packaging. Food Packaging and Shelf Life. 2020;**24**:100511

[112] Issa A, Ibrahim SA, Tahergorabi R. Impact of sweet potato starch-based nanocomposite films activated with thyme essential oil on the shelf-life of baby spinach leaves. Food. 2017;**6**:43

[113] Takala PN, Vu KD, Salmieri S, Khan RA, Lacroix M. Antibacterial effect of biodegradable active packaging on the growth of Escherichia coli, *Salmonella typhimurium* and *Listeria monocytogenes* in fresh broccoli stored at 4°C. LWT – Food Science Technology. 2013;**53**(2):499-506

[114] Chiloeches A, Cuervo-Rodríguez R, López-Fabal F, Fernández-García M, Echeverría C, Muñoz-Bonilla A. Antibacterial and compostable polymers derived from biobased itaconic acid as environmentally friendly additives for biopolymers. Polymer Testing. 2022;**109**:107541

[115] Zhai X, Zhou S, Zhang R, Wang W, Hou H. Antimicrobial starch/poly(butylene adipate-co-terephthalate) nanocomposite films loaded with a combination of silver and zinc oxide nanoparticles for food packaging. International Journal of Biological Macromolecules. 2022;**206**:298-305

[116] Available from: https://www.researchandmarkets.com/reports/5546893/food-packaging-market-global-industry-trends?gclid=Cj0KCQjwnbmaBhD-ARIsAGTPcfXV2efepbc7ZGyiGrnTNdTC1v0mgf2aWKyINGAHf-L2bzIn0qyDbWEaAqSwEALw_wcB

[117] Available from: https://www.globenewswire.com/en/news-release/2022/09/09/2513095/0/en/Antimicrobial-Packaging-Market-Size-to-Hit-Around-USD-19-7-Bn-by-2030.html

[118] Available from: https://www.alliedmarketresearch.com/antimicrobial-packaging-market-A16502

[119] Available from: https://www.dunmore.com/products/antimicrobial.html

[120] Available from: https://www.kpfilms.com/en/

[121] Zander Z, Newton D, Scaglione H, Reiber A, Agarwal P. Microbial susceptibility of various polymers and evaluation of thermoplastic elastomers with antimicrobial additives. Polymer Engineering and Science. 2021;**61**(12):3029

[122] Available from: https://www.avient.com/news/avient-launches-gls-thermoplastic-elastomers antimicrobial-technology-protects-against-microbial-growth

[123] Available from: https://www.mondigroup.com/en/home/

[124] Available from: https://prexelent.com/?utm_source=SEM&utm_medium=google&utm_campaign=prexelent&gclid=CjwKCAjwrOaBhA9EiwAUkLV4hbjSfWluLmsXvTAD3xmTC8Pb6nQRs9zzYGtlpPyAaeMQWRNelkmMhoCSbsQAvD_BwE

[125] Available from: https://aptarfoodprotection.com/

[126] Available from: https://www.basf.com/pl/pl/products/packaging-and-print.html

[127] Available from: https://www.dow.com/en-us.html

[128] Available from: https://www.biocote.com/biocote-antimicrobial-additives/

[129] Available from: https://www.microban.com/antimicrobial-solutions/environments/food-service

Chapter 11

Perspective Chapter: Development of Food Packaging Films from Microorganism-Generated Polyhydroxyalkanoates

Özcan Bulantekin and Duygu Alp

Abstract

Petroleum-based packaging (PBP) materials cause environmental pollution and toxic substance accumulation because they cannot decompose in nature for a long time. To prevent these problems, a wide variety of food packaging materials emerge as alternatives to PBP. Researchers have already discussed how polysaccharides and biopolymer-based nanocomposites are used in the development of food packaging films. This chapter, we will introduce how the microorganism-generated biopolymer, polyhydroxyalkanoates (PHAs) to be specific, is used in food packaging. PHAs, have positive social and environmental impact when compared to traditional plastics in terms of production and recycling. Considering that industrial wastes contain high quality polysaccharides, essential oils and proteins, using them in the production of biodegradable packaging will both reduce environmental problems and provide economic gain by reprocessing the wastes into products with higher added value. However, it has some disadvantages in competition with synthetic plastics and applications as biomaterials due to some properties such as poor mechanical properties, high production costs, limited functionality, incompatibility with conventional heat treatment techniques and susceptibility to thermal degradation. In this chapter, we will discuss the future and potential difficulties that may be experienced in the production or dissemination of PHA as a packaging material.

Keywords: polyhydroxyalkanoates, food packaging, biodegradable, bioplastics

1. Introduction

The packaging of foods plays an important role in the preservation of the product throughout the storage and distribution chain. The main purpose of packaging is to protect food from contamination and extend its shelf life [1]. To meet this purpose, packages are produced from different raw materials. Plastic packages are frequently used in food packaging due to their high quality, easy processing, and combination properties. However, today, the harmful effects of plastic waste on the environment have led the food industry to use biodegradable packaging materials [2].

Biodegradable packaging materials are obtained from renewable bio-based resources and are often referred to as "biopolymers" [3]. Biopolymers are strong alternatives to nonbiodegradable and nonrenewable plastic packaging materials due to their environmentally friendly nature [4]. Synthetic films used as packaging materials negatively affect the environment since they are not biodegradable, while biological-based packaging materials obtained from renewable materials draw attention to their biological compatibility, biodegradability, environmental protection, and reliability [3]. Biodegradable polymers are defined as polymers that can be broken down into simple molecules, such as carbon dioxide, methane, water, and inorganic compounds in a certain process in nature as a result of the enzymatic activity of microorganisms [5, 6]. Bioplastics are generally materials that can be produced from renewable resources such as sugar, corn, and potatoes, containing proteins, lipids, and polysaccharides or from certain microorganisms, algae, and fungi [7, 8]. Environmentally friendly biodegradable polymers have good mechanical and barrier properties and are thus seen as potential packaging materials to replace traditional plastic packaging materials [4, 6].

2. Bioplastics and their sources

Bioplastics which are biodegradable materials made from renewable resources are the new materials of the twenty-first century and are of great importance [9, 10]. Starches from corn, potatoes, wheat, rice, barley, and oats; fibers obtained from pineapple; jute, hemp, banana stems, cassava, newspaper pulp, waste paper, and citrus waste are renewable sources for bioplastics. The use of new techniques for bioplastic production that promotes sustainable solution and reduces plastic waste has been greatly promoted in recent years [11]. Bioplastics can be produced from the inedible parts of food. Food wastes, such as orange peel, pomegranate peel, banana peel, and potato peel, are used in bioplastic production. In recent years, cellulose, hemicellulose, starch, pectin, and these lignocellulosic raw materials are made useful for the bioplastic film production trend [11]. Biobased (biodegradable) packaging materials are a potential alternative to replace petroleum-based (synthetic) polymers. One of the reasons for this situation is the decrease in the demand for petroleum-based products and the movement toward renewable resources to produce plastics and the reduction of the amount of gases released into the atmosphere. Another reason is to reduce the solid waste problem by turning biodegradable materials into compostable organic residues after use [12, 13].

Starch is the most widely used source in the production of biopolymer packaging materials due to its edibility and ease of raw material supply. The main source of starch used for these purposes is usually corn the fact remains that the mechanical properties, such as strain rate and tensile and flexibility strength of films produced from starch are not sufficient. Therefore, starch can be chemically modified or mixed with other substances. Plasticizers, such as glycerol, polyether, and urea, are used to reduce the fracturability of starch [14]. Since starch shows hydrophilic properties, it is not suitable for liquid food products with high moisture content, however, starch-based films have good oxygen barrier properties [15], as well they are used as an alternative to petroleum-derived materials because it is inexpensive and biodegradable [16, 17]. The ability of starch to be hydrolyzed by microorganisms, and used as a carbon source and whether they have the ability to produce α-amylase enzyme is strain specific feature. For this reason, an external source of this enzyme is needed in order for starch to be hydrolyzed by microorganisms and used as a carbon source [18]. **Figure 1** shows bioplastics and their sources.

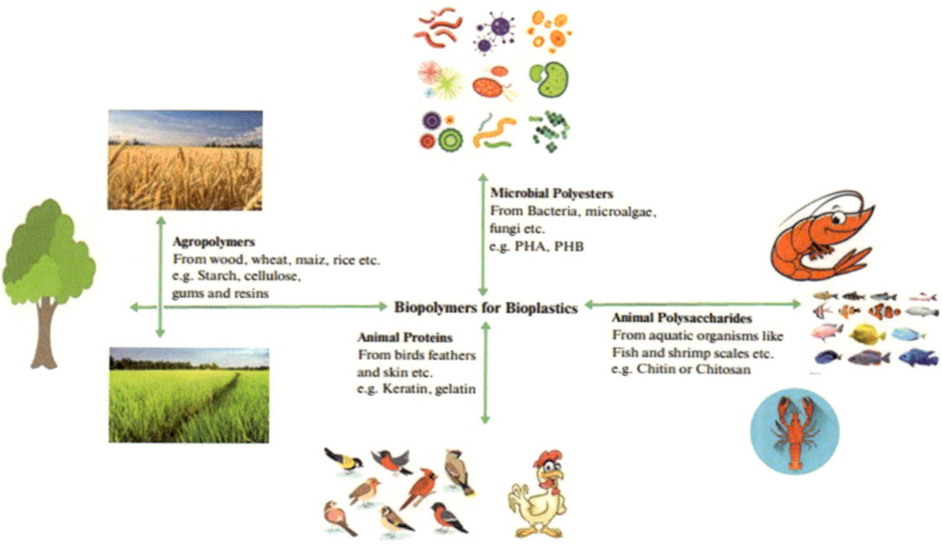

Figure 1.
Bioplastics and their sources.

Cellulose consists of glucose monomer units linked by glycosidic bonds and is an inexpensive source because it is found in all plants. However, its hydrophilic properties, low solubility, and high crystal structure create difficulties in its use in packaging, and due to the successive hydroxyl side chains, it causes low moisture barrier properties in cellulose-based packages. Also, the packaging material formed due to the high crystalline structure is brittle and has poor flexibility and tensile strength [14, 19]. For these reasons, research now focus on cellulose derivatives for packaging purposes. Cellulose-based biopolymer cellophane films, known as candy wrappers, are highly transparent and colorful. Cellophane treated by lamination, injection, and extrusion molding exhibits good film-forming properties [20]. Because it is insoluble and has excellent dimensional stability, it is also used in packaging products ranging from laminate, flower wrapping, and cheese to coffee and chocolate [19]. Researchers stated that starch/PVA, which is a composite biodegradable film, reinforced with cellulosic fiber is suitable for use in food packaging [21]. Cellulose derivatives are obtained by the reaction of cellulose in the presence of an aqueous solution of sodium hydroxide and an esterifying reagent. Cellulose derivatives, such as hydroxypropyl cellulose, hydroxypropyl methylcellulose, carboxymethylcellulose, and methylcellulose, are used for edible films/coating. Suspensions of these substances have thermogelation properties that form gels when heated and, on the contrary, regain their original consistency when cooled. Such films are poor water barriers and show poor mechanical properties due to the hydrophilic nature of the molecules [22]. The quality of the moisture barrier can be improved by adding hydrophobic compounds, such as fatty acids, to the cellulose matrix [23]. Due to the high production cost, the use of cellulose-based plastic is limited in the market.

Chitosan, the second most abundant natural polymer after cellulose, is obtained by partial deacetylation of the natural polysaccharide chitin [24]. Chitosan is an important waste of the fishing industry. It has many functions, such as antimicrobial effect against bacteria, molds, and yeasts, moisture adsorbing, precipitation, film formation, and enzyme immobilization [25]. It also exhibits good oxygen and carbon dioxide permeability, but a major drawback is that it has poor solubility in neutral solutions. Sun et al. [26] determined that chitosan films combined with apple polyphenols can be used as bioactive packaging material to increase the shelf life of foods. Pectin, which is a heteropolysaccharide found in the cell walls of plants, is frequently used to thicken jams and jellies [27]. In the industry, mostly apple pulp and citrus peels are used as a source. These types of edible films are used for certain food-related functions, such as food packaging materials, anti-browning, flavor enhancer, and antimicrobial functions, their production usually uses casting method or extrusion [28, 29]. Pectins, which are frequently used to produce biodegradable films, can be supported with nano-structured fillers to compete with commercial polymers because they show poor physical and barrier properties. The nanocrystals obtained from cellulose, a natural component, impart rigidity and strength to the films [30].

Whey, which is a by-product of the cheese and casein industry and makes up 90% of the processed milk volume, is used for humans and animals, while some is thrown into the environment. Whey, which is an inexpensive substrate and carbon source for bacterial growth, is also preferred in the production of polyhydroxyalkanoate (PHA) [31]. Wheat bran contains high protein, carbohydrates, and minerals, and they are suitable for use as waste. Various studies have been carried out to evaluate wheat bran waste. One of them is the use by Van-Thuoc et al. [32] as a carbon source for bacterial growth and PHA production. They determined that growing Halomonas boliviensis LC1 resulted in biomass production of 3.19 g/l and PHB production of 1.08 g/l. Soy proteins are generally a by-product of the soybean oil industry. Generally produced by wet casting, soybeans are preferred in edible films and coatings because of their good film-forming properties. Although their water permeability is high, their oxygen permeability is good like other protein films [33–35]. Gelatin is a naturally occurring hydrocolloid polymer derived from animal skin, bones, and related tissues. Gelatin-based bioplastic films are widely used as packaging material in the food industry [36]. Many studies showed that the physical and mechanical properties of biofilms made of gelatin can be improved with biocomposite films, such as gelatin-starch or gelatin-starch-glycerol [37]. Lipids are used as protective coatings against transfer, but besides lipids, wax and other resin materials have several disadvantages as packaging materials, such as a waxy taste and texture, oily surface, and bitterness. Among its advantages in terms of packaging, the lipid component reduces water transmission, while the hydrocolloid component acts as a gas barrier and even contributes to the strength and structural integrity [29, 38].

In the food industry, large amounts of both solid and liquid waste occur due to the production, preparation, and consumption of food. These wastes cause problems while being destroyed, but they are valuable products as biomass and nutritional components [39, 40]. Especially fruit and vegetable industry is an area that generates a lot of waste. Fruits and vegetables are consumed fresh as well as processed into juice and jam, and vegetables into canned products, and as a result of all these processes, wastes such as shell, seed and pulp with high polysaccharide, protein, and lipid content are formed. These wastes are reusable for different processes. For this reason, a "zero waste approach" is being tried to be adopted. The zero-waste approach is based on the use of organic waste generated after processing, in a different field, such as chemistry, medicine, cosmetics, or as a raw material again in food production by subjecting it to various processes. Especially fruit and vegetable industry wastes contain plenty of pectin and different essential oils. These organic substances in the structure of waste are used not only to enhance the mechanical and barrier feature of packaging materials but also for the production of biodegradable films [41–44]. In recent years, the production of biodegradable packaging materials obtained from fruit and vegetable wastes has gained importance. Biodegradable films obtained from these are unfortunately not at a level to compete with commercial polymers in terms of mechanical and barrier properties. However, it is possible to develop these properties with nanotechnology applications [45]. Considering the mechanical and barrier properties of commercial polymers, although they are suitable for use as packaging materials, interest in natural polymers is increasing because they are not sustainable and biodegradable. However, films obtained from natural polymers show poor barrier and mechanical properties. In this respect, biodegradable polymers are not yet competitive with commercial polymers. For this reason, developed biopolymers are obtained by supporting many biopolymers with organic or inorganic additives [46, 47]. By integrating nanoparticles into biopolymers, the negative mechanical and barrier properties of biodegradable films can be eliminated, and new materials with completely different properties can be developed. Polymer/clay composites improve the barrier properties of thin films [48, 49]. It is stated in the literature that bio-nanocomposites and nanoparticles have an inhibitory effect on the growth of some bacteria [50], act as a carrier of antimicrobial substances [51], and directly form an antimicrobial film [52]. Antimicrobial bio-nanocomposite films are formed by using fillers, such as chitosan [53], nano-silver [54], zinc oxide [55], and titanium dioxide. The use of food industry wastes in the production of biodegradable films has recently been one of the topics of interest in terms of the environmentalist approach. The components contained in the waste can add different properties to the packaging material, such as elasticity, strength, biodegradability, transparency, and antimicrobial activity.

Orange peels are some waste rich in pectin content, and when the films obtained by using the powder form of this waste were dried, seen that the films obtained from orange peel powder, had values close to the tensile strength of commercial polymers, such as low-density polyethylene, high-density polyethylene, polytetrafluorethylene, polypropylene, and polystyrene (16–32 MPa), were obtained [56]. Another industrial output evaluated for use in bioplastic production is soybean waste. Soybean is a raw material that is processed in large quantities and produces excessive waste. Therefore, it is obtained economically at a lower cost than other bioplastic-produced materials [57]. In a study on the evaluation of food industry waste, a film was obtained by mixing lemon peels and potato pulp at different rates. The amount of potato pulp released during the production of potato chips is between 12% and 20% of the total production. Potato pulp is rich in starch, cellulose, hemicellulose, and fermentable sugar content, and is a potential waste for biopolymer film production [58]. Lemon, on the other hand, is a product that is usually consumed fresh or processed into fruit juice, and its peel is very rich in flavonoids, pectin, and essential oils.

In addition, the antioxidant and antimicrobial properties of lemon pulp are also mentioned in the literature [59]. Pomegranate is a raw material that is generally processed into products, such as pomegranate juice or jam, after processing, approximately 55% of pulp is produced. In addition to its antioxidant and antibacterial properties, it has a pulp rich in pectin, tannin, and moisture [60]. Studies are carried out to obtain a film from pomegranate peels [41, 42]. After the banana processing process, up to 30% of the fruit is exposed. Banana peel is a waste not only rich in moisture, protein, pectin, and potassium but also rich in dietary fiber (cellulose), antioxidant, and phenolic compounds [61]. In a study conducted by adding pectin and cellulose nanocrystals obtained from banana peels to the film formulation, it was desired to improve physical and barrier properties. This study shows that nanofillers can improve the mechanical and barrier properties of biodegradable films [62]. Pectins, which are frequently used to produce biodegradable films, can be supported with nanostructured fillers to compete with commercial polymers because they show poor physical and barrier properties. The nanocrystals obtained from cellulose, a natural component, impart rigidity and strength to the films. In the production of biodegradable films, not only fruit and vegetable wastes but also the shells of nuts are used. It has been stated in the literature that shells containing a high percentage of starch are a suitable material for the production of bio-thermoplastics [63]. The same is true for walnut shells. In biodegradable films, walnut shells act as absorbent [64] and reinforcing material [65]. In a study in which starch obtained from cashew shells was used as a film matrix, cellulose obtained from walnut shells was also used as a filler. This study proved that the cellulose obtained from walnut shells shows very good barrier properties against oxygen [66].

3. Polyhydroxyalkanoate

3.1 Structure and classification of PHA

Recently, Zhang et al. [67] have reported about 150 different types of PHAs. They have various types and structures, this diversity is due to the number of carbon atoms, molecular structure, and chain lengths they have [68, 69]. However, when the carbon chain length is taken into account, three types of PHAs emerge. These are short-chain (scl-PHAs), medium-chain (mcl PHAs), and long-chain (lcl-PHAs) [18, 69, 70]. The most common and known member of the PHA family is poly-3-hydroxybutyrate (P3HB). It polymerizes to give a polymeric chain and consists of (R)-3HB repeating unit (monomer) [69–73]. Poly(3-hydroxybutyrate), poly(4-hydroxybutyrate), poly(3-hydroxyvalerate), and poly(3-hydroxybutyrate-co-3-hydroxyvalerate) are known scl-PHAs, and they have three to four carbon atoms, and they usually can be used in food packaging and disposable products [69, 74–76].

In the preparation of these PHAs containing 3-hydroxyvalerate (3HV) or 4-hydroxybutyrate (4HB) monomers, copolymers containing a mixture of four carbon chain length subunits and bacteria synthesizing these polymers with valeric acid are used. The incorporation of HV into the PHB polymers results in a less hard and brittle poly(3-hydroxybutyrate-co-3-hydroxyvalerate) [70, 77]. Medium-chain length PHAs with 6-14 carbon atoms are considered medium chain length. They consist of homopolymers such as poly(3-hydroxyhexanoate), poly(3 hydroxyoctanoate), or P(3HO) [69, 75]. They are synthesized by various bacteria with β-oxidation or novo biosynthesis pathway [78]. mcl-PHAs are flexible and elastic, having low crystallinity with low tensile strength and high elongation-to-break ratios [70]. On contrary to short-chain (scl-PHAs), they are rare, and less used in process of bioplastics [69]. The difference between the two classes is mainly due to substrate specificity. While an eutrophus can polymerize 3HAs consisting of 3–5 carbon atoms, *Pseudomonas oleovorans* can only use 3HAs consisting of 6–14 carbon atoms in PHA synthesis [79, 80].

3.2 Properties of PHA

Hydrophobicity, melting point, glass transition temperature, degree of crystallinity, and some mechanical properties differ depending on the composition of the monomer [80, 81]. Structural differences in the monomers that make up PHAs cause them to differ chemically as well. For example, poly3-hydroxybutyrate (PHB) has good moisture resistance compared to polypropylene, and barrier properties against gases. They also have a high degree of crystallinity, about 55–80%, and form fine crystals with melting points of about 175°C [82]. Considering its tensile and impact strength, UV resistance, and oxygen permeability, PHB is similar to isotactic polypropylene. This shows that it has the potential to be a packaging material [82]. Having good resistance to hydrolytic attack, PHAs are insoluble in water and resistant to UV [18]. In addition to these properties, they are biodegradable in nature [83]. The main reason why the degradation of PHAs depends on their species and composition is that they have chiral molecules [84]. The type of polymer, its composition, environmental conditions, and microorganism species are effective in the biodegradation of PHAs. It is known that microorganisms produce different PHA-depolymerase enzymes to decompose PHAs and thus they are effective on them [85].

Some thermal and mechanical properties, such as melting temperature, glass transition temperature, crystallinity, tensile strength, and percent elongation, determine the quality of PHA [69]. Some thermal properties of PHAs are crystallization and heat resistance, and they affect the polymer quality [86]. The glass transition temperature (Tg) for the amorphous phase and the melting temperature (Tm) for the crystalline phase are expressed [69]. Increasing the number of carbons in the side chain from one to seven causes Tg to decrease and Tm to increase. As a result, the melting temperature rises from 45 to 69°C [69, 74, 87]. Medium-chain length PHAs are crystalline, have more tensile strength, and have high elongation at break. Due to this feature, they show different mechanical properties as compared to short-chain length PHAs, which have a high crystallinity usually 60–80%. Short-chain length PHAs are more brittle and stiff compare to medium-chain length PHAs [69, 80]. The addition of different monomers or mixing with PHA are methods used to improve the fracture flexibility and elongation of the polymer [69, 88]. One of the purposes of these processes is to reduce the difficulties created by the lack of flexibility of PHAs in terms of food packaging uses [89]. Blending with other polymers can decrease brittleness but it is not enough to be competitive with fossil fuel-based polymers used for food packaging [89]. PHBs have similar water vapor permeability to thermoplastics, such as PVC or PET, and their properties are seen as potential for food packaging applications. The fact that they do not swell and have lower hydrophilicity is seen as an advantage compared to various biopolymers, such as starch and cellulose [89, 90]. In addition to these properties, PHAs also have good barrier properties to some organic solvents. They show relatively high permeability to the moderately polar solvents chloroform, acetone, and toluene, while they have lower permeability to methanol, n-hexane, and isopropyl ether [90, 91].

3.3 The microbial production of PHA

The first discovered PHA, P3HB, is produced by *Bacillus magaterium*, and French researcher Maurice Lemoigne isolated it between 1923 and 1927 [69, 92, 93], also more than 300 species have been reported to produce these polymers. These species include various gram-positive and gram-negative bacteria, fungi, and microalgae [80, 94, 95]. Biodegradable PHAs can be produced by many bacterial species and their different strains [69, 96]. Although many bacterial species are capable of producing a variety of biopolymers, only a few have high productivity and high production rate [69, 97, 98]. Among these bacteria; *Wauteria eutropha*, *Azotobacter* spp. *Bacillus* sp., *Pseudomonas putida*, *Pseudomonas fuorescens*, *P. oleovorans*, *Ralstonia eutropha*, *Cupriavidus necator*, *Burkholderia* sp., *Halomonas* sp., *Haloferax* sp., *Aeromonas* sp., *Thermus thermophilus*, *Hydrogenophagobacter*, *Saxogradanobacteria* de *Saxogradia Erwinia* sp., and recombinant *E. coli* [80, 97, 98]. Buhwal et al. [99] isolated bacteria that accumulate polyhydroxyalkanoates (PHA) from pulp, paper, and wastewater. The isolates *Enterococcus* sp. NAP11 and *Brevundimonas* sp. NAC1 showed maximum PHA production between 79.27% and 77.63%, and they are considered good candidates for industrial production of PHB. Preusting et al. [100] investigated to high concentration of PHA and high productivity with *P. oleovorans* by fed-batch and continuous culture, and they were reached in a continuous mode, and the culture productivity was 11.6 g/l and 0.58 g/(l h), respectively.

Another study by Guo-Qiang et al. [101] found to *Pseudomonas stutzeri* 1317 synthesized a variety of PHAs when grown in glucose and/or fatty acids. The use of recombinant *E. coli* is common in PHA production, as in many areas. Recombinant *E. coli* has been used for PHA biosynthesis, to synthesize the biopolymer to extremely high intracellular levels, and to produce the P(3HB-co-3HV) copolymer. As PHB synthesis with recombinant *E. coli*, it depends on the amount of acetyl-CoA available. Some advantages of using recombinant E. coli for PHA production are rapid growth, high cell density, ability to use a few inexpensive carbon sources [79]. Masood et al. [102] investigated various parameters on the yield of PHAs produced by bacillus cereus. They determined that B. cereus was able to produce both PHAs copolymer and tercopolymer, and it is depending on the type of substrates. Some fungi, such as *Aspergillus fumigatus*, *Saccharomyces cerevisiae,* and *Yarrowia lipolytica,* can be producers [69]. Microalgae have the ability to produce pigments, carotenoids, proteins, enzymes, sugars, fatty acids, polysaccharides, and vitamins, as well as many bioactive compounds. They are ideal for PHA production, but the number of knowns is limited. Some of those are *Nostoc muscorum*, *Chlorella minutissima,* and *Botryococcus braunii* [103]. Cyanobacteria are known to produce PHA by oxygenic photosynthesis, also various studies show that some cyanobacteria have natural capabilities to store PHAs. Although species-specific, some cyanobacteria produce predominantly PHB [104, 105]. It is known to produce PHA in archaea, which needs salt to maintain its growth and can optimally tolerate 5% NaCl (*w*/*v*). First reported in 1970 from the Dead Sea, designated *Halobacterium marismortui* [105, 106] these halophilic archaea produce PHB under nutrient-abundant carbon sources [105, 107]. Different bacteria produce different types of PHAs [108], Fluorescent *Pseudomonas* species are preferred, because they have the ability of mcl-PHA synthases, and can synthesize of PHAs with 6–14 carbon atoms [109]. **Figure 2** shows PHA biosynthesis process scheme.

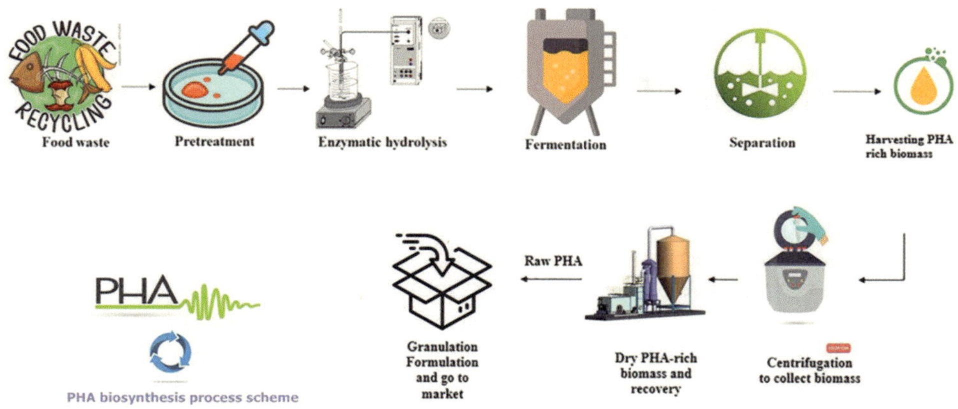

Figure 2.
PHA biosynthesis process scheme.

3.4 Carbon sources for the production of PHAs

The substrates used in the biosynthesis of PHAs, which are synthesized by bacteria through a metabolic process, are generally small molecules. This is due to the rigid cell walls of bacteria. Large molecules cannot be transported into the cell, they must undergo an extracellular transformation by the microorganism or by a chemical process in order to be used. The substrates that can be used are simple sugars (monosaccharides), triacylglycerol, and hydrocarbons. Most microorganisms use simple sugars. Triacylglycerol and hydrocarbon metabolism are less common. Different bacteria for the same substrate can produce PHAs with different compositions [110]. Monosaccharides and disaccharides do not need any hydrolysis to be used in the production of PHA, while polysaccharides cannot be fermented unless hydrolyzed first [110]. It has been reported by various researchers that *Azotobacter vinelandii*, *Alcaligenes latus,* and *Hydrogenophaga pseudoflava* can hydrolyze sucrose, consisting of a glucose unit linked to fructose, extracellularly to glucose and fructose, both of which can then be used for cell growth, and thus have the ability to produce PHA [110–114]; (Jiang et al., 2016). Lactose, a disaccharide composed of galactose and glucose, can be used by microorganisms. Whey is preferred as a source [110].

Whey, rich media that is suited for microbial growth, is obtained by precipitation and removal of milk casein during cheese-making processes [115]. Being a good PHA producer does not mean that the microorganism can produce PHA directly from whey. Vandamme and Coenye [116] stated that *C. necator*, *Wautersia eutropha,* or *Alcaligenes eutrophus* can accumulate up to 80% of its dry weight PHA. *Escherichia coli* cells, which can consume lactose as a solution for microorganisms that can produce high levels of PHA but cannot obtain it from whey, have been modified to express PHA biosynthesis genes from microorganisms that produce high PHA [115, 117]. Starch, the main component of maize, rice, and potatoes, is a polymer of D-glucose and must be enzymatically or acid hydrolyzed to fine maltose and then to glucose for the industrial production of PHAs [110]. Triacylglycerols are the main components of animal fats and plant oils [110], they were considered a cheap and viable source for the biosynthesis of PHAs [102]. In order for bacteria to use triacylglycerols in the production of PHA, they must be able to secrete lipase. The use of triacylglycerol as a carbon source was determined by Shiotani and Kobayashi in 1993 with *Aeromonas caviae* [110, 118]. Several studies have reported that some gram-negative bacteria can produce PHA using waste glycerol [119], oleic acid [120], or palm kernel oil [121].

3.5 Applications of PHA in the food industry

The packaging industry is responsible for the consumption of more than 40% of the total plastic produced worldwide. Food packaging accounts for about half of this percentage. The excessive use and unconscious disposal of disposable plastics cause large amounts of plastic waste. Therefore, packaging materials that are not easily recyclable, in particular, need to be converted from nonbiodegradable plastics to biodegradable plastics, such as PHAs [122, 123]. **Figure 3** shows the different applications of PHAs. Having good barrier properties to gases and water, PHAs are hydrophobic, thermoplastic, and nontoxic to humans and nature. Because of these material properties, PHAs become suitable for packaging materials [124, 125]. The three most commonly used bio-based plastics for packaging are PLA, starch-based plastics, and cellophane [126].

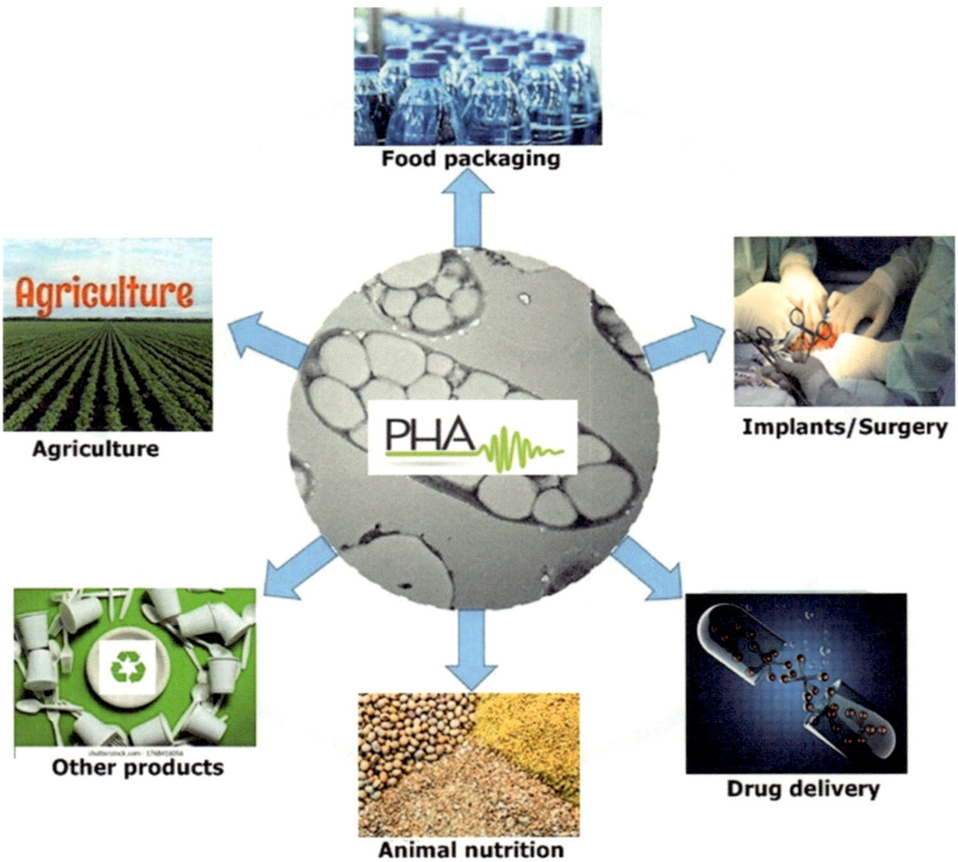

Figure 3.
Varied applications of PHAs.

PHAs are an important biodegradable packaging material used as films, foils, etc., and such barrier-coated packaging keeps oxygen and moisture inside the packaging [127, 128]. The film-forming feature of PHA provides a protective barrier against oxygen and UV rays. PHA can be used as food packaging material and has similar properties to polypropylene. PHA is insoluble in water and resistant to water and hydrophobic [129, 130]. Today, boxes, paper, boards, and cardboard are available in the market as coated with PHA [131]. PHA is used in the food industry for interior surface coatings in beverage and milk packaging. Studies have been conducted to show that they can be used instead of PP and HDPE in the packaging of foods with high-fat content, such as ready-made sauces, mayonnaise, and margarine cream [5, 132, 133]. Many companies are focusing on food packaging materials that combine conventional packaging materials with nanoparticles, thereby improving the mechanical performance of the packaging material and providing antimicrobial properties [128]. Castro-Mayorga et al. [134] synthesized an antimicrobial PHA containing silver nanoparticles and an active food that inhibits the growth of *Salmonella enterica*. They developed active food packaging.

In recent years, PHAs have become available as packaging materials for food-related products, such as straws and bottles. Danimer Scientific, who developed the first fully biodegradable plastic straw made of Nodax™ PHA in 2018. In addition, the Cove developed drink bottles, which were made completely of PHAs in 2019, California. Containers mix the soil in about 5 years; however, traditional PET plastic bottles can take around 500 years to decompose. Food and beverage manufacturing company Nestle has recently started work with Danimer Scientific to develop biodegradable plastic bottles [135, 136]. Yield10 agricultural bioscience company is the largest holding in the PHA industry with an annual production capacity of 50,000 tons [137]. Global biodegradable plastic materials are used in a variety of applications, and the packaging industry is the largest contributor, contributing 60% of global bioplastic production [138].

4. Conclusion and future outlook

The use of biodegradable packaging materials is becoming widespread day by day with the increase in environmental awareness, the desire to move away from the use of petroleum-derived packaging materials and the developing technology. The fact that natural resources, such as starch, cellulose, and protein, are the raw materials of a significant portion of biodegradable plastics increases the usability of these packaging materials. The most important reason why bioplastics cannot compete with plastics yet is high research and development costs and low production capacity. In order to reduce the production cost of PHA, it is important to use some cheap carbon sources and to make the best use of the waste. It has been revealed that biodegradable films have weaker properties in terms of barrier and mechanical properties, which are important in food packaging when compared to commercial plastics. In order to strengthen the barrier and mechanical properties of the films, montmorillonite, cellulose nanocrystals, nanoclay, and similar nanofillers can be used as well as biocomposite applications. It is clear that biodegradable packaging materials have potential for the food industry. It is thought that with the increase in crude oil prices in the future, renewable raw material sources will gain more importance and the production of environmentally friendly plastics will replace today's plastics. For this reason, it is thought that the use of environmentally friendly plastics, which do not have raw material shortages compared to petrochemical plastics, will be produced in much larger quantities with the help of new processes to be developed and more detailed studies should be done on these materials. By evaluating the food industrial wastes together, it can be achieved to develop high-performance packaging materials with stronger mechanical and barrier properties. As biodegradable packaging materials develop, it is expected that the production and use of disposable materials, such as biodegradable cups, cutlery, and plates, will become widespread.

Author details

Özcan Bulantekin[1] and Duygu Alp[2*]

1 Department of Cookery, Ağrı İbrahim Çeçen University, Dogubayazit Ahmedi Hani Vocational School, Ağrı, Turkey

2 Department of Gastronomy and Culinary Arts, Ardahan University, School of Tourism and Hospitality Management, Ardahan, Turkey

*Address all correspondence to: duygualp1@gmail.com

IntechOpen

© 2022 The Author(s). Licensee IntechOpen. This chapter is distributed under the terms of the Creative Commons Attribution License (http://creativecommons.org/licenses/by/3.0), which permits unrestricted use, distribution, and reproduction in any medium, provided the original work is properly cited. (cc) BY

References

[1] Popović SZ, Lazić VL, Hromiš NM, Šuput DZ, Bulut SN. Biopolymer packaging materials for food shelf-life prolongation. In: Biopolymers for Food Design. Academic Press; 2018. pp. 223-277

[2] Ahmadzadeh S, Khaneghah AM. Role of green polymers in food packaging. In: Choudhury I, Hashmi S, editors. Encyclopedia of Renewable and Sustainable Materials. USA: Elsevier Inc.; 2020. pp. 305-319

[3] Bajpai P. Biobased Polymers: Properties and Applications in Packaging. USA: Elsevier; 2019. pp. 129-138

[4] Shankar S, Rhim JW. Bionanocomposite films for food packaging applications. Reference Module in Food Science. 2018;**1**:1-10

[5] Peelman N, Ragaert P, De Meulenaer B, Adons D, Peeters R, Cardon L, et al. Application of bioplastics for food packaging. Trends in Food Science & Technology. 2013;**32**(2):128-141

[6] Siracusa V, Lotti N. Biobased plastics for food packaging. In: Hasmi S, Editors. Reference Module in Food Science. Amsterdam, The Netherlands: Elselvier; 2018

[7] Karana E. Characterization of 'natural'and 'high-quality'materials to improve perception of bio-plastics. Journal of Cleaner Production. 2012;**37**:316-325

[8] Sarasa J, Gracia JM, Javierre C. Study of the biodisintegration of a bioplastic material waste. Bioresource Technology. 2009;**100**(15):3764-3768

[9] Boonniteewanich J, Pitivut S, Tongjoy S, Lapnonkawow S, Suttiruengwong S. Evaluation of carbon footprint of bioplastic straw compared to petroleum based straw products. Energy Procedia. 2014;**56**:518-524

[10] Mohanty AK, Misra M, Drzal LT. Sustainable bio-composites from renewable resources: opportunities and challenges in the green materials world. Journal of Polymers and the Environment. 2002;**10**(1):19-26

[11] Shah M, Rajhans S, Pandya HA, Mankad AU. Bioplastic for future: a review then and now. World Journal of Advanced Research and Reviews. 2021;**9**(2):056-067

[12] Berthet MA, Angellier-Coussy H, Guillard V, Gontard N. Vegetal fiber-based biocomposites: which stakes for food packaging applications? Journal of Applied Polymer Science. 2016;**133**(2)

[13] Väisänen T, Haapala A, Lappalainen R, Tomppo L. Utilization of agricultural and forest industry waste and residues in natural fiber-polymer composites: a review. Waste Management. 2016;**54**:62-73

[14] Liu L. Bioplastics in food packaging: innovative technologies for biodegradable packaging. San Jose State University Packaging Engineering. 2006;**13**:1348-1368

[15] Cha DS, Chinnan MS. Biopolymer-based antimicrobial packaging: a review. Critical Reviews in Food Science and Nutrition. 2004;**44**(4):223-237

[16] Lu DR, Xiao CM, Xu SJ. Starch-based completely biodegradable polymer materials. Express Polymer Letters. 2009;**3**(6):366-375

[17] Xie F, Yu L, Liu H, Chen L. Starch modification using reactive extrusion. Starch-Stärke. 2006;**58**(3-4):131-139

[18] Raza ZA, Abid S, Banat IM. Polyhydroxyalkanoates: characteristics, production, recent developments and applications. International Biodeterioration and Biodegradation. 2018;**126**:45-56. DOI: 10.1016/j. ibiod.2017.10.001

[19] Venkatachalam H, Palaniswamy R. Bıoplastıc world: a revıew. Journal of Advanced Scientific Research. 2020;**11**(3)

[20] Zepnik S, Kesselring A, Kopitzky R, Michels C. Bioplastics Magazine. 2010;**1**:44-47

[21] Priya B, Gupta VK, Pathania D, Singha AS. Synthesis, characterization and antibacterial activity of biodegradable starch/PVA composite films reinforced with cellulosic fibre. Carbohydrate Polymers. 2014;**109**:171-179

[22] Phillips GO, Williams PA. Editors, Handbook of Hydrocolloids. USA: Elsevier; 2009

[23] Morillon V, Debeaufort F, Blond G, Capelle M, Voilley A. Factors affecting the moisture permeability of lipid-based edible films: a review. Critical Reviews in Food Science and Nutrition. 2002;**42**(1):67-89

[24] Casettari L, Vllasaliu D, Castagnino E, Stolnik S, Howdle S, Illum L. PEGylated chitosan derivatives: Synthesis, characterizations and pharmaceutical applications. Progress in Polymer Science. 2012;**37**(5):659-685

[25] Bostan K, Aldemir T, Aydın A. Chitosan and its antimicrobial activity. Journal of Turk Mikrobiyolology Cemiyeti. 2007;**37**:118-127

[26] Sun L, Sun J, Chen L, Niu P, Yang X, Guo Y. Preparation and characterization of chitosan film incorporated with thinned young apple polyphenols as an active packaging material. Carbohydrate Polymers. 2017;**163**:81-91

[27] Müller-Maatsch J, Bencivenni M, Caligiani A, Tedeschi T, Bruggeman G, Bosch M, et al. Pectin content and composition from different food waste streams. Food Chemistry. 2016;**201**:37-45

[28] Espitia PJP, Du WX, de Jesús Avena-Bustillos R, Soares NDFF, McHugh TH. Edible films from pectin: physical-mechanical and antimicrobial properties-a review. Food Hydrocolloids. 2014;**35**:287-296

[29] Janjarasskul T, Krochta JM. Edible packaging materials. Annual Review of Food Science and Technology. 2010;**1**:415-448

[30] Karakuş E, Ayhan Z. Production of environmentally friendly biodegradable packaging materials from food waste. Journal of Food. 2019;**44**(6):1008-1019

[31] Koller M, Bona R, Chiellini E, Fernandes EG, Horvat P, Kutschera C, et al. Polyhydroxyalkanoate production from whey by Pseudomonas hydrogenovora. Bioresource Technology. 2008;**99**(11):4854-4863

[32] Van-Thuoc D, Quillaguaman J, Mamo G, Mattiasson B. Utilization of agricultural residues for poly (3-hydroxybutyrate) production by Halomonas boliviensis LC1. Journal of Applied Microbiology. 2008;**104**(2):420-428

[33] Coltelli MB, Wild F, Bugnicourt E, Cinelli P, Lindner M, Schmid M, et al. State of the art in the development and properties of protein-based films and coatings and their applicability to cellulose based products: an extensive review. Coatings. 2016;**6**(1):1

[34] Khwaldia K, Arab-Tehrany E, Desobry S. Biopolymer coatings on paper packaging materials. Comprehensive Reviews in Food Science and Food Safety. 2010;**9**(1):82-91

[35] Zink J, Wyrobnik T, Prinz T, Schmid M. Physical, chemical and biochemical modifications of protein-based films and coatings: an extensive review. International Journal of Molecular Sciences. 2016;**17**(9):1376

[36] Hanani ZN, Roos YH, Kerry JP. Use and application of gelatin as potential biodegradable packaging materials for food products. International Journal of Biological Macromolecules. 2014;**71**:94-102

[37] Podshivalov A, Zakharova M, Glazacheva E, Uspenskaya M. Gelatin/potato starch edible biocomposite films: correlation between morphology and physical properties. Carbohydrate Polymers. 2017;**157**:1162-1172

[38] Aulin C, Karabulut E, Tran A, Wågberg L, Lindström T. Transparent nanocellulosic multilayer thin films on polylactic acid with tunable gas barrier properties. ACS Applied Materials & Interfaces. 2013;**5**:7352-7359

[39] Deniz E, Yeşilören G, İşçi NÖ. Biomass and biofuel potential of food industry in Turkey. Food. 2015;**40**(1):47-54

[40] Kanmaz EÖ, Saral Ö. The relationship between antioxidant activities and phenolic compounds in subcritical water extracts from orange peel. Journal of Food. 2017;**42**(5):485-493

[41] Oliveira TÍS, Rosa MF, Cavalcante FL, Pereira PHF, Moates GK, Wellner N, et al. Optimization of pectin extraction from banana peels with citric acid by using response surface methodology. Food Chemistry. 2016a;**198**:113-118

[42] Oliveira TÍS, Zea-Redondo L, Moates GK, Wellner N, Cross K, Waldron KW, et al. Pomegranate peel pectin films as affected by montmorillonite. Food Chemistry. 2016b;**198**:107-112

[43] Pereira PHF, Oliveira TÍS, Rosa MF, Cavalcante FL, Moates GK, Wellner N, et al. Pectin extraction from pomegranate peels with citric acid. International Journal of Biological Macromolecules. 2016;**88**:373-379

[44] Wikiera A, Mika M, Starzyńska-Janiszewska A, Stodolak B. Endo-xylanase and endo-cellulase-assisted extraction of pectin from apple pomace. Carbohydrate Polymers. 2016;**142**:199-205

[45] Da Silva ISV, Neto WPF, Silvério HA, Pasquini D, Zeni Andrade M, Otaguro H. Mechanical, thermal and barrier properties of pectin/cellulose nanocrystal nanocomposite films and their effect on the storability of strawberries (Fragaria ananassa). Polymers for Advanced Technologies. 2017;**28**(8):1005-1012

[46] Grigoriadi K, Giannakas A, Ladavos AK, Barkoula NM. Interplay between processing and performance in chitosan-based clay nanocomposite films. Polymer Bulletin. 2015;**72**(5):1145-1161

[47] Romero-Bastida CA, Tapia-Blácido DR, Méndez-Montealvo G, Bello-Pérez LA, Velázquez G, Alvarez-Ramirez J. Effect of amylose content and nanoclay incorporation order in physicochemical properties of starch/montmorillonite composites. Carbohydrate Polymers. 2016;**152**:351-360

[48] Staroszczyk H, Malinowska-Pańczyk E, Gottfried K, Kołodziejska I. Fish gelatin-nanoclay films. Part I: effect of a kind of nanoclays and glycerol concentration on mechanical and water barrier properties of nanocomposites. Journal of Food Processing & Preservation. 2017;**41**(5):e13211

[49] Trifol J, Plackett D, Sillard C, Szabo P, Bras J, Daugaard AE. Hybrid

poly (lactic acid)/nanocellulose/nanoclay composites with synergistically enhanced barrier properties and improved thermomechanical resistance. Polymer International. 2016;**65**(8):988-995

[50] Ahmed J, Hiremath N, Jacob H. Antimicrobial efficacies of essential oils/nanoparticles incorporated polylactide films against L. monocytogenes and S. typhimurium on contaminated cheese. International Journal of Food Properties. 2017;**20**:53-67

[51] Siripatrawan U, Vitchayakitti W. Improving functional properties of chitosan films as active food packaging by incorporating with propolis. Food Hydrocolloids. 2016;**61**:695-702

[52] Reesha KV, Panda SK, Bindu J, Varghese TO. Development and characterization of an LDPE/chitosan composite antimicrobial film for chilled fish storage. International Journal of Biological Macromolecules. 2015;**79**:934-942

[53] Al-Naamani L, Dobretsov S, Dutta J. Chitosan-zinc oxide nanoparticle composite coating for active food packaging applications. Innovative Food Science & Emerging Technologies. 2016;**38**:231-237

[54] Li L, Zhao C, Zhang Y, Yao J, Yang W, Hu Q, et al. Effect of stable antimicrobial nano-silver packaging on inhibiting mildew and in storage of rice. Food Chemistry. 2017;**215**:477-482

[55] Sanuja S, Agalya A, Umapathy MJ. Studies on magnesium oxide reinforced chitosan bionanocomposite incorporated with clove oil for active food packaging application. International Journal of Polymeric Materials and Polymeric Biomaterials. 2014;**63**:733-740

[56] Bátori V, Jabbari M, Åkesson D, Lennartsson PR, Taherzadeh MJ, Zamani A. Production of pectin-cellulose biofilms: a new approach for citrus waste recycling. International Journal of Polymer Science. 2017:1-9. Article ID: 9732329. DOI: 10.1155/2017/9732329

[57] Kowalczyk M, Piorkowska E. Mechanisms of plastic deformation in biodegradable polylactide/poly (1, 4-cis-isoprene) blends. Journal of Applied Polymer Science. 2012;**124**(6):4579-4589

[58] Rogols S, Sirovatka DM, Widmaier RG. U.S. Patent No. 6,524,639. Washington, DC: U.S. Patent and Trademark Office; 2003

[59] Chutia M, Bhuyan PD, Pathak MG, Sarma TC, Boruah P. Antifungal activity and chemical composition of Citrus reticulata Blanco essential oil against phytopathogens from North East India. LWT- Food Science and Technology. 2009;**42**(3):777-780

[60] Canbolat Ö, Kamalak A, Kara H. The effects of urea supplementation on pomegranate pulp (*Punica granatum* L.) silage fermentation, aerobic stability and in vitro gas production. Journal of Ankara Üniversty Veterinary Faculty. 2014;**61**:217-223

[61] González-Montelongo R, Lobo MG, González M. Antioxidant activity in banana peel extracts: testing extraction conditions and related bioactive compounds. Food Chemistry. 2010;**119**(3):1030-1039

[62] Oliveira TÍS, Rosa MF, Ridout MJ, Cross K, Brito ES, Silva LM, et al. Bionanocomposite films based on polysaccharides from banana peels. International Journal of Biological Macromolecules. 2017;**101**:1-8

[63] Yuliana M, Huynh LH, Ho QP, Truong CT, Ju YH. Defatted cashew nut shell starch as renewable polymeric

material: isolation and characterization. Carbohydrate Polymers. 2012;**87**(4):2576-2581

[64] Ding D, Zhao Y, Yang S, Shi W, Zhang Z, Lei Z, et al. Adsorption of cesium from aqueous solution using agricultural residue–walnut shell: equilibrium, kinetic and thermodynamic modeling studies. Water Research. 2013;**47**(7):2563-2571

[65] Sarsari NA, Pourmousa S, Tajdini A. Physical and mechanical properties of walnut shell flour-filled thermoplastic starch composites. BioResources. 2016;**11**(3):6968-6983

[66] Harini K, Mohan CC, Ramya K, Karthikeyan S, Sukumar M. Effect of Punica granatum peel extracts on antimicrobial properties in walnut shell cellulose reinforced bio-thermoplastic starch films from cashew nut shells. Carbohydrate Polymers. 2018;**184**:231-242

[67] Zhang B, Carlson R, Srienc F. Engineering the monomer composition of polyhydroxyalkanoates synthesized in *Saccharomyces cerevisiae*. Applied and Environmental Microbiology. 2006;*72*:536-543. DOI: 10.1128/AEM.72.1.536-543.2006

[68] Pötter M, Steinbüchel A. Biogenesis and structure of polyhydroxyalkanoate granules. Microbiol Monogr. 2006. DOI: 10.1007/7171_005

[69] Muneer F, Rasul I, Azeem F, Siddique MH, Zubair M, Nadeem H. Microbialpolyhydroxyalkanoates(PHAs): Efficient replacement of synthetic polymers. Journal of Polymers and the Environment. 2020;**28**(9):2301-2323

[70] Sharma PK, Munir RI, Blunt W, Dartiailh C, Cheng J, Charles TC, et al. Synthesis and physical properties of polyhydroxyalkanoate polymers with different monomer compositions by recombinant Pseudomonas putida LS46 expressing a novel PHA synthase (PhaC116) enzyme. Applied Sciences. 2017;**7**(3):242

[71] Amache R, Sukan A, Safari M, Roy I, Keshavarz T. Advances in PHAs production. Chemical Engineering and Technology. 2009;**32**:931-936

[72] Koller M, Salerno A, Dias M, Reiterer A, Braunegg G. Modern biotechnological polymer synthesis: a review. Food Technology and Biotechnology. 2010;**48**:255-269

[73] Sudesh K, Abe H, Doi Y. Synthesis, structure and properties of polyhydroxyalkanoates: biological polyesters. Progress in Polymer Science. 2000;**25**:1503-1555. DOI: 10.1016/S0079-6700(00)00035-6

[74] Grigore ME, Grigorescu RM, Iancu L, Ion RM, Zaharia C, Andrei ER. Methods of synthesis, properties and biomedical applications of polyhydroxyalkanoates: a review. Journal of Biomaterials Science. Polymer Edition. 2019;**30**:695-712. DOI: 10.1080/09205 063.2019.1605866

[75] Hazer B, Steinbüchel A. Increased diversifcation of polyhydroxyalkanoates by modifcation reactions for industrial and medical applications. Applied Microbiology and Biotechnology. 2007;**74**:1-12. DOI: 10.1007/s00253-006-0732-842

[76] Johnston B, Radecka I, Hill D, Chiellini E, Ilieva VI, Sikorska W, et al. The microbial production of polyhydroxyalkanoates from waste polystyrene fragments attained using oxidative degradation. Polymers (Basel). 2018;**10**:950. DOI: 10.3390/polym10090957

[77] Pachekoski WM, Agnelli JAM, Belem LP. Thermal, mechanical and morphological properties

of poly(hydroxybutyrate) and polypropylene blends after processing. Materials Research. 2009;**12**:159-164

[78] Choi SY, Cho IJ, Lee Y, Kim YJ, Kim KJ, Lee SY. Microbial polyhydroxyalkanoates and nonnatural polyesters. Advanced Materials. 2020;**32**(35):1907138

[79] Khanna S, Srivastava AK. Recent advances in microbial polyhydroxyalkanoates. Process Biochemistry. 2005;**40**:607-619

[80] Anjum A, Zuber M, Zia KM, Noreen A, Anjum MN, Tabasum S. Microbial production of polyhydroxyalkanoates (PHAs) and its copolymers: A review of recent advancements. International Journal of Biological Macromolecules. 2016;**89**:161-174

[81] Crank M, Patel M, Marscheider-Weidemann F, Schleich J, Husing B, Angerer G. Techno-economic feasibility of large-scale production of bio-based polymers in Europe (PRO-BIP), Final Report Prepared for the European Commission's Institute for Prospective Technological Studies (IPTS). 2004

[82] Winnacker M. Polyhydroxyalkanoates: recent advances in their synthesis and applications. European Journal of Lipid Science and Technology. 2019;**121**(11):1900101

[83] Bugnicourt E, Cinelli P, Lazzeri A, Alvarez V. Polyhydroxyalkanoate (PHA): review of synthesis, characteristics, processing and potential applications in packaging. Express Polymer Letters. 2014;**8**:791-808. DOI: org/10.3144/expresspolymlett.2014.82

[84] Boyandin AN, Prudnikova SV, Karpov VA, Ivonin VN, Do NL, Nguyen TH, et al. Microbial degradation of polyhydroxyalkanoates in tropical soils. International Biodeterioration and Biodegradation. 2013;**83**:77-84. DOI: 10.1016/j.ibiod.2013.04.014

[85] Masood F, Yasin T, Hameed A. Comparative oxo-biodegradation study of poly3-hydroxybutyrate-co-3-hydroxyvalerate/polypropylene blend in controlled environments. International Biodeterioration and Biodegradation. 2014;**87**:1-8. DOI: 10.1016/j.ibiod.2013.09.023

[86] Nduko JM, Taguchi S. Microbial production of biodegradable lactate-based polymers and oligomeric building blocks from renewable and waste resources. Frontiers in Bioengineering and Biotechnology. 2021;**8**:618077

[87] Gopi S, Kontopoulou M, Ramsay BA, Ramsay JA. Manipulating the structure of medium-chain-length polyhydroxyalkanoate (mcl-PHA) to enhance thermal properties and crystallization kinetics. International Journal of Biological Macromolecules. 2018;**119**:1248-1255. DOI: 10.1016/j.ijbiomac.2018.08.016

[88] Sanhueza F, Acevedo S, Rocha P, Villegas M, Seeger R. Navia, polyhydroxyalkanoates as biomaterial for electrospun scaffolds. International Journal of Biological Macromolecules. 2019;**124**:102-110. DOI: 10.1016/j.ijbiomac.2018.11.068

[89] Plackett D, Siró I. Polyhydroxyalkanoates (PHAs) for food packaging. In: Multifunctional and Nanoreinforced Polymers for Food Packaging. Woodhead Publishing. 2011:498-526

[90] Miguel O, Fernandez-Berridi MJ, Iruin JJ. Survey on transport properties of liquids, vapors, and gases in biodegradable poly(3-hydroxybutyrate) (PHB). Journal of Applied Polymer Science. 1997;**64**:1849-1859

[91] Padermshoke A, Katsumoto Y, Sato H, Ekgasit S, Noda I, Ozaki Y. Melting behavior of poly(3-hydroxybutyrate) investigated by two-dimensional infrared correlation spectroscopy. Spectrochim Acta-PartA Mol. Biomol. Spectrosc. 2005;**61**:541-550. DOI: 10.1016/j.saa.2004.05.004

[92] Koller M. Switching from petro-plastics to microbial polyhydroxyalkanoates (PHA): the biotechnological escape route of choice out of the plastic predicament? EuroBiotech Journal. 2019;**3**:32-44. DOI: 10.2478/ebtj-2019-0004

[93] Laycock B, Halley P, Pratt S, Werker A, Lant P. The chemomechanical properties of microbial polyhydroxyalkanoates. Progress in Polymer Science. 2013;**38**:536-583. DOI: 10.1016/j.progpolymsci.2012.06.003

[94] Chanprateep SJ. Journal of Bioscience and Bioengineering. 2010;**110**:621-632

[95] Olivera E, Carnicero D, Jodra R, Minambres B, Garcia B, Abraham G. Environmental Microbiology. 2001;**3**:612-618

[96] Chodak I. Polyhydroxyalkanoates: origin, properties and applications. Monomers, Polymers and Composite from Renewable Resources. 2008:451-477. DOI: 10.1016/B978-0-08-045316-3.00022-3

[97] Koller M. Production of Polyhydroxyalkanoate (PHA) biopolyesters by extremophiles? MOJ Polymer Science. 2017;2:69-85. DOI: 10.15406/mojps.2017.01.00011

[98] Nahar S, Jeong MH, Hur JS. Lichen-associated bacterium, a novel bioresource of polyhydroxyalkanoate (PHA) production and simultaneous degradation of naphthalene and anthracene. Journal of Microbiology and Biotechnology. 2019;**29**:79-90. DOI: 10.4014/jmb.1808.08037 64

[99] Bhuwal AK, Singh G, Aggarwal NK, Goyal V, Yadav A. Isolation and screening of polyhydroxyalkanoates producing bacteria from pulp, paper, and cardboard industry wastes. International Journal of Biomaterials. 2013

[100] Preusting H, Hazenberg W, Witholt B. Continuous production of poly(3-hydroxyalkanoates) by *Pseudomonas oleovorans* in a high cell density, two-liquid-phase chemostat. Enzyme and Microbial Technology. 1993;**15**:311-316

[101] Guo-Qiang C, Jun X, Qiong W, Zengming Z, Kwok-Ping H. Synthesis of copolyesters consisting of medium-chain-length - hydroxyalkanoates by Pseudomonas stutzeri 1317. Reactive and Functional Polymers. 2001;**48**:107-112

[102] Masood F, Abdul-Salam M, Yasin T, Hameed A. Effect of glucose and olive oil as potential carbon sources on production of PHAs copolymer and tercopolymer by Bacillus cereus FA11. 3 Biotech. 2017;7:87. DOI: 10.1007/s13205-017-0712-y

[103] Afreen R, Tyagi S, Singh GP, Singh M. Challenges and perspectives of polyhydroxyalkanoate production from microalgae/cyanobacteria and bacteria as microbial factories: An assessment of hybrid biological system. Frontiers in Bioengineering and Biotechnology. 2021;**9**:624885

[104] de Philippis R, Ena A, Guastini M, Sili C, Vincenzini M. Factors affecting poly-β-hydroxybutyrate accumulation in cyanobacteria and in purple non-sulfur bacteria. FEMS Microbiology Reviews. 1992;**103**:187-194

[105] Saharan SB, Grewal A, Kumar P. Biotechnological production of

polyhydroxyalkanoates: A review on trends and latest developments. Chinese Journal of Biology. 2014

[106] Kirk RG, Ginzburg M. Ultrastructure of two species of halobacterium. Journal of Ultrastructure Research. 1972;**41**:80-94

[107] Fernandez-Castillo R, Rodriguez-Valera F, Gonzalez Ramos J, Ruiz-Berraquero F. Accumulation of poly (*β*-hydroxybutyrate) by Halobacteria. Applied and Environmental Microbiology. 1986;**51**:214-216

[108] Rehm B. Microbial Production of Biopolymers and Polymer Precursors: Applications and Perspectives. Norfolk, UK: Caister Academic Press; 2009

[109] Kim DY, Kim YB, Rhee YH. Evaluation of various carbon substrates for the biosynthesis of polyhydroxyalkanoates bearing functional groups by *Pseudomonas putida*. International Journal of Biological Macromolecules. 2000;**28**:23-29. DOI: 10.1016/S0141-8130(00)00150-1

[110] Jiang G, Hill DJ, Kowalczuk M, Johnston B, Adamus G, Irorere V, et al. Carbon sources for polyhydroxyalkanoates and an integrated biorefinery. International Journal of Molecular Sciences. 2016;**17**(7):1157

[111] Grothe E, Moo-Young M, Chisti Y. Fermentation optimization for the production of poly(β-hydroxybutyric acid) microbial thermoplastic. Enzyme and Microbial Technology. 1999;**25**:132-141

[112] Page WJ. Production of poly-β-hydroxybutyrate by *Azotobacter vinelandii* strains UWD during growth in molasses and other complex carbon sources. Applied Microbiology and Biotechnology. 1989;**31**:329-333

[113] Reid SJ, Abratt VR. Sucrose utilisation in bacteria: genetic organisation and regulation. Applied and Environmental Microbiology. 2005;**67**:312-321

[114] Tanamool V, Imai T, Danvirutai P, Kaewkannetra P. Screening, identification and production of polyhydroxyalkanoates (PHAs) by sucrose utilizing microbes isolated from soil environments. Journal of Biomechanics. 2010;**150**:361

[115] Amaro TM, Rosa D, Comi G, Iacumin L. Prospects for the use of whey for polyhydroxyalkanoate (PHA) production. Frontiers in Microbiology. 2019;**10**:992

[116] Vandamme P, Coenye T. Taxonomy of the genus Cupriavidus: a tale of lost and found. International Journal of Systematic and Evolutionary Microbiology. 2004;**54**:2285-2289. DOI: 10.1099/ijs.0.63247-0

[117] Ahn WS, Park SJ, Lee SY. Production of poly (3-hydroxybutyrate) by fed-batch culture of recombinant *Escherichia coli* with a highly concentrated whey solution. Applied and Environmental Microbiology. 2000;**66**:3624-3627. DOI: 10.1128/aem.66.8.3624-3627.2000

[118] Shiotani T, Kobayashi G. Alternating Copolymer and Its Production Method. Japanese Patent Application 93049,16. 1993;(April)

[119] Cavalheiro JMBT, Raposo RS, de Almeida MCMD. Effect of cultivation parameters on the production of poly (3-hydroxybutyrate-co-4-hydroxybutyrate) and poly (3-hydroxybutyrate-4-hydroxybutyrate-3-hydroxyvalerate) by *Cupriavidus necator* using waste glycerol. Bioresource Technology. 2012;**111**:391-397. DOI: 10.1016/j.biortech.2012.01.176

[120] Cheema S, Bassas-Galia M, Sarma PM. Exploiting metagenomic diversity for novel polyhydroxyalkanoate

synthases: production of a terpolymer poly (3-hydroxybutyrate-co3-hydroxyhexanoate-co-3-hydroxyoctanoate) with a recombinant *Pseudomonas putida* strain. Bioresource Technology. 2012;**103**:322-328. DOI: 10.1016/j.biortech.2011.09.098

[121] Bhubalan K, Rathi DN, Abe H. Improved synthesis of P(3HB-co-3HV-co-3HHx) terpolymers by mutant *Cupriavidus necator* using the PHA synthase gene of Chromobacterium sp. USM2 with high affinity towards 3HV. Polymer Degradation and Stability. 2010;**95**:1436-1442. DOI: 10.1016/j.polymdegradstab. 2009.12.018

[122] Bucci DZ, Tavares LBB, Sell I. PHB packaging for the storage of food products. Polymer Testing. 2005;**24**(5):564-571

[123] Geyer R, Jambeck JR, Law KL. Production, use, and fate of all plastics ever made. Science Advances. 2017;**3**(7):e1700782

[124] Grumezescu AM, Oprea AE, editors. Nanotechnology Applications in Food: Flavor, Stability, Nutrition and Safety. Academic Press; 2017

[125] Koller M. Poly (hydroxyalkanoates) for food packaging: application and attempts towards implementation. Applied Food Biotechnology. 2014;**1**:3-15

[126] Rahman R, Sood M, Gupta N, Bandral JD, Hameed F, Ashraf S. Bioplastics for food packaging: a review. International Journal of Current Microbiology and Applied Sciences. 2019;**8**(3):2311-2321

[127] Basumatary IB, Mukherjee A, Katiyar V, Kumar S. Biopolymer-based nanocomposite films and coatings: recent advances in shelf-life improvement of fruits and vegetables. Critical Reviews in Food Science and Nutrition. 2020;**62**(7):1912-1935. DOI: 10.1080/10408398.2020.1848789

[128] Montazer M, Harifi T. New approaches and future aspects of antibacterial food packaging: from nanoparticles coating to nanofibers and nanocomposites, with foresight to address the regulatory uncertainty. In: Grumrzescu AM, Editors. Food Packaging. USA: Academic Press; 2017. pp. 533-565

[129] Guillard V, Gaucel S, Fornaciari C, Angellier-Coussy H, Buche P, Gontard N. The next generation of sustainable food packaging to preserve our environment in a circular economy context. Frontiers in Nutrition. 2018;5:121. DOI: 10.3389/fnut.2018.00121

[130] Rastogi VK, Samyn P. Bio-based coatings for paper applications. Coatings. 2015;**5**(4):887-930

[131] Sharma V, Sehgal R, Gupta R. Polyhydroxyalkanoate (PHA): properties and modifications. Polymer. 2021;**212**:123161

[132] Masood F. Polyhydroxyalkanoates in the food packaging industry. In: Nanotechnology Applications in Food. Academic Press. 2017. pp. 153-177

[133] Singh RS, Kaur N. Microbial biopolymers for edible film and coating applications. Advances in Biotechnology. 2015;**12**:187-216

[134] Castro-Mayorga JL, Fabra MJ, Cabedo L, Lagaron JM. On the use of the electrospinning coating technique to produce antimicrobial polyhydroxyalkanoate materials containing in situ-stabilized silver nanoparticles. Nanomaterials. 2016;**7**(1):4

[135] Boucher J. Development of biodegradable water bottle. 2021.

Available from : https://www.foodpackagingforum.org/news/development-of-biodegradable-waterbottle [Accessed: November 2021]

[136] Cov. The drink bottle made entirely of PHA. Available from: https://drinkcove.com [Accessed: 14 May 2021]

[137] Obulisamy PK, Mehariya S. Polyhydroxyalkanoates from extremophiles: a review. Bioresource Technology. 2021;**325**:124653

[138] Muthusamy MS, Pramasivam S. Bioplastics – an eco-friendly alternative to petrochemical plastics. Current World Environment: An International Research Journal of Environmental Sciences. 2019;**14**(1):49

Chapter 12

The Place of Packaging System in Advancing Food Preservation for Promoting Food Products' Market Share

Thomas Amarachukwu Uzombah

Abstract

The place of packaging system in advancing food preservation for promoting food products' market share was studied. It was observed that packaging system is a version of food preservation, used to prepare food products for transportation, distribution, storage, marketing, luring consumers and usage. It makes freshly/processed food products last beyond natural shelf-life while maintaining an acceptable quality, nutritional and sensory characteristics by shielding them from interacting with natural environmental atmosphere. Consumers' increasing demand for fresh-like/mildly food with valuable quality, nutritional, safety and sensorial properties instigated emergence of novel packaging system having better preservative potentials that guarantee meeting their expectations. By advancing food preservation, packaging elongates products' marketability and enlivens organizational marketing strategies for increasing products' sales/their associated market shares. It achieves this through the functions it performs which include; containment, protection/preservation, communication, convenience, tamper indication, and traceability. It is recognized as a salesman, an advertising agent and a sales promoter; and a beacon of relaunching of products to revive/heighten consumers' interests in the products. It is part of marketing strategies, often used to add values to products and for brand promotion. Packaging system advances food preservation and creates opportunities for products' sales and the associated gain in market share.

Keywords: food packaging, advancing food preservation, food products, market share

1. Introduction

Packaging mediates between food industry and the market and, determines the stake of food products in the market and hence, the rewards of efforts put in by producers in food processing. Processing of foods is as important as ensuring that the processed foods get to the consumers in good condition, hence, the indispensable role of packaging in food processing operations. Marsh and Bugusu [1] considered packaging as instrument of maintaining the benefits of food processing by facilitating safe movement of processed food products from the point of production through channels of distribution to the point of consumption. Food packaging, by maintaining the wholesomeness of products, undoubtedly, creates preservative environment for enhancing food quality and safety as well as reducing postharvest wastes which constitute the key objectives of sustainable food system [2]. Food packaging configures atmosphere surrounding food products and shields the products from interacting with natural environmental atmosphere by either reducing the presence of oxygen and increasing carbon dioxide or vice versa depending on one adjudged to favor keeping quality of the given product(s). The preservative potential of packaging was highlighted by Himanshu and Bindu [3] in their report that acknowledged packaging as an essential system for preserving food quality, minimizing food wastage as well as reducing the use of chemical food preservatives and their associated effects. The foregoing, indicates that food packaging is a version of food preservation system, an instrument that could be used in advancing food preservation for efficient and effective distribution, marketing and making the products get to the consumers in an acceptable condition. The packaging system purveys the advancement of preservation of food products, enables increased consumers to increasingly patronize the products which usually translate to more sales for the organization with its attendant more market share. The efforts of food industry to provide sufficient food for people could only be successful by packaging system which, according to Himanshu and Bindu [3] serves several purposes which include; protecting products from the point of production through distribution channels to the consumers, or after purchase for the consumer, increasing ease of utilization of the products, improving consumers' interest or acceptability of the products, supporting what they further recognized as self-service, consumer affluence, company and brand recognition, and innovation opportunity. Although, the primary target of packaging is to advance the preservation of foods; it has been observed to have indirectly become a major factor in gaining customers. The report of Draskovic [4] attested to the link between food packaging and gaining consumers' attention and, stated that it is a key to motivating consumer to purchase products. The forgoing, not only, determines preferences of consumers with respect to packaging designs but their behaviors toward purchasing of food products. The logistical and functional roles packaging play in the food system makes it outstanding in every activity associated with the system such as preservation/containment, enhancing longevity of shelf-life, distribution or movement of products from point of production to the table of the consumers. The importance of food preservation in promoting marketability and market share of products/organizations in food industry cannot be over emphasized. As one of the key drivers of marketing, it pushes the food industry to flow with dynamic changing consumption habits of consumers occasioned by busier life styles, small household unit and general insufficient time. Packaging system as a version or integral part of preservation system, undoubtedly elongates keeping quality, nutritional and sensory values of fresh and processed food products to assure their availability, at right conditions, both for the increased sales of products by producers/marketers cum gain in the products' market shares. There is insufficient information on the importance of packaging system on preservation or elongation of shelf-life of fresh/processed food products which enables selling more of the products and gaining market shares. Hence, the objective of this paper was to discuss the place of packaging system in advancing food preservation for promoting food products' market shares.

1.1 Definition of concepts

1.1.1 Food packaging

This is defined as an enclosure of fresh or processed foods in a container called a package with an intention to hold and keep the foods for a later use. This definition explains the containment, protection or preservation functions of food packaging indicated in the literature. Coles [5] alluded to its marketing orientation to define it as 'a coordinated system of preparing food for transport, distribution, storage, retailing, and end-use to satisfy the ultimate consumer with optimal cost'. It can be equally defined as a post-processing operation, which although not part of raw materials, is designed to improve or further the values of fresh or processed food products to food industry. Shin and Selke [6], aligned with that above definition by stating that it is a follow-up to the completion of food processing operations for maintaining the benefits of processing efforts, carried out to enable food products travel safely for long distances from their point of origin to the table of consumers at a wholesome condition. They averred that food packaging emanated from the demands of modern or advancement in food system that revolves around ensuring that food products move through distribution channels to the consumer in safe and acceptable quality conditions. Also, food packaging, may be considered as enclosure of fresh or processed food products for passing the relevant information to expected consumers to encourage their patronage. This definition is in tandem with the report of Marsh and Bugusu [1] that stated that packaging is the face of food products designed to expose the products to consumers and or induce/influence the expected consumers' patronage. Similarly, food packaging is defined as preparing or arranging or configuring processed or fresh food products for easier or convenient utilization of the products. Packaging role in making food products easier for consumers to consume is reflected in designing some packages to have dispensing and resealing features, ability-to-cook and eat-the-product within its specific packaging features, easier handling features etc. [4].

The definitions of Food packaging indicate that it is a conglomerate of activities designed to ensure that food products are handled to maximally benefit those that have interest in the products such as the producers, the marketers and the consumers. Hence, the definition spans through enclosure to protecting/preserving the product, containing for safe distribution of the products, easing difficulties in the utilization of the products and; to communicating to draw attentions of the potential consumers in order to encourage the marketability of the products. It mediates between the producers of fresh or processed food products and the consumers, ensuring movement of the product through distribution channels with untampered values.

1.1.2 Organizational market share

Organizational market share is the portion of market sales accruable to a given organization by the sales of its product(s) measured within a given period of time. Some school of thought considers market share as that percentage of total customers' purchases of a given product that goes to a given individual, organization or nation. It is also defined as the measure of the relative size of a business with respect to other participants in the same market, but is not a measure of a relative size elsewhere in the value chain [7]. **Table 1** shows the contribution of packaging to the values of market share gained by products due to the types of packaging materials used.

Materials	Market share (%)		Compound annual
	2011	2016	Growth rate (%)
Flexible packaging	20.4	21.6	4.8
Flexible plastic	13.6	14.9	5.4
Flexible foil	1.7	1.7	3.3
Flexible paper	5.1	5.0	3.4
Rigid plastic packaging	22.1	24.4	5.7
Board packaging	30.6	29.4	2.9
Metal packaging	15.6	14.3	1.9
Glass packaging	6.7	6.4	2.7
Other packaging	4.7	3.9	−0.4

Source: Su et al. [8].

Table 1.
Packaging market trends.

Market share is a measure of consumers' preference for product over other similar products or consumers' preference for product from a given organization over other products from other organizations, indicating that higher market share emanate from making greater sales, having less efforts to sell more and having strong barrier to entry for competitors [9]. The types of market shares mentioned by the literature are value market share and volume market share. The value market share is based on the total share of a company out of the total segment sales. Volumes refers to the actual numbers of units that a company sells out of a total units sold in the market. In marketing literature, market share is considered as a demonstration of the effort of an organization to compete in product-market place. This view is in agreement with the report of Farris *et al.* [9] that considered market share as a reflection of successful efforts an organization unleashed in selling its product by advertising and promotion, product/service offering quality and price, channel and customer relationships, and selling activities.

1.1.3 Advancing food preservation

The concept of advancing food preservation is all about taking, the elongation of wholesomeness (quality and edibility) of fresh and processed food products, to greater levels by applying scientific knowledge using available technologies and procedures to discourage the activities of agents of food deterioration/spoilage and foodborne diseases. Preservation attempts at elongating the quality or usefulness of the products by elimination or slowing down microorganisms, oxidation of available fats that instigate rancidity or activities of enzymes responsible for chemical deteriorations [10]. The literature indicates that the elimination of the spoilage/deterioration agents is usually, taken care of, during processing (conversion of raw food materials to desired products), as the heat energy applied for processing eliminates or reduces the microbial loads and or inactivate enzymes. Advancing food preservation goes beyond ensuring maintenance of food shelf-life and safety to seeing that mildly processed food products have an elongated shelf-life with the right quality, nutritional values, and safety, sensory attributes (taste, appearance, color etc.). Ayoub [11], associated advancing preservation with the emergence of novel preservation process which shifted from the use of conventional methods of processing like thermization, pasteurization, and in-container sterilization; to novel food processing technologies. His report showed that these technologies are grouped into non-thermal processes which uses physical means as inactivation factors like pressure, electromagnetic field, sounds etc. and they include high pressure processing (HPP), pulse electric fields, irradiation, ultrasound, cold plasma, ozone, supercritical; and thermal which mainly uses energy generated by microwave and radio frequency and they include microwave, radio frequency, ohmic heating and inductive heating. Advancing preservation by preventing recontamination entails creating barriers that circumvent interactions between food products and outside environmental atmosphere using packaging system that encloses the product with different atmospheric conditions unfavorable to the activities of deteriorating/spoilage agents. Also, advancing food preservation may, also, be considered as increasingly prolonging keeping quality and enhancing the economic orientation of food products with an insignificant or no effect on the physiochemical conditions of food. Hassan et al. [12] adduced that advancing food preservation is, also, achieved through application of advanced techniques involving the use of natural ingredients, example is the use of natural antioxidants in preservation of foods. Hence, advancing food preservation is keeping or maintaining the quality of food products, from the point of production through the channels of distribution, with all the conditions that coincide with the expectations of the consumers. The same reasons that instigated the evolvement of novel food processing technologies like production of safer, long shelf-life, higher quality, good nutritional valuable, and healthy promoting food products with acceptable sensorial parameters [13] caused the emergence of novel packaging technologies, as mildly processed food products, which for instance, are increasingly desired by consumers are to be ensured to be in acceptable condition from the point of production through to the time of consumption [14]. Packaging systems are facilitators of unitization of advancement in food preservation. The novel packaging technologies outlined by [13] to include; active packaging, intelligent packaging, eco-friendly packaging, bioactive packaging, nanotechnology, modified atmosphere packaging, controlled atmosphere packaging and others, are advancing food preservation technology.

2. Developments in the food packaging

The development of the packaging started with the traditional packaging systems. The accounts of archeologists and historians, according to the literature, tied the inception of food packaging with people becoming more nomadic and the need for them to protect, store and transport food and other goods. At that time, they made use of untreated natural resources they could lay hand on, such as animal skins, gourds, fashioned leaves and other plant materials. The advancement in people's ways of life translated to the enhancing the use of these resources in new fashioned way by making clay pots, weaving baskets using plant fibers and making bags from animal skins, which enabled storage and transportation of food materials between settlements. As people grew in population and advanced in lifestyles, innovation and creativity; sophisticated packaging system evolved to cater for their prevalent expectations. The packaging industry, thus, developed from the use of locally available natural resources in the traditional setting through to the use of sophisticated materials concerned with achieving four major functions of containment, protection, convenience and communication.

Traditional packaging materials, most of the times, target at providing containment, some protections and convenience with little or no regard to the communication aspect of packaging function [15]. Thus materials could be said to be readily available to meet up the supply needs of the industry, but fall short of some important ones. For example, plant materials like leaves used for packaging either as a direct wrapping or weaving/forming them into containers and baskets provided good packaging for containment with an insignificant contribution to preservation at long term and communication [16]. The shortcomings of the traditional packaging systems and man's quest for better packaging materials evolved the use of plant materials to produce paper and paper boards, which almost provided all the packaging functions of containment, protection, communication and convenience. Sarkar and Aparna [17] reported the changes packaging system passed through to attain the present status today in which materials such as paper and paperboard, metals, glasses, plastics and other materials designed by the blends of two or more of these aforementioned items for better packaging functions. According to literature, it took over 150 years for food packaging to undergo changes and finally emerge in its current form [18].

However, it is worthy to mention that the pattern of involvement of the materials in food packaging indicated that paper which took over from traditional materials had shortcomings of low strength, being opaque and high permeability to moisture, gases and vapors and hence, are unsuitable for adequate protection and preservation of food products. The glass developed to address these lapses, though provided the advantages of transparency, imperviousness, and inertness; but their fragility and heavy weight to strength ratio became limiting factors [1]. The first evidence of pottery and glass being made was about 7000 B.C., yet industrialization of the process by the Egyptians was not seen until about 1500 B.C. [19]. It is interesting to note that the primary materials used to make glass at that time, limestone, soda, sand, and silica, are the same materials that are used today, although many additives have been developed to color glass and give it varying properties [17].

The metal containers came into use in the 1700s for food packaging, and have excellent strength but their shape limitation was their main disadvantage [18]. Metal cans were initially manufactured for snuff, for which they provided an excellent barrier to products either from losing moisture, flavor, odors or absorbing same from their external environment [17]. The metal containers were later used in the canning operation that was discovered by Nicholas Appert when he answered a challenge from French Emperor Napoleon Bonaparte to develop a method to preserve food for his army [20]. The use of metal in packaging materials are still ongoing as they provide the expectations of man with regard to having ideal containment, protection, convenience and communication.

The involvement of plastics in food packaging was as a result exploration for better packaging materials. The ability of plastics to be made into thin films and containers is one of the most important advantages of involvement of plastics in packaging industry [2]. The benefits of involving plastics as packaging materials outweigh their demerits. Their use reduces packaging weight, volume, costs and ease of transportation of packaged food products. The involvement of plastics as packaging materials, facilitated evolution of technologically advanced packaging systems such as modified atmosphere or controlled atmosphere packaging, active packaging, intelligent packaging and others [11]. Also recent developments in food packaging have shown the involvement of ecofriendly materials like biodegradable or edible materials in food packaging system, in an effort to minimize problems of increasing environmental wastes occasioned by packaging materials. The development of food packaging from traditional through the conventional packaging to novel packaging systems showed a progressive improvement in the advancement in the contributions of packaging to

Parameter	Traditional packaging	Conventional packaging	Novel packaging
Food quality	Slightly maintained	maintained	Maintained
Shelf-life	Slightly maintained	maintained	Maintained
Sensorial properties	Not maintained	Slightly maintained	Maintained
Nutritional value	Lost at long term	Good	better nutritional values
Safety	Not assured	Good	Good
Fresh-like food	Not available	Not available	Available

Sources: Buckow and Bull [21]; Hameed et al. [13]; and Pal et al. [14]

Table 2.
Traditional, conventional, and novel packaging materials and their impacts on food parameters.

food industry as indicated in the **Table 2** below. Buckow and Bull [21], Pal et al. [14] and Hameed et al. [13] mentioned areas of improvement in the maintenance of quality, nutritional values, sensorial parameters and shelf-life, with longer storage period, as the packaging systems developed from traditional through the conventional to novel packaging, especially for the fresh-like or mildly processed food products being increasingly demanded by consumers.

2.1 Some of the pictures of the packaging materials

The pictures showing examples of the various types of conventional packaging materials afore-mentioned in **Figures 1–7** below.

Figure 1.
Flexible paper packaging.

Figure 2.
Rigid paper packaging.

Figure 3.
Rigid plastic packaging.

Figure 4.
Flexible plastic packaging.

Figure 5.
Flexible Aluminums foil food packaging.

Figure 6.
Metal food packaging.

Figure 7.
Glass packaging.

3. The contributions of packaging in the development of food industry

Food industry is a conglomerate of activities involved in handling agricultural products, which runs, possibly from the farm gate through to the utilization or consumption of the products at an expected condition of consumers' adjudged food value. Food packaging is a process of making fresh or processed food product to be utilized beyond the point of production, which facilitate the link between the producers and the consumers, by ensuring that the products are transported through the distribution channels to the consumers, with unchanged quality, nutritional status and physical/sensory parameters. The literature recognized food packaging as the champion of identification, anticipation and satisfaction of customers and increasing patronage of food products for productive marketing and gaining of market share. The support of food packaging to the enterprising of food industry is founded on the forgoing as well as in ensuring diversifying patronage of the products through their unitization process. The contributions of food packaging to the development or success of food industry can be considered in terms of primary and secondary roles [22]. The literature mentioned four principal or primary and other ways food packaging promote food industry to include containment, protection/preservation, communication and convenience constituting primary functions and others like temper indication, traceability, portion control/unitization. Some of these functions are discussed below.

3.1 Containment

This role entails holding or containing food products to prevent them from throwing away or scattering during logistic or handling operations. For instance, containment makes movement of the products from the point of production to the point of storage/usage possible or convenient. This indicates that containment is the key factor in either the handling or the usage of the products. According to Shin and Selke [6], containment is the key factor for food packaging functions because products must be contained in a container to get to the consumers. Also, containment facilitates handling of the products in discrete units as an added value to the products. By unitizing the products, containment eases the marketability or distribution of the products and hence, improves revenues realizable from the products. Containment can equally be considered as a fundamental basis for protecting or preserving the products from unnecessary interactions between the products and the outside environmental constituents during handling operations.

3.2 Protection/preservation

Packaging is the beacon of protection or preservation of food products from physical damage arising from shock, vibration, compressive forces, poor handling (abrasion) etc. during distribution and transportation of the products. It also safeguards the products from other environmental damages that occur due to the influence of water, air, light, environmental odors, microorganisms [6]. The protective function of packaging equally involves shielding products from unacceptable interactions or destructive tendencies of environmental factors such microorganisms, insects/rodents and others. Packaging process creates conducive micro-climate or environment that enhance shelf-life around food products by putting in place a different formation of oxygen, carbon dioxide, water vapor etc. other than that obtained in natural atmospheric environment. Marsh and Bugusu [1] stated the protection/preservation function of food packaging can be considered to be in three dimensions, physical, chemical and biological with the chemical minimizing compositional changes instigated by environmental influences like exposure to gases, moisture and, light (visible, infrared, and ultraviolet); biological, preventing interactions of microorganisms(pathogenic and spoilage agents), insects, rodents and other animals to circumvent spoilage and incidence of diseases.; physical aspects catering for shock, vibration, compressive forces, poor handling (abrasion) etc. during distribution and transportation of the products.

The choice of containers for food packaging, according to the literature, depend on many factors which include the physical condition (liquid, semi-solid and solid), chemical composition of food, physical strength of the material, material' resistance to permeability of liquid, vapor, environmental odor etc. usually, the packaging literature is replete with information on various containers applied in food system as packaging materials. The report indicated the use of natural resources like leaves and other plant materials, animal skins, cement paper bags, jute bags, basket, bamboo, cane basket, and pottery, discarded bottles and jars, old stock of paper prints, broad leaves, empty fluted gourds, fruit shell, coconut shells, maize-sheath, glass-sided boxes, jute sacks, poly sacks, polyethylene bags grouped as traditional packaging materials [15, 22, 23]; that are still used amongst Igbos, Yorubas, Hausas and other tribes of Nigeria in their localities. These material are being used not necessarily for their protective/preservative potential but for their availability at a very low cost, as according to Evivie et al. [15] they are unsuitable for protecting agricultural products or for efficient logistical roles and transportation and thus are inapplicable in industrial settings. The attempt to improve on the functions of traditional packaging evolved modern packaging materials which include paper and paperboard, glass, metals, and plastics in conventional packaging system. Paper and paperboards are used in wide range of forms in food packaging for dried food products like powdered foods, flours, cereals, cereals etc. as well as in production of corrugated boxes, milk cartons, sacks, and paper plates [2]. The protective/preservative properties of plain paper are poor because of its low strength and high permeability to moisture, gases and water vapors [1]. Also, glass containers, which according to Grayhurst and Girling [24], attributed to being chemically inert and odorless, are used to package wide range of food products like food beverages such as beers, wines, spirits, liqueurs, soft drinks and mineral water and others, whether in solid or liquid form. Another aspect of containers used in food packaging are metals. The report of Page et al. [25] indicates that metal containers are strong and impervious to water/liquid, gas, food vapors, environmental odors and are used to package wide variety of foods. The two metals mostly involved in food packaging are aluminum and steel [1]. While steel is used in making three-piece cans for packaging liquid milk, drinks and processed foods such as beans and peas, aluminum is used to produce two-piece cans (for carbonated drink and seafood), foils and laminated paper [2]. Furthermore, plastic containers appear to dominate other ones afore-mentioned in food packaging because they are valuable in terms of being moldable, heat sealable, inclusive to production processes by having the nature of being produced, filled and sealed as part of production operation [1]. The use of plastic in packaging has facilitated advancement in food packaging that guarantees better maintenance of nutritional and sensorial properties of food products [2]. The emergence of modified atmosphere packaging, active packaging, intelligent and other recently developed novel packaging systems which target at meeting the consumer' desires for fresh-like food products, with unchanging valuable nutrients and sensorial attributes, health promoting potential; was facilitated by packaging containers constituted of plastic and or biodegradable materials.

3.3 Communication role

Packaging passes information about the products, providing what the consumers need to know in terms of the chemical constituents, nutritional characteristics, mode of utilization, shelf-life etc. This information is necessary to draw attentions of consumers or guide them in the choice of products [6]. The communication function of packaging not only includes the information provided by the written texts, but the elements of the packaging designs such as package shape, color, recognized symbols or brands. In this status, packaging performs a role of salesmen employing all its characteristics to entice expected consumers who may instantly recognize the products through appetizing pictures or distinctive brands on the package. The packaging, if transparent, expresses the same nature to lure consumers by giving them opportunity to see the products contained inside the package. Invariably, packaging by its characteristics, passes information to consumers, helps in the marketing and, may be considered a purveyor of food products' retail marketing mix that influence conventional marketing variables such as products, price, promotion, and place.

The communication aspect of the functions also encompasses how food packaging facilitates marketing of specific or well-known or branded food products. It distinguishes products from different organizations to enable the consumers choose amongst similar products present in the market. For instance, it provides information on cocoa food drinks from different companies like Cadbury and Nestle Foods, in Nigeria to enable consumers choose according their interest or appeal. Thus, food producers that introduce food products enclosed with an innovative and exquisite packages that are alluring to consumers will always outsell their competitors in product marketing. This is because the ability to present an outstanding packaging, for products, gives organizations an edge in brand competition. It promotes sales of products by providing prerequisite information consumers need every place and on the shelf, thereby acting as an advertising instrument [1]. For example, such promotional information like free extra products, uniqueness of the products, gifts, reduction in price with bulk purchases are provided to draw attention or command interests of consumers.

3.4 Convenience

The desire of to avoid burdens or inconveniences in life is equally translated to application of packaging as part of food processing operation. The use of packaging to reduce burden associated with the handling and utilization of processed food products provided element of convenience in food packaging which seem to influence the interests of the users of the products. Packaging makes the use of food products easier and simple for middlemen and consumers. Promoting convenience in the use of food products through packaging relies significantly on the physical characteristics of packaging, which can unavoidably be linked with the concept of communicative dimension of packaging (e.g. shape, size, color, material, graphic applications. Marsh and Bugusu [1] mentioned elements of food packaging that promote convenience to include ease of access, handling, and disposal; product visibility; resealability and microwavability. According to them the forgoing elements unburdens preparation and serving of food, add value and competitive advantages to food products, though may influence the amount and types of package waste requiring disposal.

The convenience aspect of packaging may be using lighter weighted materials as packaging to reduce the burdens of transportation during distribution. Such materials, though are light, must withstand physical pressures, vibrations or shock that may be occasioned by handling, especially the transportation process. The use of PET instead of glass in packaging soft drinks, recently introduced in the beverage industry in Nigeria, typifies packaging for convenience in transportation of food products. The use of lightweight materials as food package which guarantees provision of protective/preservative critical functions reduces weight eases the burden and cost of transportation of food products. Hence, considering convenience, food product's type vis-a vis the potential of the material to resist breakage and permeation of gases, water vapor, environmental odors etc. are mandatory.

3.5 Tamper indication

Food packaging could also be designed to be tamper resistant to be able to pass information of whether the product enclosed is fondled with or not in the course of movement of the products through distribution channels, marketing to the consumers. According to the literature, this role is registered in food system as tamper evidence or tamper indicator and defined as the characteristic that disclose every unauthorized attempt made by people to access or fiddle with the products. The U. S. Food and Drug Administration [26] defined "tamper- evidence or indication package as one having one or more indicators or barriers to entry which, if breached or missing, can reasonably provide visible evidence to consumers that tampering has occurred" Marsh and Bugusu [1] reported that tamper indication is majorly designed to prevent or reduce adulteration/tampering with the following features such as banding, special membranes, breakaway closures, and special printing on bottle liners or composite cans such as graphics or text that irreversibly change upon opening. The information passed by tamper-indication packaging, not only, assures consumers of the genuineness of the products and hence, contribute immensely in influencing the consumers' decision to choose amongst available similar products and in marketing of the products; but assist to keep organizational brands from being faked. Tamper-indication, thus, is a sure way of discouraging adulteration of products and ensuring that products in the market originated from the organizations that produced them. It is therefore necessary to educate consumers on identification of tamper alert by the manufacturers. This is tandem with the report of Rosette [27] that suggested that consumers should be knowledgeable of the characteristics of tamper-packaging and its indications for the products concerned. Tamper-evidence, is designed to provide, what Kumbhar et al. [28] called visual indication of package integrity of good or reasonable logistic attention given to products from manufacturing through distribution to retailing activities.

3.6 Traceability

Traceability is a characteristic of food packaging applicable in monitoring and assuring unchangeable quality of food products from processing operation through distribution channels to when the products are on the consumers' table. The literature defines food traceability as the system that trace the flow of food products through supply chain whose activity covers production, processing and distribution [29]. It facilitates the tracking of products at any stage of the supply chain. The tracking enables producers to detect defective and unsafe products for easier withdrawal from the market. In addition to tracking products, Golan et al. [30] mentioned other objectives of traceability to include improvement in supply management and differentiating and marketing of food products with subtle or undetectable quality attributes. The tracking is through technology that inscribes unique readable codes on package labels of products. Some of these technologies according to the report of Daoud and Trigui [31] include Barcodes, Radio Frequency Identification, Wireless sensor networks, Quick Response; who further cited the use of quick response code, in particular, a traceability label to access information of food products using smart phone. Traceability provides up-to-date information or histories of all batches products from starting materials to complete final products. Hence, while to the producers it enables the monitoring, identifying and withdrawal of defective and unsafe food products, it assists consumers to ascertain the originality and nature of the products. Amongst consumers that are acquainted with traceability functions, it determines their choices and disposition toward information to purchase products [32]. It indeed aids consumers in making decision on the products to purchase through provision of accurate documentation, and hence, heightens the marketability and possible economic benefits [33]. Its contributions to the advancement of food packaging system are attested to by the literature. Daoud and Trigui [31] mentioned its status in the recently developed innovative packaging like smart packaging and intelligent packaging; designed, according to the literature to maintain, improve and monitor food quality and safety [34].

4. Food packaging as unit of food preservation technique

Food packaging is undoubtedly, the engine of food industry instigating innovations in food preservation in an attempt to maintain the quality, nutritional values and sensory parameters of fresh or processed food products from the point of production to through the distribution channels to the dining tables of consumers. Although packaging of food started with containment of food to facilitate its safe distribution, it progressed with creating environment that discourage deteriorative and spoilage agents, hence the literature classified it as part of food preservation process. Food packaging is indeed a food preservation process since it contributes in the preservation of food products' quality and, guarantees food safety within the shelf-life [35]. Hence, the literature includes food preservation as one of the four major functions of food packaging [36]. Consideration of food packaging system as a preservative technique is based on the fact that packages function to protect products from spoilage and damage occasioned by environmental factors such as microbes, insects, light, heat, oxygen, water vapor, odors, dirt, dusts, etc. [37]. Ganguly [38] convincingly aligned food packaging with food preservation process by stating that it delays deterioration/spoilage of the products, retains the beneficial effects of processing, extends shelf-life, and maintains or increases the quality and safety of food. To accomplish the forgoing, packaging shields food products from uncontrolled interaction with three major external elements which include; chemical, biological and physical environment. Chemical protection targets preventing or reducing compositional changes caused by environmental influences like exposure to gases (typically oxygen), moisture (gain or loss), or light (visible, IR or UV); physical protection ensures logistical efficiency as distortion of food products due to vibration, shock, compressive forces, poor handling (abrasion) etc. during distribution and transportation of the products. Also biological protection not only shields the products from incursion of microorganisms (pathogens and spoiling agents), insects, rodents, and other animals, thereby preventing disease and spoilage; it in addition create condition that regulate senescence (ripening and aging).

The potential of food packaging to preserve fresh or processed foods is explicable with the fact that the process creates different atmospheric environment around the products that discourage activities of agents of deterioration or spoilage. Food packaging as part of food preservation functions by preventing occurrences of undesirable changes in the wholesomeness, nutritive values or sensory quality of food products by shielding the food products from the undesirable activities of biological, chemical and physical agents of food deterioration/spoilage.

Food packaging facilitates storage or preservation of food products in small units and hence, increases the number of potential consumers that can patronize the products. Other instruments of preservation like the use of high or low temperature, for instance, are applicable through the packaging system/medium, as the direct contact of the products and the temperature's source may adversely affect the quality of the products.

Also, innovations in food preservations required to satisfy the dynamic change in the demands of consumers are achievable majorly through food packaging. Preservation of freshly harvested foods to take care of consumers desiring fresh foods, especially fruits and vegetables is being achieved by use of intelligent and active packaging technologies. This assertion is in agreement with the report of Tanja et al. [39] that considered food packaging as a way of improving food preservation system through elongation shelf-life of food. Major studies of improving food preservation are centred on developing new packaging technologies. Abdullahi [40] mentioned the development of smart packaging for optimization of shelf-life using nanotechnologies to prove predominating position of food packaging in efforts to improve food preservation. These new packaging technologies involving the use of biodegradable materials in the packaging system has enabled improvement in preservation of foods, especially, fruits and vegetables for sustainable environment [39]. Sarkar and Aparna [17] also allied advancement in food packaging; which threw up active packaging, aseptic packaging, smart packaging, bioactive packaging, edible packaging; with improvement in food preservation. They further stated that, by maintaining food standard at the highest possible degree, which may help in satisfying the needs of consumers throughout the food supply chain, food packaging incidentally prevents food spoilage/wastage.

5. Food packaging as a salesman

Food packaging through its communication function serve as marketing strategy, passing necessary information all potential consumers desired to know about the product. The information is designed to provide the state of chemical constituents, nutritional and sensory characteristics, mode of utilization, shelf-life and other aspect of logistics required for the consumers' utilization of the product. Such information is necessary to stir up interest of consumers or influence their choice for the product. In this way, food packaging performs a role of salesmen, employing all its characteristics to entice expected consumers who may instantly recognize the products through appetizing pictures or distinctive brands on the package. Also, the increasing change in lifestyle and introduction of self-service retail system, has invariably substituted food packaging as a communicator and mediator between producers and final consumers, by having information designed for consumers printed on the package [4]. Other nature of packaging like transparency which enables consumers to see and ascertain the conditions of the product while in the package, also influences patronage of consumers' choices for products. The forgoing positions packaging as a purveyor of food product retail marketing mix that influences conventional marketing variables such as products, price, promotion, and place. The forgoing may have instigated Wells et al. [41] to associate packaging with marketing mix; even as Smith and Taylor [42] cited by Draskovic [4] regarded packaging as a silent salesperson that helping to inform consumers of a given brand, highlighting unique selling propositions/unique benefits (USPs), and providing friendly tips on usage. This function guides consumers to choose alright amidst increasing availability of similar brands of products in the market.

The power of branding in the marketing of food products, is seemingly the workings of food packaging designed to distinguish products from different organizations in the food industry. Fill [43], stated that packaging is a means through which buyers, especially in consumers' market, can make brand choice decision, while Draskovic [4] considered food packaging as a brand communicator. It distinguishes products according to different industries producing foods products for the patronage of consumers, though the products are the similar. For instance, it provides information on cocoa food drinks from different companies like Cadbury and Nestle Foods to enable consumers choose according their interest or appeal. Thus, any competitor having more innovative packaging that is increasingly appealing to consumers will control or determine the swing of the market. The ability to present an outstanding packaging gives organization an edge in brand competition. This assertion agrees with the reports of Löfgren [44] and Löfgren et al. [45] that stated that the tendency of a consumer to choose one brand over the other at the point of purchase significantly rely on packaging and its ability to persuade the consumers. It promotes sales of products by providing prerequisite information consumers need about the products, every place and on the shelf, thereby acting as an advertising instrument. For example, such promotional information like free extra products, uniqueness of the products, gifts, reduction in price with bulk purchases are provided through food packaging to persuade consumers to patronize the products.

6. Market share: the kernel of organizational growth

Market share which determines the take of organization from its efforts exerted in selling products in given market is a key to organizational profitability. It not only depicts the size of the company, but stand as a metric in demonstrating competitiveness and dominance of the organization in a given field. The reports of the literature indicate that market share constitute the percentage or portions of total purchases of product or service by customers that goes to a given organization. In other word, market share can be considered as portion of sales of a product of an organization from the sales of this and all similar products of other organizations brought into the market. The reports of Chaudhuri and Holbrook [46] and Varadarajan [47] showed that it is an indicator of the success of a firm's efforts to compete in a product-marketplace. It is thus connected with sales of products and all benefits associated with organizational sales such as increased profit, revenue, returns on investment, research and development and other mix of activities targeted at improving organizational productivity and hence, growth. Hence, it is boosted by organizational marketing effort, which Farris et al. [8] mentioned to includes efforts toward advertising and promotion, the quality and price of food product being offered to the consumers, channel and customer relationships, and selling activities. Marketing strategies deployed by organizations to improve competitiveness and sales for organizational growth, are incidentally purveyors of market share. The report of Akinseye et al. [48] x-rayed virtues of marketing strategy and linked it with every effort designed to boost sales, organizational competitiveness, return on investment, revenues and consequential organizational growth and development. The consideration of market share by the literature as a measure of the consumers' preference for a food product over other similar products proves its link with organizational sales, the major source or kernel of organizational growth. And gaining the attentions of consumers or achieving their preference for organizational products is by provision and maintenance of quality products at affordable price, adding values to the product through a package designs and other distinguishing characteristics that give an edge of organizational product over other similar products from competing organizations. This assures increasing market share that is associated with improved profitability. Pursuing market share through productivity improvement effort, which minimizes wastes encourage steady improved profitability and other parameters of organizational growth and development.

7. Enhancing organizational market shares through food packaging

As afore-mentioned, market share is an important consideration in promoting growth and development of organization. The desire of organizations to remain in business is tied to processes that instigate market share and, food packaging is one of the most important operations that influence organizational market share. The position of packaging system in promoting market share of food products was elucidated in the report of Sarka and Aparna [17] which stated that packaging is not only a means of presenting a product but a way of inducing customers and influencing their decision to patronize the products. The manner in which packaging induces and or attract customers to patronize products, according to Selke [49] include the design of appetizing pictures on the package, use of distinctive brands and the transparency of packaging materials which allow them view the product inside.

The use containers usable by customers for other purposes after utilization of the product, is recognized in the literature as one of the ways packaging convinces consumers to buy food products. The review of the roles of food packaging in promoting productivity of food industry indicated that it promotes sales of food products and the volume of market shares that incidentally follows the sales.

In protecting the products, the package assists both producers and the middlemen in making sales associated with distributing the product to consumers in many locations; thereby promoting products/organizational market share. The report of

Himanshu and Bindu [3] agreed with the forgoing by stating that packaging performs many function like protecting the product from the point production through to the point of consumption. The protective function also enhances sales of the products and the market share by increasing the product's shelf-life to give producers and middlemen more time to sell the products as well as to market more of the product. By enhancing longevity of the products' shelf-life, packaging system encourages more sales which favors increase in the market share of the products.

Similarly, packaging not only reflect organizational ingenuity/creativity and innovative potentials but creates a new image to differentiate/distinguish the products from similar products in the market. This referred to in the literature as branding influences inclination of consumers toward products from already known sources, which possibly results in increasing sales of the products and the attendant market shares. The report of Himanshu and Bindu [3] indicated the potential of food packaging in promoting sales through many ways including self-supporting service, consumer affluence, company and brand recognition and innovation opportunity. Products that characterize new package design and new innovation tend to flow or align with dynamic change in the demands of consumers.

Also, the contribution of food packaging in increasing sales of food products and market share is indicated in the literature of how packaging entices consumers to purchase product. Hence, it is regarded as the 'silent salesman' by Judd et al. [50]. According to Himanshu and Bindu [3], the understanding that packaging as silent salesman, has led to its use as a marketing tool deployed to create convenience for consumers and promotional values for the producers and the middlemen. Its increasing significance in marketing, is the reason for its adoption as a major factor in gaining customers to boost sales and hence, market shares. The expectations of consumers could be met through well designed package and this boost their increasing interest in a product and the marketing of such product. Fill [43] alluded to the forgoing by reporting that packaging is an instrument through which buyers makes significant brand choice decisions.

Furthermore, packaging remains a major avenue to communicate important messages to consumers or potential customers at the point of purchase of a given product. It provides potential customers information on their expectations from the product. Underwood [51] reported that the marketing messages are made through structural and visual elements, including a combination of brand logo, colors, fonts, package materials, pictures, product descriptions, shapes and other elements that provide brand associations. The marketing messages are well designed to encourage sales and its aligned market share in a given marketing environment. Draskovic [4] mentioned some of the ways packaging passes marketing message to include packaging shape or design; packaging size; packaging color; packaging material; carried information; graphic applications (brand logo, fonts, images. Through this ways, packaging increases market shares for organizations through increased sales of products.

The attempt to ease the burden of utilization of food products for the consumers through packaging, has indirectly increasingly determined their interests in the products. This act considered as convenience function of packaging system by the literature, is enhancing sales of products and possibly assisting organizations to compete for market share using creative innovative package designs. Design architecture, material constituents, the communication of information potentials and other features that entice consumers, are according to Draskovic [4] ways packaging contribute to sales and hence, gaining of market shares. He further mentioned few of the packaging feature that allure consumers for increased sales to include resealing feature which enables consumers to use the product as and when needed and, good closure systems that simplifies utilization of products. According to the literature, convenience or the ease with which a product, is removed from the package and utilized, is one of the variables that dictates consumer behavior toward making purchasing choices, and hence, enhances marketing or sales of products [4]. A well designed package that makes it easier for more segment of population, the elderly, youth and children, to utilize the products also boost sales and guarantees the organizations increasing portions of market shares from the food products. The example of 'easier opening packaging system' is the 'tear strip' at the end of biscuit pack. While Bota and Petkovic [52] mentioned typical packaging system that has 'tear strip' opening system designed to enable people with limited motion and strength to utilize products with ease; Fain [53] reported similar packaging system that makes opening easier for people with arthritis. The literature alluded to the potential of convenience packaging to facilitate increasing interests of all segment of population, on products especially when such packaging system guarantees tamper proof. Usually, the more the segments of population are inclined to a given product, the more they purchase the products thereby giving opportunity to the manufacturer to have an edge over the competitors, not only on the sales but also in gaining market share. The report of the literature indicated that the strip should be durable enough to resist tendency to collapse during transportation and logistic runs, and to ensure a 100% opening success rate.

Similarly, the efforts of food producers to communicate with the consumers through packaging system have been observed to help in increasing sales of food products and the market share accruable to the products. This communication disposition of packaging system encompasses marketing strategies of organizations. The role of packaging in marketing is referred to as marketing function of packaging by Dascovik [4] while Fill [43] in referring to it as marketing communication; mentioned color, shape, package size and the inscribed information; as dimensions or elements that communicate to consumer. The communication dimension of packaging passes information of all packaging characteristics or features designed to draw the attentions of consumers. The successful passage of these information promotes sales and incidentally helps organizations in gaining market share. Food packages with alluring features designed into them, distinguish product on a crowded shelf, pass instant information to consumers to decide on their choice in few minutes, connect consumers emotionally; to boost sales and enable organizations have edges over their competitors. Hence, the overall efforts to design packages to perform communication function, are seemingly attempts to increase sales, enhance marketing and promote the organizational market share in the prevailing market. The use of features such as brand logo, colors, fonts, package materials, pictures, product descriptions, shapes and other elements that provide brand associations as communicators; distinguishes organizational products and boosts their sales to enable the organization gain market share.

Furthermore, food packaging can be considered as culmination of organizational creative and innovative efforts toward making products available to potential consumers or a purveyor of organizational efforts to gain market share.

The nature or design of packages for food products is, most of times, prodded by instinct to market the products. This is the reason the package is designed according to the expectations of the potential consumers or as a reflection of organizational marketing strategy. This assertion is agreement with definition of food packaging, by Coles [5], as 'a means of safely and cost effectively delivering products to the consumer in accordance with the marketing strategy of the organization'. The forging indicates that food packaging is an instrument of marketing of freshly or processed food products and a constituent of organizational strategies to outdo competitors to gain more in the market share. Thus packaging system, according to the literature, aligns with clearly defined marketing and manufacturing strategies that are consistent with the corporate strategy or mission of the business [5].

Also, it is unarguable that packaging system is an instrument of improving sales and heightening products cum organizational market share. The designing of packaging to cover multi-dimensional roles is an attempt to meet the dynamic demands of consumers and, a marketing strategy to increasingly gain market share. This assertion aligns with the report of Coles [5] indicating packaging as an important instrument the manufacturers of food products and or the retailers use to pursue competitive advantage. The relaunching of food products, most of the times, is carried out by redesigning of existing package in an attempt to meet the dynamic demands of potential consumers. Coles [5] also mentioned the forgoing as one of the essence of packaging and proceeded to add the idea of its use as instrument of adding value to the existing products to woo consumers. The marketing orientation of packaging is buttressed by the report of the literature that it stands as the face of product, exposing product and convincing consumers to purchase. Hence, distinctively and innovatively designed package according to Marsh and Bugusu [1], enhances and differentiate product from competing with similar products in the market to boost sales in competitive environment; and possibly increase organizational market share. Usually, designing packages to cater for the need and wants of the consumer, opening up new distribution channels, providing a better quality of presentation, enabling lower costs, increasing margins, enhancing product/brand differentiation, and improving the logistics service to customers, fulfill marketing demands of packaging for products [5]; which incidentally enhance the take of organizations from market share.

8. Novel packaging materials and their environmental consequences

Novel packaging system evolved to cater for food consumers' dynamic change in tastes of food products occasioned by change in life styles that increasingly inclined them to demand for mildly processed and convenience food products with enhanced shelf-life. Such packaging designed with innovation or application of new ideas and knowledge in packaging process/methods; according to Majid et al. [54]; 'act by prolonging the shelf life, enhancing or maintaining the quality, providing indication and to regulate freshness of food product'. The literature mentioned the types novel packaging and materials in involved to include active packaging, intelligent packaging, bioactive packaging, eco-friendly packaging, and others [40]. Apart from the reasons already adduced for evolution of novel packaging system, the literature mentioned solving the problems of environmental pollution/destructions associated with the traditional and conventional packaging systems as part of the reasons. For instance, Salgado et al. [55] reported the enormous environmental problems caused by increasing use of plastics in place of other materials in the conventional food packaging systems; because the plastics being mostly products of petroleum, are non-degradable. The need to protect the environment and promote economic values of food packaging resulted into researches and subsequent discovery of materials that are biodegradable and renewable and hence, are eco-friendly. This statement is in agreement with the observation of Cazon et al. [56] that 'biodegradable and renewable materials represent a great alternative to protect the environment and give economic values to underutilized products and industrial waste materials'. The use of bioplastics, defined as 'plastic materials that are either biobased (partly or entirely) or biodegradable or feature both properties' [55], predominates in novel packaging system. Whereas biobased indicates that the materials sourced from biomass, biodegradable shows that it can be biologically degraded to the constituent substances such as water, carbon dioxide, methane, basic elements, and biomass by living organisms that are available in the environment [57]. Salgado et al. [55] further elucidated on the classes bioplastics and mentioned three types which includes; (i) 'biobased but non-biodegradable plastics such as bio-polyethylene (Bio-PE), bio-polyamide (Bio-PA), bio-polyethylene terephthalate (Bio-PET), bio-polytrimethylene terephthalate (Bio-PTT), biopolyurethanes (Bio-PU), biopolypropylene (Bio-PP); (ii) plastics that are biodegradable and based on fossil resources, such as poly(butylene adipate-co-terephthalate) (PBAT), poly(butylene succinate-co-butylene adipate) (PBSA), polyvinyl alcohol (PVA), polyglycolic acid (PGA), polycaprolactone (PCL); and (iii) plastics that are both biobased and biodegradable'. Petersen et al., [57] further stated that the 'last group include: polymers directly extracted from biomass such as polysaccharides (e.g., starch, cellulose, chitin, etc.) and proteins (e.g., collagen, gelatin, casein, whey, soy protein, zein, wheat gluten, etc.); whose further modification can produce additional valuable biobased materials such as cellulose acetate, cellulose acetate butyrate, cellulose acetate propionate, cellulose nitrate, regenerated cellulose, carboxymethyl cellulose, lignocellulosic products, chitosan, etc.); polymers produced by chemical synthesis using renewable biobased monomers, such as polylactic acid (PLA), a bio-polyester polymerized from lactic acid monomers produced by fermentation of carbohydrate feedstock; and polymers produced by microorganisms or genetically modified bacteria such as polyhydroxyalkonoates (PHA) like polyhydroxybutyrate (PHB) and polyhydroxyvalerate (PHV) and bacterial cellulose'. Khoo et al. [58] compared environmental impacts of bioplastic packaging materials with those of the conventional plastics and stated that bioplastics are eco-friendlier. And Salgado et al. [55] threw more light on the forgoing by reporting that 'biobased plastics, which are generally drop-in products for their petroleum-based counterparts, are adequate for material recycling and/or energy recovery, whereas biodegradable plastics are intended for organic recycling'. Also, the report of European Bioplastics Organization cited by Salgado [55] indicated that bioplastics minimizes their impacts on environment through proper bio-wastes' management, separation and collection which eventually produces valuable compost manures usable in growing crops that initiate the life cycle.

9. Conclusion

Food packaging, an aspect of post-processing operation is aversion of food preservation system. It enhances the elongation of shelf-life of freshly harvested or processed food products, and ensures that the food products get to the consumers in right quality, in an acceptable nutritive value and sensory parameters. The nature of the preservative potential of food packaging system, which assist in meeting the dynamic change in demands of consumers, in the terms of having convenient foods with guaranteed safety standard, has made food packaging an influencer of consumer decision in the choice of food products. The forgoing is the reason organizations consider food packaging as a marketing strategic factor for presenting food products to the consumers in a manner that attracts their patronage, with a view to boosting the sales of the products. The developments in food packaging systems that have resulted in the emergence of active packaging, intelligent packaging, eco-friendly packaging and others have increasingly advanced food preservation. The elongation of shelf-life, nutritive values, sensory and safety/hygienic attributes of food products, by packaging system, creates marketing opportunities for producers and middlemen in trading the products, increasing retail business activities, which unarguably translates into more sales of food products and the gain in the products' market shares.

Author details

Thomas Amarachukwu Uzombah
Department of Food Technology, Federal Polytechnic, Nekede, Owerri, Nigeria

*Address all correspondence to: tomuzomba@yahoo.com

IntechOpen

© 2022 The Author(s). Licensee IntechOpen. This chapter is distributed under the terms of the Creative Commons Attribution License (http://creativecommons.org/licenses/by/3.0), which permits unrestricted use, distribution, and reproduction in any medium, provided the original work is properly cited.

References

[1] Marsh K, Bugusu B. Food packaging? Roles, materials, and environmental issues. Journal of Food Science. 2007;**72**(3):39-55

[2] Opara UL, Mditshwa A. A review on the role of packaging in securing food system: Adding value to food products and reducing losses and wastes. African Journal of Agricultural Research. 2013;**8**(22):2621-2630. DOI: 10.5897/AJAR2013.6931, Available from: http://www.academic journals.org

[3] Himanshu MM, Bindu G. The Role of Food Packaging, Researchgate. 2019. Available from: https://www.researchgate.net/publication/344124833 [Accessed: 19 May 2022]

[4] Draskovic N. Packaging convenience: Consumer packaging feature or marketing tool? The International Journal of Management Cases. 2016 Available from: https://www.researchgate.net/publication/263058254 [Accessed: 4 November 2021]

[5] Coles R. Introduction. In: Coles R, McDowell D, Kirwan MJ, editors. Food Packaging Technology. London, UK: Blackwell Publishing, CRC Press; 2003. pp. 1-31

[6] Shin J, Selke SEM. Food Packaging, 2014, In: Clark S, Jung S, Lamsal B, Food Processing: Principles and Applications. 2nd editors. Singapore: John Wiley and Sons. p: 249-273

[7] Edward GN. Market Share, Profitability and Implications for Strategy Formulation, Unpublished Thesis submitted to the Graduate School of Business Administration, University of the Witwatersrand, Johannesburg, South Africa in fulfilment of the Requirements for the Degree of Doctor of Philosophy

[8] Su Y, Yang B, Liu J, Sun B, Cao C, Zou X, et al. Prospects for replacement of some plastics with lignocellulose materials: A brief review. BioResources. 2018;**13**(2):1-27

[9] Farris PW, Neil T, Bendle PE, David JR. Marketing Metrics: The Definitive Guide to Measuring Mar Keting Performance. Upper Saddle River, NJ: Pearson Education; 2010. pp. 1-41

[10] Prokopov T, Tanchev S, Methods of Food Preservation. 2007 Available from: https://www.researchgate.net/publication/226177792 [Accessed: 10 August 2022]

[11] Ayoub A, Hameed F, Julie D. Food packaging technology and its emerging trends: A review. International Journal of Current Microbiology and Applied Sciences. 2018;*7*(10):3363-3378

[12] Hassan A, Hussain A, Jabeen F, Manzoor M, Nasreen Z. Advanced techniques for food preservation: A review. Journal of Food Safety and Hygiene. 2016;**2**(3-4):47-53

[13] Hameed F, Ayoub A, Gupta A. Novel food processing technologies: An overview. National Journal of Chemical Studies. 2018;**6**(6):770-776

[14] Pal M, Devrani M, Hadush M. Recent Development in Food Packaging Technologies, Beverage and Food world. 2019;**46**(1):21-22. Available from: https://www.researchgate.net/publication/330555330

[15] Evivie SE, Ogwu MC, Ebabhamiegbebho P, Abel E. Nigerial and nigerian food industry, challenges and

opportunities. In: Ogunlade C, Adeleke K, Oladejo M, editors. Food, Technology and Culture in Africa. Ibadan, Nigeria: Reamsworth Publishing Limited. 2021. p. 28-100. Available from: https://www.researchgate.net/publication/342732825, [Accessed: 30 July 2022]

[16] Oladepo OW, Taiwo KA, Ilori MO. Environmental policy implications of packaging waste disposal practices in Nigerian food sector. In: Evivie SE, Ogwu MC, Ebabhamiegbebho P, Abel E, editors. Nigerial and Nigerian Food Industry, Challenges and Opportunities. In: Ogunlade C, Adeleke K, Oladejo M, editors. Food, Technology and Culture in Africa. Ibadan, Nigeria: Reamsworth Publishing Limited. 2020. p. 28-100. Available from: https://www.researchgate.net/publication/342732825 [Accessed: 30 July 2022]

[17] Sarka S, Aparna K. Food Packaging and Storage. In: Prakash S, editors. Research Trends in Home Science and Extension. Vol. 3. New Delhi, India: AkiNik Publications; 2020. DOI: 10.22271/ed.book.959. Available from: https://www.researchgate.net/publication/345063737 [Accessed: 18 May 2022]

[18] Sen C, Das M. Trends in Food Packaging Technology. In: Meghawal M, Goyal MR, editors. Food Process Engineering: Emerging Trends in Research and Their Applications. 1st ed. New York: Apple Academic Press; 2016. Available from: https://www.researchgate.net/publication/313109461 [Accessed: 9 July 2022]

[19] Berger K, Welt B. A Brief History of Packaging. In: Sarka S, Aparna K, Prakash S, editors. Food Packaging and Storage. Research Trends in Home Science and Extension. Vol. 3. New Delhi, India: AkiNik Publications; 2005. Available from: http://edis.ifas.ufl.edu/pdffiles/AE/AE20600.pdf; [Accessed: 4 January 2009]

[20] Francis FJ. Wiley Encyclopedia of Food Science. 2nd ed. Vol. 4. New York: Wiley; 2000. p. 2306

[21] Buckow R, Bull M. Advanced Food Preservation Technologies, Microbiology Australia. 2013:108-111. DOI: 10.1071/MA13037. Available from: https//:wwww.researchgate.net>2578, [Accessed: 5 Iuly 2022]

[22] Adejumo BA, Ola FA. The appraisal of local food packaging materials in Nigeria. Continental Journal of Engineering Sciences. 2008;**3**:13-20

[23] Sarpong SO, Exploring the Concept of Standout in Structural Packaging-Challenges and Prospects in Ghana, Department of Communication Design in the Faculty of Art, Kwame Nkrumah University of Science and Technology. 2015, (Master of Communication Design thesis) Available from: http://ir.knust.edu.gh/bitstream/123456789/6907/3/SAMUEL%20OSEI%20SARPONG% 203.pdf Nigerian food sector, [Accessed: 5 July 2022]

[24] Grayhurst P, Girling PI. Packaging of Food in Glass Containers, In: Coles R, Kirwan MJ, editors. Food and Beverage Packaging Technology. Second Edition; Oxford, UK: Blackwell Publishing; 2011. pp. 137-156, DOI: 10.1002/97814444392180.ch6 [Accessed: 8 August 2022]

[25] Page B, Edward M, May N. Metal Packaging, In: Coles R, Kirwan MJ, editors. Food and Beverage Packaging Technology. Second Edition, Oxford, UK: Blackwell Publishing; 2011. pp. 137-156, DOI: 10.1002/97814444392180.ch5

[26] U.S. Food and Drug Administration. CFR—Code of Federal Regulations Title 21, Federal Register of the Government of USA; 2019. Available from: https://www.accessdata.fda.gov./scripts/cdrhcdocs/fr=211.132

[27] Rosette JL. Packaging Cosmetics and Pharmaceuticals, Encyclopedia of Chemical Technology, updated by Staff Published Online, John Wiley & Sons, 15 JUL 2005

[28] Kumbhar MS, Choudhary NH, Dighe DA, Singh MC. Tamper-evident pharmaceutical packaging—Needs and advances. International Journal of Pharmaceutical Sciences Review and Research. 2012;**13**(2):141-152

[29] Codex Alimentarius Commission. Report of the 20th session of the Codex committee on general principles. Joint FAO/WHO food standards Programme. In: Joint FAO/WHO Food Standards Programme. Paris, France; 2004; Rome, Italy: Viale delle Teme di Caracaalla, 00153

[30] Golan E, Krissoff B, Kuchler F, Calvin L, Nelson K, Price G. Traceability in the U.S. Food Supply: Economic Theory and Industry Studies. Agricultural Economic Report No. 830. Washington, DC: Economic Research Service, U.S. Dept. of Agriculture; 2004. p. 48

[31] Daoud MK, Trigui IT. Smart packaging: Consumer's perception and diagnostic of traceability information. In: Proc. 4th International Conference on Digital Economy, ICDEc. Springer. 2019;**358**:352-370. DOI: 10.1007/978-3-030-30874-2_28

[32] Alba JW, Hutchinson JW. Dimensions of consumer expertise. Journal of Consumer Research. 1987;**13**:411-454

[33] Stanisławek MD, Miarka D, Kowalska H, Kowalska J. Traceability to ensure food safety and consumer protection as typified by case studies of three meat processing plants. South African Journal of Animal Science. 2021;**51**(2):241-249, Available from: http://www.sasas.co.za [Accessed: 22 June 2022]

[34] Ahmed I, Lin H, Zou L, Li Z, Brody AL, Qazi IM, et al. An overview of smart packaging technologies for monitoring safety and quality of meat and meat products. Packaging Technology and Science. 2018;**31**(7):449-471

[35] Ghaani M, Cozzolino CA, Castelli G, Farris S. An overview of the intelligent packaging technologies in the food sector. Trends in Food Science and Technology. 2016;**51**:1-11. DOI: 10.1016/j.tifs.2016.02.008

[36] Biji KB, Ravishankar CN, Mohan CO, Gopal TKS. Smart packaging systems for food applications: A review. Journal of Food Science and Technology. 2015;**52**:6125-6135. DOI: 10.1007/s13197-015-1766-7

[37] Fuertes G, Soto I, Carrasco R, Vargas M, Sabattin J, Lagos C. Intelligent packaging systems: Sensors and nanosensors to monitor food quality and safety. Journal of Sensors. 2016:4046061. DOI: 10.1155/2016/4046061

[38] Ganguly S. Recent technological advancements in food packaging: A review. World Journal of Biology and medical Sciences. 2014;**1**(3):21-23

[39] Tanja IR, Ivan SR, Branka MP, Aleksandra RN. Antimicrobial nanomaterials for food packaging applications. Food and Feed Research. 2016;**43**(2):119-126

[40] Abdullahi N. Advances in food packaging technology: A review. Journal of Postharvest Technology. 2018;**06**(4):55-64

[41] Wells LE, Farley H, Armstrong GA. The importance of packaging design for own label food brands. International Journal of Retail & Distribution Management. 2007;**35**(9):677-690

[42] Smith PR, Taylor J. Marketing communication: An integrated approach

2004. In: Draskovic N, editor. Packaging Convenience: Consumer Packaging Feature or Marketing Tool?. 2016;**12**(2):267-275 Available from: https://www.researchgate.net/publication/263058254. [Accessed: 4 November 2020]

[43] Fill C. Marketing Communications: Interactivity, Communities and Content. 5th ed. Harlow: FT, Prentice Hall, Pearson Education; 2009 Available from: https://www.worldcat.org.title.oclc

[44] Löfgren M. Winning at the first and second moments of truth: An exploratory study. Managing Service Quality. 2005;**15**(1):102-115

[45] Löfgren M, Witell L, Gustafsson A. Customer satisfaction in the first and second moments of truth. Journal of Product & Brand Management. 2008;**17**(7):463-474

[46] Chaudhuri A, Morris BH. The chain of effects from brand trust and brand affect to brand performance: The role of brand loyalty. Journal of Marketing. 2001;**65**(2):81-93

[47] Varadarajan R. Customer information resources advantage, marketing strategy and business performance: A market resources based view. Industrial MarketingManagement. 2020;**89**:89-97

[48] Akinseye EM, Onimole SO, Ekundayo OA, Adebusoye AB. Impact of marketing strategies on organizational growth: A study of selected industries in Lagos State Nigeria. International Journal of Business and Economics Research. 2022;**11**(1):1-7. DOI: 10.11648/j.ijber.20221101.11

[49] Selke S. Green packaging. In: Boye J, Arcand Y, editors. Green Technologies in Food Production and Processing. New York: Springer; 2012. pp. 443-470

[50] Judd D, Aalders B, Melis T. The Silent Salesman: Primer on Design, Production and Marketing of Finished Package Goods. Singapore: Octogram Design PTE Limited; 1989. p. 26

[51] Underwood RL. The communicative power of product packaging: Creating brand identity via lived and mediated experience. Journal of Marketing Theory and Practice. 2003;**11**(1):62-76

[52] Bota J, Petkovic G. Evaluation of zipper tear strip design structure for paperboard packaging, proceedings of 10th international symposium on graphic engineering and design, In: Conference Paper. 2020. Available from: https://www.researchgate.net/publication/346261194. DOI: 10.24867/GRID-2020-p32

[53] Fain B. Food Packaging Design Accessibility Guidelines, Arthritis Australia's Accessible Design Division. Available from: https://arthritisaustralia.com.au/wordpress/wp-content/uploads/2018/01/Food-Packaging-Design-

[54] Majid I, Nayik GA, Dar SM, Nanda V. Novel food packaging technologies: Innovations and future prospects. Journal of Saudi Society and Agricultural Sciences. 2018;**17**:454-462

[55] Salgado PR, Di Giorgio L, Musso YS, Mauri AN. Recent development in smart food packaging focused biobased and biodegradable polymers. Frontier Sustainable Food System. 2021;**5**(630393):1-30

[56] Cazon P, Velazquez G, Ramirez JA, Vasquez M. Polysaccharide based films and coatings for food packaging: A review. Food Hydrocall. *2017;68:136-148*

[57] Petersen K, Væggemose Nielsen P, Bertelsen G, Lawther M, Olsen MB, Nilsson NH, et al. Potential of biobased materials for food packaging. Trends in Food Science & Technology.

1999;**10**:52-68. DOI: 10.1016/S0924-2244(99)00019-9

[58] Khoo HH, Tan RBH, Chng KWL. Environmental impacts of conventional plastic and bio-based carrier bags part 1: Life cycle production. International Journal of Life Cycle Assessment. 2010;**15**:284-293

Chapter 13

Application of Edible Packaging in Dairy and Food Industry

Bhushan Devidas Meshram, Vaibhav Kisanrao Lule, Shivani Vyawahare and Rekha Rani

Abstract

There are number of food packaging materials such as glass, paper and cardboard, metals and plastic are available. However, the plastic is a mostly used non-biodegradable packaging material which causes environmental pollution. To overcome these problems, the biodegradable/edible food packaging is currently into focus for use. Edible packaging can be used in film as well as coating form. The materials are used for preparation of edible packaging varies in their function according to their sources. Some examples of edible film (packaging) are starch-based, collagen-based, zein-based, gluten-based, etc. Additives are added during the formation of film to enhance their positive role for packed food. Each additive has their unique role when combined with film material. These types of films have various functions, which would help to increase shelf life of food by acting barrier between food and external environment. The main advantage of edible packaging over synthetic packaging is that this may be safely eaten as a part of food product and thus, may reduce packaging waste and pollution. Edible film is physically and nutritionally better that the synthetic food packaging. Edible film used in food packaging should be passed by FDA as GRAS, then it can be used in food packaging. Edible packaging has several applications in dairy, food, confectionary, meat and also in pharmaceutical industry.

Keywords: classification, manufacturing and coating technology, functional properties, applications

1. Introduction

Packaging is the science, art and technology of enclosing or protecting products for distribution, storage, sale and use. There are several packaging materials present in the market such as plastic, paper, cardboard, PET and also new technologies such as Active, intelligent vacuum, aseptic packaging, etc. Edible packaging is one of them. According to Food Production Daily, a new type of edible food packaging that does not affect the environment and that can be eaten with food that is inside the package has been invented. Harvard Professor and biomedical engineer, David Edwards, developed edible packaging, also called 'future of food packaging' Zoe [1]. Edible packaging is categorised into films, coating, pouches and sheets. The edible coating (EC) is prepared directly on the food, whereas edible films (EF) and layers having

IntechOpen

thickness 10 mm or less than 254 μm and more than 254 μm, respectively, are separately prepared, and then food is packed in it, in the pouches form or placed between the food layers [2, 3]. As packaging material contains various additives (flavourings, colourings, sweeteners) [4]. Lipid, carbohydrates, protein (casein), tomato skin are used to make edible packaging [5]. In the recent year, use of edible film and coating is emphasised due to its function of food protection from negative environmental effect and also helps to increase shelf life of food. Materials from which EF and EC are made are decided by their functions. Edible packaging materials have exceptional properties such as barrier and mechanical properties, enhance sensory properties and optical properties, making them attractive alternative for food packaging [2]. Based on the type of food and storage conditions, components of edible packaging are selected [6]. Currently, edible packaging gives replacement to other packaging because it inhibits loss of gas, aroma and moisture of the packed food. Krochta and De [3] added nanofillers to film and coating for improved quality and acceptability. Various methods are used to manufacture film. Casting method is generally used. By comparing EF with synthetic packaging, EF has benefits over it because EF is eaten with contained food as a component of same food, and if not eaten, it has biodegradability [4, 7, 8]. EF satisfies industry requirements by keeping quality of food, fulfils consumer desires and reduces environmental pollution [9].

2. Need of edible packaging

Food packaging is very crucial for various purposes such as labelling, protecting the contained food, caring for the food, measuring, attracting the consumers, etc. Mostly, plastic is used as food packaging material. Globally, factories produce 400 MT of plastic per year approximately [10]. About 12.7 metric tons of plastic waste enters the ocean every year which affects the life of marine lives ([11] https://www.condorferries.co.uk/plastic-in-the-ocean-statistics). From total plastic waste about 79% is dumped in land, 12% incinerated and 9% is recycled. This plastic waste can take up to 400 years to break down in the landfill harming animal life, polluting cities or devastating landfills [12]. Dumped plastic affects (i.e. reduces) the moisture and oxygen transfer rate of soil and deteriorates the quality of the land. **Figure 1** shows state-wise plastic waste generation of India. It seems that Maharashtra state produces plastic waste in higher percentage. To overwhelm the harmful environmental effects as well as health effects of plastic and improvement of food, many companies are trying to replace excess plastic packaging with such packaging material which will be degraded, which does not produce much waste and does not have any negative effect on human health, food as well as an environment like edible packaging materials which do not produce packaging waste, and if may consumer throw the edible packaging, then it is degraded which required the same period as its contained food.

(https://cpcb.nic.in/uploads/plasticwaste/Annual_Report_2019-20_PWM.pdf)

2.1 Impact of edible packaging on environment

Some packaging material has exclusive use, and that may be thrown away after it's used rather than recycled or reused [13]. Such plastic waste goes into rather in landfill, incinerate or in ocean which polluted the soil (land), air or water, respectively. As consumer demanded for the plastic-free food due to its negative health effect, companies are affording to get solution on it. The best way to contend the plastic in food

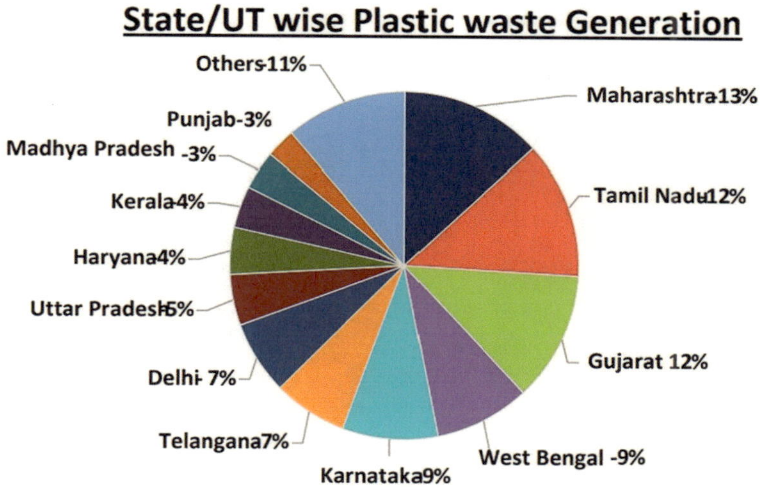

Figure 1.
State-wise plastic waste generation. Source: https://www.business-standard.com/article/markets/plastic-ban-maharashtra-gujarat-among-top-indian-plastic-waste-producers-122070400736_1.html.

packaging and also feasible to get zero plastic waste from food industry is by encouraging the customer to eat food along with its wrapper [14]. But this is only possible when that wrapper is GRAS for eating. Edible packaging is the finest solution to the plastic for reducing the waste which affects the environment and health of consumer. Nowadays, companies produce edible packaging in the form of pouches, wrappers, sachets, containers, plates, etc.

Commonly, edible packaging can be eaten, and in some amount, it is not eaten by consumers, but there is no need to reject the waste or no need to recycle it because edible packaging has biodegradability as it is composed of edible and environmentally safe components which biodegrade in less period of time [15]. The demand of edible packaging could rise by 6.9% yearly till the 2024, and market worth could be almost $2 billion, informed by global research firm TMR (Transparency Market Research) [14].

2.2 History

Edible coating and films have been used for centuries to protect foods. The history of packaging is given in **Table 1**, the idea is derived from the natural protective coating on some foods such as the skin of fruits and vegetables [16]. In the twelfth and thirteenth centuries, China used a wax coating to decrease water loss, on lemon and oranges. In 1856, the first cellulose EF was developed, and in 1907, phenol-formaldehyde (Bakelite) resin was used. This was the starting point of a series of developments and inventions giving birth to a great range of packaging materials that nowadays are employed [17].

2.3 Classification of edible films and coating

Edible film and coating are classified based on their raw material sources, i.e. protein, polysaccharide, lipids and composites. They possess various functions such as retarding loss of moisture and volatile compounds, acting as a barrier for fat and

Year	Country	Findings
Twelfth century	China	Waxes were applied to oranges and lemons
Fifteenth century	Japan	Yuba (1st free-standing edible film) from soymilk
Sixteenth century	England	Food products were coated with lard (control moisture loss)
1930s	US	Hot-melt paraffin waxes were have been used to coat citrus fruits
1950	—	Carnauba wax and oil–water emulsion—for coating fresh fruits and veggies
Twentieth century	—	Casing for sausages and chocolate coating for nuts and fruits
Source: Robertson [16].		

Table 1.
History of packaging.

oil and having a high selective gas permeability ratio of CO_2/O_2 as compared with conventional packaging [18].

2.3.1 Protein-based edible films

Technological and functional properties of protein may improve by changing the structure of protein for preparing edible film and coating [19]. Protein-based edible films have additional stimulating mechanical and barrier properties than polysaccharides [20]. They have the capacity to slab the gases due to its structure in which hydrogen-bonded network is packed tightly [21]. Food quality is declined mostly due to oxidation of lipid which can be preserved by using protein-based edible film having ability to inhibit oxygen permeation [22]. Many protein materials have been tried: collagen, corn zein, wheat gluten, soy protein isolates, fish proteins, ovalbumin, whey protein isolate, casein, etc [23]. In addition to their nutritional value, milk protein such as casein has several key physical characteristics for active performance in edible films such as emulsification and water solubility [24]. Sesame seed protein was mostly used in manufacturing of edible film [25]. Protein-based edible films transfer various additives such as plasticisers, antioxidants, essential oil, antimicrobial agents, etc. The diffusion of additives components from the surface to the interior is controlled by keeping the film on the food surface [15]. Nuts, cashew nuts and beans require special packaging which is fulfilled by the protein-based film [26].

2.3.1.1 Collagen

Collagen is a protein obtained from animal which is rich in glycine, proline and hydroxyproline and hydrophilic in nature [27]. Among all protein-based edible film, collagen is most commercial and successfully used film. For production of homogeneous surface film, high concentration of hydrolysed collagen is used [28]. Collagen powder and fibres were revealed to be suitable for the production of the bio-composite film in which fibres act as fillers and boost the effect [29].

2.3.1.2 Gelatin

Partial acid or alkali hydrolysis of collagen is used to produce gelatin at high temperature in the presence of water. Due to unique properties of gelatin, it is widely used in cosmetics, food, pharmaceuticals, industry [30]. Edible film obtained from gelatin has impermeability to CO_2 and flexibility due to its random arrangement of polypeptide chain in water solution [31]. Also gelatin possesses antioxidant property. Antimicrobial activity of gelatin is studied by Gomez-Guillen et al. [32]. But still, the relation between antimicrobial activity and peptide characteristics of gelatin is not revealed. Edible packaging materials act as carriers for additives, gelatin is one the carriers which carries bioactive components [33]. Gelatin has advantages such as having ability of good film forming, low in price, non-toxicity, biodegradability and gas and oil resistance property but has poor thermal stability, water resistance and mechanical property [34]. Gomez-Guillen et al. [32] stated that formation of an active packaging and extending the functional properties of biodegradable films were only possible due to usage of natural antioxidants and antimicrobial compounds. Gelatin-based edible films have poor water vapour barrier property which can be overcome by adding surfactant, i.e. lecithin, to it [35].

2.3.1.3 Casein

Casein is a milk protein. It has coil-like structure [36] due to its structure, it can be processed easily. Casein is dissolved in water, but after being dipped in water, it gains about 50% weight [37]. By treating the aqueous solution of casein molecule, it forms films, which are flexible, tasteless and transparent [31, 33]. Casein film helps to retard migration of aroma, CO_2 and O_2, because casein contains more number of polar groups which have excellent adhering property [38]. By treating the casein film with buffer at its isoelectric point may enhance the mechanical property and reduce solubility of casein film [39]. Its drawback is its high price.

2.3.1.4 Gluten

Gluten has cohesiveness and elasticity which help to improve mechanical property of film [40]. Wheat gluten chiefly contains glutenins and gliadins, insoluble and soluble in aqueous alcohol, respectively [41]. Edible film made from wheat gluten has water barrier property [42], films are transparent, homogeneous and strong in nature [43]. By using high-pressure treatment to gluten film, the texture can be changed from smooth to rubber-like texture [44].

2.3.1.5 Zein

Zein is a maize protein which is hydrophobic in nature. It is by-product of the oil industry and bio-ethanol [45]. It can be used in manufacturing of edible film, coating and pouches [22]. It is mostly used in confectionery industry as a coating material [46]. It consists of protein which is alcohol-soluble [47]. Zein film is insoluble in water due to more number of non-polar amino acids in it, and it also helps to enhance its water vapour barrier property [48]. Concentration of alcohol in alcoholic zein solution may changes its physico-chemical properties which affect the film properties.

For producing functional film from zein, antioxidant and antimicrobial compounds are added to zein coating and films [49]. By treating the zein film forming solution with gamma irradiation helps to enhance its appearance, water barrier property and colour [50]. Zein also combines with polysaccharide and other protein-based film such as glucomannan to improve its property [51]. Zein has ability to assemble by itself due to its hydrophobic and hydrophilic nature [52].

2.3.2 Polysaccharides based films

Polysaccharides such as starch, cellulose, chitin and chitosan, etc. are used for manufacturing of EF and EC. These ingredients are selected according to their suitability for mechanical strength, functional properties, etc. [19]. Polysaccharides are long chain of polymers which are made up of the repeating units of mono or disaccharides which are joined together by bonds called glycosidic bonds. In film characteristics and formation of it, H-bond has important role [15]. Polysaccharide films are made by coacervation process. During the coacervation process, the interaction of long-chain polymer gets changed, and this leads to forming new intermolecular hydrophilic and H-bonding after evaporating the solvent [2]. Polysaccharide coating has barrier properties such as O_2, oil, and aroma, but it has poor moisture barrier property [19]. And it also has good structural and strength ability. Use of polysaccharide edible film helps to delay ripening of fruit and vegetables and hence, increases its shelf life too [53].

2.3.2.1 Starch and derivatives

Starch is chiefly present in tubers, roots and seed of plants. Maize, wheat, potato, etc. are mainly used in the industry as starch source [54]. Starch produces semi-crystalline and amorphous layer due to formation of linear amylose and amylopectine (branched) by glucose polymerisation process [55]. Amylase is a polymer having linear structure which affects the amorphous nature of starch granules [54]. The concentration of amylase and amylopectine in starch is 20–30% and 70–80%, respectively [56]. Amylopectine and amylase are naturally present in granular form having size about 1–100 μm [54]. Mechanical property, barrier property of starch are dependent on it sources, concentration of amylase and amylopectine [57]. The application of starch-based films in food packaging is capable because of their environmental appeal, low cost, flexibility and transparency [58, 59]. Corn starch is a high amylose starch which is good source for formation of edible starch-based film. Film can be made from aqueous solution of starch and drying it for free-standing film [60]. Starch-based edible films help to prevent change in taste, appearance and flavour of contained food because film is tasteless, transparent and odourless in nature [57]. Starch film is highly hydrophilic in nature; due to this hydrophobicity, this film has poor barrier property for water. Main advantage of starch film is that it has excellent barrier property of CO_2 and O_2 [61].

2.3.2.2 Cellulose and derivatives

Cellulose is a common natural polymer present in nature. Repeating units of D-glucose linked by β-1,4, glycosidic bonds formed cellulose [62]. Cellulose has crystalline structure and tight packing of polymer chain which helps to resist salvation in aqueous media of edible film [60]. Due to the presence of intermolecular hydrogen bond cellulose having insolubility in water to overcome it, cellulose treated with alkali helps to make it soluble in water [56]. Etherification of cellulose results in the formation of water-soluble ethers: methyl cellulose (MC), carboxy methyl cellulose (CMC), hydroxyl propyl methyl cellulose (HPMC) and hydroxyl propyl cellulose (HPC), these form good-quality edible film [63]. Edible film based on cellulose derivatives has various characteristics/properties such as resistant to CO_2 and O_2, transparent, odourless, flexible, water soluble, etc. The hydrophobic:hydrophilic ratio of these film components decided its WVP [31, 64]. Šuput et al. [61] stated that MC is less hydrophilic in nature, hence has water barrier property. Gas and barrier property of cellulose is mainly dependent on its molecular weight [3]. Cellulose-based film acts as anti-rancidic agent due to the presence of large surface area and bipolar in structure [65].

2.3.2.3 Chitin and chitosan

Chitin is a biopolymer found in abundance in nature after cellulose [66]. It is found in cell wall of fungi, exoskeleton of crustaceans and many other biological materials [67]. Chitin contains poly (β-(1–4)-2-acetamido-D-glucose), and chitosan is obtained in the presence of alkalin from N-deacetylation [68]. Chitosan has anti-microbial property, which is very effective over yeast and mould, followed by Gram-positive and Gram-negative bacteria [15]. In mechanism of chitosan's antimicrobial action, it results in harmful leakage of microbial proteinaceouse and intercellular components when positive charge chitosan and negative charge microbial cell membrane are interacted. Chitosan helps to prevent toxic production and microbial growth in product because it contains nearly amount of chelate trace metals [69]. From shrimp shells, chitin is obtained about 30%. In the process of obtaining chitin, shells are treated with NaoH at 85–110°C, and after obtaining it, for removing the $CaCO_3$, weak HCl solution was used at room temperature having about 1–10% concentration [70]. The physicochemical properties of Chitosan is dependent on the method used to obtained chitosan and used apparatuses [56]. It has been extensively used in films and coatings due to its ability to inhibit bacterial and fungal pathogens' growth [71]. Santos et al. [72] stated that the antimicrobial property of chitosan film is enhanced by the availability of fatty acids in it. Chitosan forms coating which is semi-permeable in nature, which helps in delaying the ripening rate of fruits and vegetables by modifying the internal atmosphere. For obtaining tough, flexible and clear film, chitosan is made from aqueous solution [73, 74].

2.3.2.4 Alginate

Alginate is a sodium salt of alginic acid which has the ability to form film [70]. Alginate polysaccharide is chiefly extracted from seaweed which is in brown colour [75]. Divalent cations such as Mg, Mn, Ca, Fe are used for making alginate coating material which possess gelling character [76]. D-mannuronic acid (M) and L-gulurinic acid (G) are present in alginate having different proportion, arrangement [77], and it affects its physical property [78]. If the ratio of M/G is less than 1, it is means alginate contained higher amount of guluronic acid and hence formed strong bond while, if the ration of M/G is more than 1, it shows less amount of guluronic acid in alginate which may result in flexible structure [79]. The alginic acid obtained from different species of brown algar may contain different polymers such as mannumaric acid unit as main or guluronic acid as a main or both/partial mannuronic and gulurinic acid [80]. Alginate has colloidal nature with stabilising, thickening and suspending property which makes it suitable for film forming [81, 82]. Alginate-based edible film or coating has poor moisture barrier due to its hydrophilic nature [76], and also it has some desirable properties such as reduction in shrinkage and barrier for colour and odour [83]. Alginate forms coating/film by evaporation of solvent method which is done with or without gelation of it. For retarding dehydration and for protecting the oxidation of lipid in meat product, alginate coating/film is used [84]. For preparing the alginate, pieces of seaweed are dipped in sodium carbonate. After 2 h. slurry of sodium alginate was obtained which has glycellulose (undissolved part of seaweed). Obtained slurry was diluted with water and filtered through filter cloth, and then pressing is carried out. Then precipitation is done to obtain calcium alginate [81, 82].

2.3.3 Lipid-based film

From antique time, lipid was used as an edible coating for fruits. Mainly waxes were used for coating the citrus fruits [19]. The factors on which efficiency of lipid material are decided for formation of edible film or coating are its structure, nature of lipid used, its hydrophilic nature, its state and its interaction with other components of films, etc. [85]. To increase resistance of water penetration of film, lipids are generally combined with polysaccharides or proteins as multilayer coating [86]. Natural wax, surfactants and acetylated mono-glycerides are examples of lipid compounds which act as protective coating for fruits and vegetables [60]. Some examples of lipids, waxes and resins used for making edible film and coating are sunflower oil, cocoa butter, palm oil, etc.; paraffin, carnauba wax, candelilla wax, etc.; and tragacanth gum, gum arabic, etc., respectively [19].

2.3.3.1 Waxes and paraffin

The non-polar substances are produced by naturally as well as synthetically called as wax [61]. Waxes are soluble in organic solvent and insoluble in bulk water because waxes are hydrophobic in nature [87]. Wax micro-emulsion is manufactured by adding the water to molten wax in the presence of base and the fatty acids; this method is called as water-to-wax method. Present base plays the role of inverting the emulsion to wax-in-water [88]. Waxes are applied in thin layer on fruits. Thin layer on fruits is considered as edible. but if that layer is thick like on the cheese, it should be removed before eating. For getting humidity barrier, mostly used waxes are paraffin, bee wax and carnauba wax [60].

2.3.3.2 Shellac resins

The secretion of insect Lacciferalacca is called shellac resin which is the complex mixture of polymer that is aliphatic alicyclic hydroxyl acid. Shellac resin in not safe for eating, it is not permitted by GRAS. This is used only for adhesion and coating. And it is soluble in alkaline as well as alcohol solution. Hernandez [89] reported that the maximum use of resin is done in pharmaceutical industry than the food industry as a coating material. For coating citrus and other fruits, resin and its derivatives are used mainly as they have good barrier property which maintains their quality. But from this, coating gases are passed because coating has different gas permeability [90]. The internal environment of coated fruit with shellac and wood resin and wax coating are different, i.e. has high CO_2 and low O_2 and high in ethanol content and has low ethanol content, respectively [91–93]. Coating based on wood and shellac resin helps to enhance the prevalence of post-harvest pitting [94, 95].

2.3.4 Composite films

The film which is made from combinations of edible substances to make it stronger than before to overcome its drawbacks is called composite film [3]. The main motto of manufacturing composite film is utilisation of synergistic changes in film to overcome its individual lacks in some properties [19]. For making composite film, two different polymers are combined, such as lipid and protein; carbohydrates and protein; carbohydrates and lipid or/and synthetic as well as natural polymer. These composite coatings are applied in the form of solution, successive layer of film or coating, an emulsion, etc. [60].

2.3.4.1 Polysaccharide-protein edible coatings

The protein-polysaccharide-based composite film has good mechanical and water-vapour permeability property because interaction between protein-polysaccharide results in mono-phase film [18]. This composite film is proposed to be used as edible film for packaging in food industry [96]. Combination of carboxy methyl cellulose with soy protein improves the structure and properties of edible film [97]. By decreasing gas transfer rate in fruits, shelf life is increased; this is possible by using edible film made from collagen-galactomannan mixture [98].

2.3.4.2 Lipid-protein edible coatings

Protein has large number of polar group and is hydrophilic in nature. Due to its hydrophilicity, it has poor water vapour barrier, although it has good mechanical and oxygen property. Lipid is hydrophobic in nature. By combining these two makes film with stronger and good barrier properties as well as organoleptic property enhances and increases its market value [99]. Combined film also improves its water solubility [100]. To improve optical, mechanical and barrier properties of sodium caseinate–oleic acid–beeswax film, calcium caseinate is added to it [101].

2.4 Film/coating additives

Various material/substances are added to edible film to improve its properties such as mechanical, handling and structural and/or for improving active functions to coating or films [53]. For examples:

2.4.1 Plasticiser

For improving mechanical property of edible film, plasticisers are added to film solution during formation/manufacturing of film. Plasticiser has small molecular weight and is hydrophilic in nature. These small molecules are situated between their polymeric networks and make them stronger [2]. Commonly used plasticisers in edible packaging are polyols (sorbitol, glycerol), mono-, di- or oligosaccharides (glucose, sucrose), lipid and its derivatives (fatty acids, surfactants). Generally, the selection of plasticisers requires considering plasticiser's compatibility, efficiency, permanence and economics [102].

2.4.2 Antimicrobials

Antimicrobial compounds are additives used to control biological deterioration and to inhibit the growth of microorganisms, including pathogenic microorganisms. Antimicrobial agent can be incorporated directly into the food during manufacturing or may be incorporated into food packaging materials [30]. There are several groups of antimicrobial compounds potentially incorporated into edible films, including chemical agents, natural extracts and probiotics [103]. Janjarasskul and Krochta [2] stated that for controlling growth of microorganisms, both natural and synthetic antimicrobial agents are added in edible film, and this is used as an alternative for it.

2.4.3 Natural extracts

2.4.3.1 Plant/spice extract

Different plant/spices, such as seeds, roots, bark, buds, flowers and leaves, are used to create the extracts. The phenolic chemicals, such as catechin, tannin, ferulic acid, caffeic acid, gallic acid and carvacrol, which are found in various portions of plants and spices, are mainly responsible for their antimicrobial properties [103]. Extraction of essential oil from plants (cinnamon, clove, onion, garlic, radish, etc.) comprises phenolic compounds such as phenolic acid and flavonoids which have biological activities such as antimicrobial and antioxidants [104].

2.4.3.2 Enzyme

The most employed antimicrobial enzyme is lysozyme which is made up of hydrophilic monopeptide chains [30]. Lysozyme is a nutraceutical and is produced from egg white, milk and blood [105]. It is shown to be more effective against Gram-positive bacteria. It separates *N*-acetylmuramic acid and *N*-acetylglucosamine bond of the peptidoglycan in the cell wall of bacteria [106]. Gram-negative bacteria have lipid-based outer cover to their cell walls; due to this, lysozyme is less effective on it [15]. By hydrolysing the peptidoglycan, lysozyme causes bacterial death by destroying the cell wall of bacteria [30].

2.4.3.3 Bacteriocins

Bacteriocins are macromolecules which contain protein and produced from the different varieties of bacteria and have different mode of action, chemical property. Bacteriocins are naturally occurring antimicrobial substances. They are small-molecular-weight peptides produced by microorganisms and effectively inhibit the growth of food spoilage bacteria, mainly Gram-positive bacteria [105]. The most employed antimicrobial bacteria are nisin and pediosin [103]. Antimicrobial efficiency of bacteriocins is influenced by their concentrations and number and species of microorganisms, using condition, interaction or inactivation by food elements and temperature and pH of the product [105].

2.4.3.4 Probiotics

Live bacteria known as probiotics can improve health when ingested in adequate quantities [107]. After 12 days of storage testing, Bekhit et al. [108] found that the film of hydroxyl propyl methylcellulose (HPMC) contains microencapsulation of *Lactococcuslactis subsp. Lactis* was successful in reducing the growth of *Listeria monocytogenes* by a five-log cycle when compared with control film. According to Beristain-Bauza et al. [109], whey protein isolate films containing *Lactobacillus rhamnosus* cell-free supernatant (12 or 18 mg/ml in film-forming solution) maintain inhibitory action against Gram-positive (*Salmonella typhimurium* and *E. coli*) and Gram-negative (*E. coli*) bacteria (*L. monocytogenes and S. aureus*). Probiotics can be incorporated into the edible polymer matrix to be used in the food packaging industry because of their safety, operative properties and useful qualities.

2.4.3.5 Emulsifiers

Emulsifier has both polarity and non-polarity; due to this, it acts as a surface active agent which has the ability to mix the two immiscible substances such as oil and water, by changing or modifying interfacial energy of these immiscible substances [2]. Emulsifiers are very important because they help to achieve proper and sufficient wetability to product which is essential for proper surface area and adhesion to the wrapping material [53]. Many proteins have emulsifying properties owing to their amphiphilic nature [2].

2.4.3.6 Chemical agents, organic acids and salts

Organic acids and their salts are mostly used as chemical antimicrobial agents for food products due to their efficacy and cost [110]. They are produced by chemical synthesis or chemical modification of natural acids [111]. The most widely used organic acids in film packaging are acetic acid, lactic acid, sorbic acid and citric acid. Films containing organic acids have been developed as a consequence of numerous scientific studies. For instance, Uranga et al. [112] reported that 20% (w/w) citric-acid-contained gelatin/chitosan films decreased *Escherichia coli* in liquid culture. Furthermore, Rocha et al. [113] established films made of anchovy protein that are antifungal and contain 1.50% (w/v) sorbic acid or benzoic acid.

3. Edible film manufacturing and coating application method

The edible film is commonly wrapped around a food product surface as a solid matrix and can act as primary packaging deprived of any sensory or nutritional appeal. Edible film can be utilised in the shape of pouches or sachets specifically for energy drinks and meal replacement shakes [8]. Aqueous solutions are transformed into edible films during the coating operation using specialised equipment. Only two basic converting methods – casting on steel belt conveyors and casting on a disposable substrate (such as release paper) on a coating line – are commonly employed to make edible films. However, a wide range of technologies are available for making thin coatings and films [114].

3.1 Casting

Casting (**Figure 1**) is a manufacturing process by which a liquid material is usually poured into a mould and then allowed to solidify [115]. The most popular method for forming films, known as solvent casting, is typically used at laboratory scales. Three fundamental steps are involved in casting a biopolymer-based movie: the biopolymer is dissolved in a suitable solvent, the solution is cast in a mould and the casted solution is dried [116]. To dissolve soy protein isolate polymer, the chosen polymer is dissolved or dispersed in an appropriate solvent (ethanol) [117]; this process is known as solubilisation. The resulting solution is poured into a glass plate with Teflon coating or a predetermined mould. The drying procedure gives the solvent enough time to evaporate, resulting in a polymer coating that adheres to the mould. For the casting of films to facilitate solvent removal and easy peeling of the film, air driers such as hot air ovens, tray dryers, microwaves and vacuum driers are used [118]. To improve the intramolecular interaction between the polymer chains and achieve a proper microstructure for the film, the air-drying process for casting edible film is crucial [119]. Quick-drying casting techniques have had a negative impact on the physical and structural qualities of the film [120]. The primary benefit of the casting method of film formation is its low cost and ease of production without the need for specialist equipment [121]. However, the film formation in solvent casting depends on the solubility of the polymer rather than melting [122].

In the casting process, following are the key drawbacks: (i) Limiting the forms (the simplest forms that frequently arise are simple sheets and tubes). (ii) Possibility of harmful solvent being trapped inside the polymer. (iii) Denaturation of proteins and other molecules that are incorporated into polymers using solvents [121]; vacuum drying of films can be used to remove the hazardous solvent [123]. (iv) A cap on the quantity of films that can be made [124]. (v) Films with various features can be created when evaporation levels and temperatures are variable [125]. (vi) Casting requires long drying time, which is not possible for commercial production [117].

3.2 Steel belt conveyors

Wherein the solutions are cast or spread uniformly on a continuous steel, then the moisture is removed by passing through a drying chamber. After being removed from the steel belt, the dry film is wound into mill rolls for subsequent processing. Additionally, the steel belt revolves around sizable drums that are located at either end of the line. With the aid of traditional coating equipment, the solution is evenly applied at one end of the line before being dried in a chamber. To avoid sticking or blocking, a thin, secondary dusting of starch powder may be applied to the dried film as it leaves the drying chamber. It is also decorated or marked with edible inks; given a variety of other treatments; and then taken from the steel belt and wound into enormous master rolls. In general, the steel belt conveyor lines (**Figure 2**) are 50–100 feet long when measured from the centre of the two drums. The widths of steel belts range from 20 to 60 inches. One of the highly desired characteristics of steel belt conveyors is the ability to directly cast aqueous solutions on the belt surface. This reduces the cost of a separate carrier web-like polyester film or coated paper while increasing uniformity, heat transfer and drying efficiency. However, some coating formulas could adhere to the steel belt too firmly. The coated substrate is then stripped from the belt and wound into a master roll [114].

3.3 Extrusion method

Extrusion is alternative technique used for producing polymeric films, and a pictorial view is given in **Figure 3**. This technique is preferred over a casting technique because it requires less energy and takes less time to remove water for making film [128].

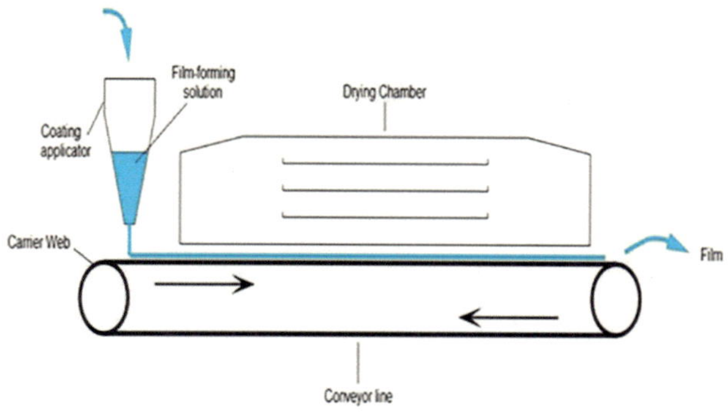

Figure 2.
Steel belt conveyor. Source: Gamboni et al. [126].

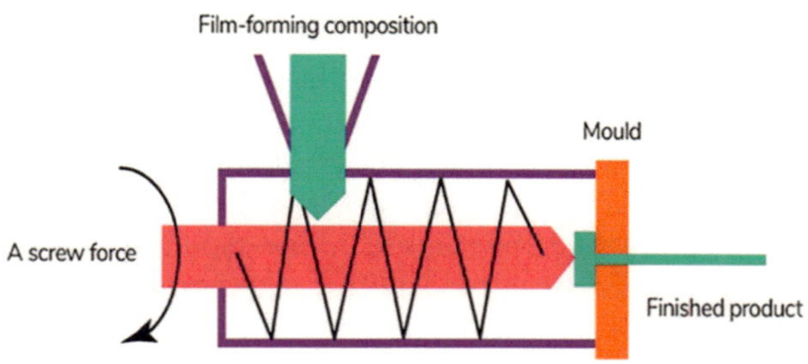

Figure 3.
Extrusion machine for polymeric films. Source: Suhag et al. [127].

It is one of the foremost polymer processing techniques currently in use at a commercial scale [129]. The extrusion process, commonly, is divided into three zones: (i) the feeding zone, (ii) the kneading zone and (iii) the heating zone at the final part/exit from the machine [130]. This technique best works with a minimum content of water or solvents; therefore, it is also called a dry process. However, to increase film flexibility, plasticisers are needed [131]. The mechanical (specific mechanical energy) and thermal energies (extruder barrel temperature) are involved in this method to yield an extruder-based edible film [132]. Certain parameters such as moisture content of the feed, screw speed, temperature of the barrel, the diameter of die, pressure at the die, energy input, etc. are critical for the extrusion process to influence the final products [127]. If high temperature is produced during the process, it affects the sensorial and nutritional properties of biopolymer of edible film which restrict its usage for high temperature with low moisture content FFS [133]. Co-extrusion is a method that can be used to create multi-layer films and gives flexibility in determining the required film properties. In addition to enhancing the produced film's functionality and processing capabilities, the multilayer also benefits from the inventive structure of the multilayer film [134]. For preparing the edible film from pectin/starch blends added with pasticiser and glycerol, Fishman [128] used extrusion method in which extruder has nine heating zones and two screws. Liu [135] give optimal parameters/conditions for preparation of pectin film that must be 225 rpm and temperature should be 125°C in third zone and 110°C in fourth zone. These conditions were fixed on the basis of film's physical and mechanical properties such as elongation, puncture strength, colour, thickness, etc. [136]. A short processing time with low energy consumption, better mechanical and optical qualities, such as elongation and transparency of edible film, are the key benefits of extrusion film formation over casting technique [129, 137]. It is a high-performance, inexpensive and efficient method utilised in the commercial food manufacturing industry [138, 139].

3.4 Electrospinning

3.4.1 Principle

The process is chiefly carried out in three phases such as (1) jet initiation, (2) elongation and (3) solidification of solution (Masoud Aman [140]). An electrospinning (**Figure 4**) having components such as syringe or capillary tube which transfer solution to which high voltage is given by using high-voltage battery followed by producing the nanofiber and solution is collected with the help of collector [142].

A widely used technique for processing biopolymer-based film-forming solutions worldwide is electrospinning [143]. It is an economical method that can create thin films that could increase a material's solubility and improve an application [144, 145]. In the electrohydrodynamic process used in electrospinning, a liquid droplet is electrified to create a jet, which is then stretched and lengthened to create fibres [146, 147]. For producing thin film, strong electric field is applied with small size orifice to spinneret (usually, a hypodermic needle with blunt tip). The diameter size of electrospun polymer fibre made by this technique generally is from 10 to 1000 nm and hence performs better electrical, mechanical and thermal properties than the synthetic packaging ([148] BG book). This electrospun fibre acts as an adhesion; hence, this can be used to combine two layers of biopolymer which increases the barrier properties of EF [149]. Electrospining technique helps to tightly hold two layers which maintain the thickness of film, and due to nanometric size of fibre, it enhances the mechanical property without affecting the optical property of the film [150]. The modified spinning process has plate die in place of spinneret which can produce flat film from casein [151]. Direct current (DC) or alternating current (AC) power supplies are both acceptable (AC). Surface tension causes the liquid to protrude from the spinneret and form a pendent droplet during electrospinning. When a droplet is electrified, the electrostatic attraction between surface charges with the same sign causes them to repel one another, transforming the droplet into a Taylor cone from

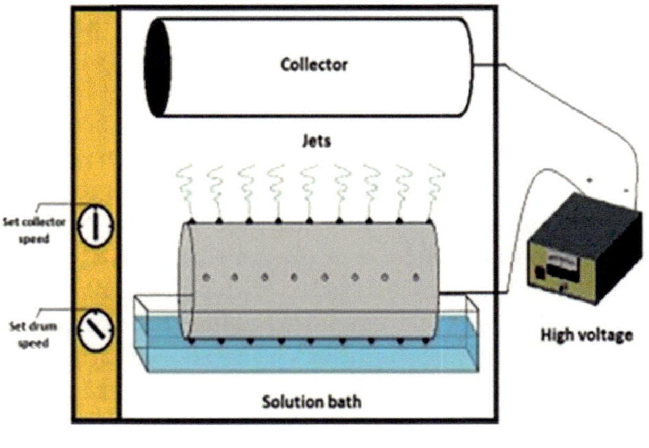

Figure 4.
Electrospinning machine for making biopolymer-based film. Source: Ebrahimi et al. [141].

which a charged jet is released. Due to bending instabilities, the jet first extends in a straight line before undergoing ferocious whipping motions. The jet is then compressed to smaller diameters, where it quickly hardens and deposits solid fibres on the grounded collector [152, 153]. Cui et al. [154] produced the nanofibers based on chitosan, gelatin and clove oil by electrospinning. The experiment of wrapping cheese by the obtained nanofiber for inhibiting microbial contamination proved that electrospinning was a creative strategy to fabricate nanofibers applied onto food [155]. Ebrahimi et al. [141] introduced a nozzle-less electrospinning device as a favourable technology to produce food-grade nanofibers on large scale with a high production yield in comparison to a nozzle-based instrument.

3.5 Thermoplastic method

The thermoplastic method uses shear pressures, high temperatures and little water to continuously shape the materials, allowing for industrial-scale large-scale manufacture of films [156]. It has been claimed that this technique can expand applications and get beyond the drawbacks of conventional techniques such as casting [157]. For chitosan films and gelatin films, the thermoplastic method – also known as the 'dry process' – involves extrusion strategy [158], blown approach [159], compression moulding [160] and the aforementioned combination [138]. The thermoplastic method, which is typically used for synthetic polymers, bio-plastics made from proteins, polysaccharides and other biopolymers, can be carried out as a continuous unit operation using control of temperature, size, shape and moisture [161]. For the most part, thermoplastic processing is an effective way to create chitosan or gelatin films for commercial uses in a wide variety of food products. However, there is no information available regarding thermoplastically manufactured chitosan-gelatin composite films [155].

4. Coating methods/technology

There are various methods used for the application of edible coting on the surface. For successful coating application food's surface property plays an important role. It helps to give an overall equal thickness of coating material. Typical methods for forming a coating include panning, fluidised-bed processing, spray coating and dipping [114].

The coating is done in four steps:

I. Application of edible coating material or solution on the food surface using various coating methods.

II. Adhesion of edible coating material on the food surface.

III. The formation of a film on the surface of food is called "Coalescence'.

IV. Stabilisation of coating layer on food surface by heating, cooling, drying.

4.1 Enrobing

In enrobing, the product to be coated is dipped in an edible coating solution or molten lipid (**Figure 5**). After a particular period, the fresh and frozen product

Figure 5.
Enrobing technique. Source: https://biztaskplus.com/chocolate-enrobing-how-it-works-using-a-chocolate-enrober/.

produces a bland taste, moisture loss takes place and more oil absorption during frying, and to overcome this problem, coating with edible batter may improve palatability and flavour of it. This method is largely used in the chocolate industry. The principle behind using enrobing in the chocolate industry is to cover the centre of confectionery with tempered chocolate [114].

4.2 Pan coating

Pan coating or panning is used to apply thick or thick layers. Mainly this method is used for the hard and spherical products (**Figure 6**). It is conducted in the batch process, not in a continuous process [114]. It involved a perforated stainless steel pan in which coating solution is spread over the product with the help of a spray gun. The speed of the pan depends on the size of the product.

There are three types of panning methods used:

- Hard panning – In this method, the hard coating of sugar syrup is applied by repeating the coating process. Applied sugar syrup coat is dried to form crystals on the surface.

- Soft panning – In this method, a soft and thick layer is formed by applying the mixture of corn syrup and sugar to the product, and it is dried by applying dry sugar on it.

- Chocolate panning – A fat-based coating is applied around the centre product. This coating could be of chocolate, white/coloured confectionery, etc. [163].

4.3 Drum coating

Drum coating is one of the methods through which a thin or thick layer can be applied to the hard or solid product (nut). **Figure 7** illustrates the drum coating machine, it is run continuously. It is used in oiling and salting of nuts which enhance the palatability and flavour and helps to delay oxidation. In chocolate, this method of coating helps to delay the absorption of moisture.

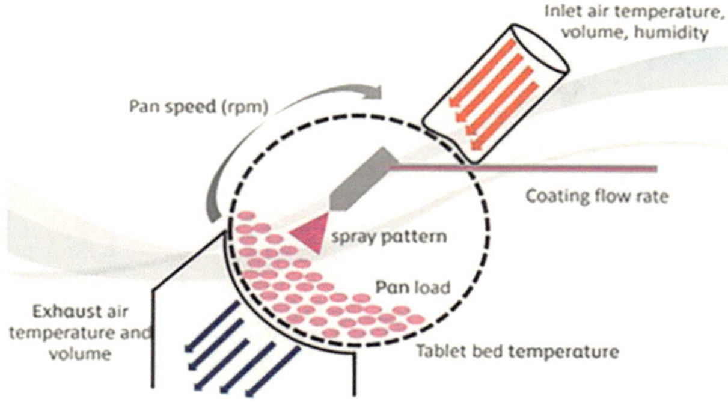

Figure 6.
Pan coating technique. Source: Agrawal and Pandey [162].

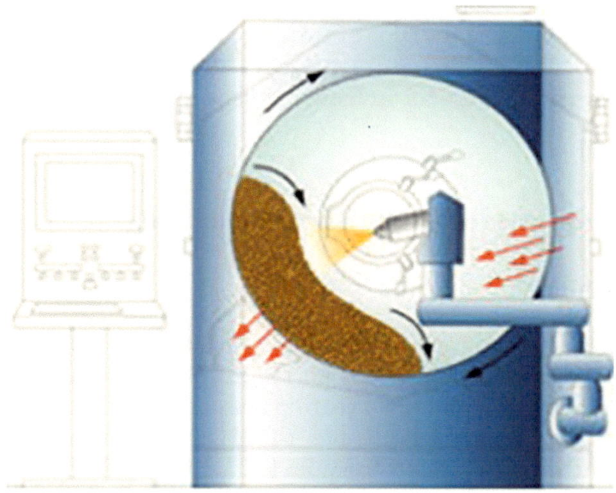

Figure 7.
Drum coating machine. Source: Suhag et al. [127].

4.4 Screw coating and fluidised-bed coating

Screw coating is used to apply thin layer coating material on solid as well as firm food (**Figure 8**). It is a continuous process. To improve anti-caking of cheese, this coating method is used. For applying very thin layer of edible coating on dry particles having smaller size and very low density, this coating method is used mostly. Coating the powder by this method prevents agglomeration and enhances the dispersion and solubility of coating material [114].

4.5 Spray coating

Coating by spraying method helps to reduce in waste of coating solution by means of spraying system [80]. Spray coating is mainly preferred for that food which has large surface area [48]. The spray coating method is used mostly in combination with another method such as pan coating. But we can use it alone too. With the help of the spray coating method possible to spray molten chocolate, molten waxes, an aqueous

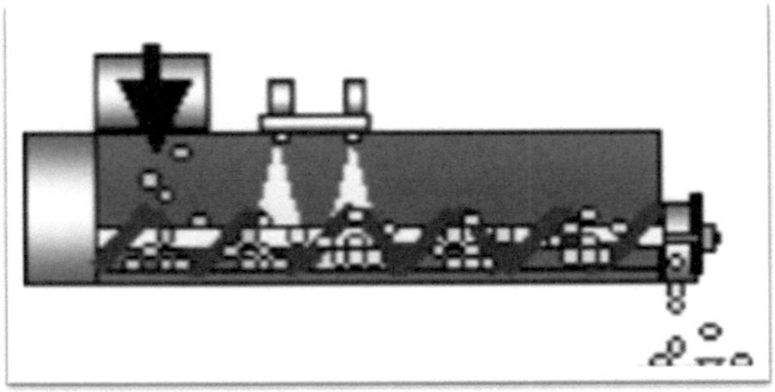

Figure 8.
Screw coating. Source: Gamboni et al. [126].

solution or molten lipid in both thin as well as thick layers. For spraying, high-pressure nozzles are used but also the type of product which is coated. Spraying may be affected by fluid viscosity, temperature, pressure, the shape of the nozzle [114]. The apparatus for spray coating of wax on fruit and vegetables is patented by Cuning and Caulkins [164].

5. Functional properties of edible films and coating

5.1 Mechanical property

The edible bio-based film can resist normal external force created during handling and must have the proper mechanical property [155]. Mechanical property means film should be most effective in case of tearing resistance, stiffness, tensile strength, puncture resistance, bursting strength, etc. [61, 87]. Adequate mechanical strength ensures the integrity of a film and its freedom from minor defects, such as a pin hole, which ruin the barrier property. Coatings can also lessen damages to foods during handling and transportation. Sometimes, edible coatings and films may be used to change handling properties of materials. For example, edible films can encapsulate liquids or powders to improve handling. Simple protein films are stronger and more stretchable than composite films containing lipids. Lipid content was linearly associated to strength reduction. A WPC protein emulsion film with lipid of higher melting point exhibited better mechanical strength than that with lipid of a lower melting point [165]. Polysaccharide has lesser mechanical property than protein-based film. In the mechanical properties of the edible film, essential oil plays a vital role. Plasticiser helps to decrease tensile strength and increase the elongation property of edible film [112].

5.2 Colour property

Colour property is mostly used for selecting proper food packaging material [51]. Chitosan has its natural colour; due to this, chitosan edible film has higher greenness and yellowness and has lower lightness than commercial films [166]. Due to the presence of polyphenol in extract of chitosan-gelatin, the composite film became yellow [167]. The colour of the developed edible film can change the overall appearance of the food inside the packaging films. The edible film not only improves sensory quality but also affects the food quality due to that comes in contact with outer light.

Therefore, the colour property requires proper and careful design to avoid this and for higher consumer acceptance [155].

5.3 Barrier property

One of the utmost important functions of edible packaging is barrier property. It cuts the contact of food with the external environment [168]. It includes oxygen barrier property, water barrier and light barrier property [169]. These properties are closely related to oxidation, microbial entrance and growth and spoilage in the food.

5.3.1 Gas barrier property

Oxygen and carbon dioxide affect the respiration of postharvest fruit and vegetables and also speed up the oxidation of lipid containing food [170]. Gas permeation and WVP have same mass transfer principle [171]. Oxygen barrier property is efficiently measured by parameter, i.e. oxygen permeability or rate of oxygen transmission through it. These are inversely proportional to each other [172]. By comparing chitosan and gelatin film, the chitosan-based film has good oxygen property than the gelatin-based film [173]. Although the gelatin-based film has a poor oxygen barrier, chitosan-gelatin-based composite edible film has excellent oxygen barrier properties. One of the main benefits of whey protein films over polysaccharide films is lower gas permeability. Gas barrier properties of milk protein films are also better than those of several commonly used in edible, synthetic films. This feature can be used to form integrated packages of milk protein films and synthetic polymer films [174]. Hydrocolloid films provide good oxygen barrier properties in the absence of moisture. Gelatin films can be used to cover candy and dry foods, microencapsulate flavours, and keep frozen meats from rotting. Moisture improves oxygen permeability by greatly improving the transferability of the macromolecule chains. Composite films composed of carboxymethyl cellulose and fatty acid sucroesters have adequate oxygen barrier properties while yet being somewhat CO_2 permeable [175]. On cooled bananas, this type of coating reduced CO_2 exchange by only around half while reducing oxygen transfer by five times. This result is influenced by the kind, variety and temperature of the fruit.

5.3.2 Water vapour permeability (WVP)/barrier property

For preserving food and increasing the shelf life of food, water vapour barrier property is considered mostly and importantly too [176]. The edible film should have low WVP because those films have low WVP and are mainly ideal for food that contains more moisture; it helps to avoid moisture transmission between food and environment which results in a better quality of food [171]. WVP can be reduced by adding an antimicrobial agent to it. Chitosan films show higher water vapour permeability properties than gelatin-based films [170]. The chitosan-gelatin composite film possesses hydrophilic nature, and it forms a compact network; due to this, it has higher water vapour barrier properties [170]. Whey protein films effectively bound the water vapour condensation in fruit and vegetable packaging, thus restricting microbial spoilage. For same reasons, the RH and plasticiser type considerably affect the moisture permeability properties of protein films [177]. Combining linear and globular proteins results in a decrease in WVP [178]. WVP was lowered by gelatin and defatted soy proteins from 8.45 to 5.55 g mm kPa-1 h^{-1} m^{-2} [178]. Plastisisers have an impact on

WVP as well; it has been noted that utilising glycerol instead of PEG-400 and sorbitol results in greater WVP values. Sorbitol will provide the film with higher moisture barrier qualities because it is anticipated that edible films will have lower WVP [25].

5.3.3 Thermal property

The thermal property of edible film is related the resistance of that film to temperature which helps to protect food when stored in different temperatures. Thermo-mechanical property of film can be determined by scanning the calorimetry and thermo-gravimetric method [179, 180].

5.3.4 Antimicrobial property

The antimicrobial property of edible film is improved by adding the antimicrobial compound to it. The edible film also has carrier property so it carries antimicrobial agents, which play their role upon food when comes in contact with it. Antimicrobial activity of edible film containing antimicrobial agent is very effective against fungi and bacteria [171]. In chitosan-based edible film, mainly curcumin, apple peel polyphenols, etc are present [181, 182]. The Agar disc diffusion method is used to measure the antimicrobial activity of antimicrobial agent added edible film. Previously microorganisms inoculated film cuts are placed on agar plate and incubated in suitable conditions, and then the inhibition zone is observed around the disc films [183].

5.3.5 Moisture barrier properties

Films with appropriate moisture barrier properties are required for a great number of applications. Many lipid compounds, such as animal and vegetable fats, aceto-glycerides, surfactants and waxes [184], have been used in the formation of edible films and coatings because of their excellent moisture barrier properties. Waxy coatings on fresh fruit and vegetables thus reduce weight loss due to dehydration during storage by 40–75% [185]. A multicomponent film was established by Guilbert in 1986 [186] and is made of casein or gelatin, carnauba wax, glycerol monopalmitate and monostearate. This film demonstrated good water vapour barrier qualities when applied as an emulsion and subsequently acidified with lactic acid after drying. Krochta et al. [184, 187] also reported that the composite films of casein and aceto-glycerides or wax placed as an emulsion.

5.4 Edibility and biodegradability

The film is made from components that are totally edible, and if that is not eaten by consumers, also it has biodegradability property because it contains (environmentally safe) constituents or ingredients [15].

5.5 Carrier properties

Antioxidants, antimicrobial, flavouring compounds, pigments and nutrients are added to the film during the blending process of raw material. In such cases, a functional group of the film is bonded with these additives and makes the film stronger and has good carrier property [168].

6. Applications of edible film and coating in the dairy and food industry

6.1 Dairy industry

6.1.1 Paneer

Cinnamon essential oil–added edible film has excellent antioxidant as well as antimicrobial activity against spoilage and pathogenic microorganisms. Paneer packed in alginate-calcium edible film increases the quality of the panner during storage than the control sample of panner. It helps to increase its shelf life from 5–6 days to 13 days [188].

6.1.2 Cheese

The literature study suggests that cheese packaging is one of the potential application areas for edible packaging given that antimicrobial film in cheese has been found to have a significant impact on its shelf life. According to Fajardo et al. [189], saloio cheese's storage stability was increased by employing chitosan-based film as a natamycin carrier; they also discovered that the product was stable for 7 days in ambient storage settings. Cheddar cheese was covered in whey protein films, and they discovered that the coatings preserved the products' features and sensory attributes normally [190].

The goal of an edible whey-protein-based coating packed with antimicrobial agents is an efficient approach to stop the development of harmful germs and increase in shelf life of the product. Ricotta cheese coated with chitosan/whey protein edible film and stored at 4°C for 30 days exhibited a significant reduction in mesophilic and psychrotrophic counts compared with control [191]. Henriques et al. [192] stated that edible whey protein concentrate coatings prepared by heat denaturation or UV irradiation and merged with antimicrobials (lactic acid and natamycin) seem to be a potential alternative for commercial coatings in ripened cheese.

For prolonging the shelf life of low-fat cut cheese, it is coated by edible material added with oregano essential oil, which acts as antimicrobial as a mandarin fibre, which is based on nano-emulsion by retarding the growth of microorganisms during storage of low-fat cut cheese [193]. Zein-based blend coating helps to lower the weight loss by 30% and also helps to avoid microbial contamination of mode cheese or short ripening period cheese for more than 50 days as normal unpacked cheese sample gets contaminated after 21 days [194]. Shelf life of Mongolian cheese coated with water chestnut starch-chitosan film is determined at 8°C. This film contains perilla oil which acts as an antimicrobial agent and also helps to reduce weight loss [195]. The active coating of mozzarella cheese helps to retard the growth of microorganisms, and also it improves its sensory quality of it. It includes the addition of potassium sorbate (PS), sodium benzoate (SB), calcium lactate (CL) and calcium ascorbate (CA). Potassium sorbate gives the best effect than others. It increases its shelf life too [196].

6.1.3 Ice-cream holding edible cone

Ice-cream cone is mainly made up of flour and sugar, and it contains acetylated monoglycerides and has good moisture barrier property, which holds the ice-cream for longer as well as it is edible too. The chocolate coating of cone is used as a barrier which maintained the crispiness of that cone [197].

6.2 Food industry

6.2.1 Edible packaging of red chilli from mango kernel starch (MKS)

Red chilli is packed in mango kernel starch (MKS) as a sample and commercially PE packed red chilli as control. By comparing this for 6 months at 40°C, MKS packed chilli powder gives better results in pungency and colour. MKS film is produced by casting method which contains glycerol and sorbitol in a 1:1 ratio. This MKS film also helps to increase the shelf life of red chilli powder [198].

6.2.2 Edible coatings for fresh fruits and vegetables

The film is formed by adding fruit purees which interact with biopolymer and active compound of film material, which helps to increase shelf life of perishable foods [70]. Rangel-Marrón et al. [199] studied that papaya puree is added to the edible coating, increases the shelf life of minimally processed pumpkin, carrots, papaya, etc. For enhancing the shelf life of cut mango, composite coating is made from mango puree, gaur gum, sesame protein and calcium chloride. This composite coating aids to decrease the degradation of ascorbic acid, carotenoids and phenolic content lever of fruit [25]. In fresh-cut fruits and vegetables, protein-based edible coatings can act as moisture or gas barriers, which may reduce moisture loss and/or reduce oxygen intake from the environment and later reduce the respiration rate [200]. Edible whey protein coatings in apple and potato slices act as oxygen barriers and delay browning reactions [201]. Tien et al. [202] reported that whey protein isolate seems to possess superior antioxidant capacity than calcium caseinate. Whey protein concentrate and bees-wax-based coatings with 1% ascorbic acid or 0.5% cysteine are revealed to be the most effective means of preventing browning in apples [203]. Whey protein isolate coatings on freeze-dried strawberry pieces in milk revealed substantial reduction of rehydration ratio, resolving the problems of 'rapid rehydration velocity' and 'loss of freeze-dried strawberry texture' [204]. By reducing moisture loss and preventing microbial growth, edible coatings made with whey protein/pectin in the presence of TGase significantly delay the onset of spoilage in fresh-cut apples, carrots and potatoes for up to 10 days. However, their antioxidant properties, hardiness and chewiness were unaffected. According to Ochoa et al. [205], edible layers made of natural were derived from Euphorbia antisyphilitica with the potent antioxidant 0.01% elegiac acid which can significantly increase the quality and shelf life of exquisite golden apples (EA). Lower oxidative and hydrolytic rancidity and better sensory quality were found in walnuts and pine nuts covered with a homogenised coating solution of whey protein isolate and carnauba wax [86].

6.2.3 Egg and meat-based products

According to Caner [206], the whey protein film significantly extended the fresh egg quality's shelf life when kept at room temperature. Eggs covered with whey protein lost weight while kept at room temperature for around 4 weeks, but control eggs gained weight by 5.66%. (uncoated eggs). Additionally, coated eggs showed better albumen quality and a lower pH than control eggs. Even after 4 weeks, the coated eggs' yolk index values of 0.26–0.9 showed good quality.

Whey protein films significantly substantiated the decline in lipid oxidation and inhibition of the growth of spoilage and pathogenic microorganisms in meat products [207]. Shon and Chin [208] informed that the whey protein packaging mixed with natural antioxidant extracts revealed the reduction in the moisture loss and showed lower thiobarbituric acid-reactive substances and peroxide values (PV) in sausage and cooked meatballs when stored for 8 weeks at 4°C. Fernandez-Pan et al. examined the effects of a number of essential oils from oregano, clove, coriander, laurel, mastic thyme, rosemary, sage and tea tree on WPI films used for coating fresh skinless chicken breast (2012). The findings demonstrate that films containing essential oils reduced *Listeria innocua*, *Pseudomonas fragi*, *Staphylococcus aureus* and *Staphylococcus enteritidis*, with oregano oil demonstrating the strongest suppression. Similar to this, turkey meat's shelf life was increased by Ferulago Angulate Essential Oil (FAEO) (0.05%) added to a gelatin-chitosan-based film (Naseri et al.).

Fresh meat has such biological composition that makes it perishable. Fresh meat contains 12–20% protein, 0–6% carbohydrates and 3–45% fat and muscle tissue has 42–80% water, approximately 75.5% [209]. As fresh meat has much amount of moisture in it, this condition is susceptible for microbial growth. To avoid such condition, packing of fresh meat in edible film has antimicrobial activity, which retards the growth of microorganisms as well as enhances the flavour of meat [171].

6.2.4 Edible coating for deep fat-frying products

Mashed potato balls were coated with corn zein, hydroxyl propyl methyl cellulose (HPMC) or methyl cellulose (MC) film-forming solution and uncoated considered as control. By comparing the control sample with a coated sample of potato balls, in coated ball observed that there is reduction in loss of moisture 14.9%, 21.9% and 31.1% in CZ, HPMC and MC coated ball, respectively. Also reduction in fat uptake by the ball is observed, 59.0%, 61.4% and 83.6% in CZ, HPMC and MC respectively. MC has the most effective barrier properties [210].

6.2.5 Seafoods

The majority of seafood, which includes fish and fish products, has a short shelf life due to the rapid proliferation of germs, which can endanger the health of consumers and cause a financial loss. Castro et al. [211] demonstrated the effectiveness of a film made of whey protein concentrate and green tea extract when applied on fresh salmon. They discovered that this combination successfully postponed the oxidation of the salmon's lipids until 14 days after storage. Red sea bream's shelf life was dramatically increased after ginger essential oil, fish sarcoplasmic protein and chitosan were combined with the fish. These three ingredients significantly decreased oxidation and protected against microbial decay [212]. The majority of the fat in salmon is easily digestible unsaturated fatty acids.

6.2.6 Nuts

When combined with high heat treatment during the roasting process, peanuts' high oil and unsaturated fatty acid content makes them highly susceptible to oxidative rancidity. The main cause of oxidative rancidity in roasted peanuts is autoxidation. Depending on the oxygen levels in storage, lipid oxidation is the most frequent reason for peanuts to degrade [26]. In the presence or absence of vitamin, native and heat-denatured WPI coating postponed oxidation and increased the shelf life of peanut to 31 weeks at 40, 50 and 60°C [213]. According to Maté et al. [214], whey protein isolate

(WPI) film or coating created with ascorbic acid (AA-WPI) considerably slowed down the oxidation of lipids in peanuts kept at 23, 35 and 50°C. All of the aforementioned temperatures saw the AA-WPI covered.

7. Advantages and disadvantages of edible film and coating over commercial synthetic packaging

7.1 Advantages

- Edibility and biodegradability film and coating can be consumed directly with contained product, which does not produce any waste for decomposition, and if some consumer dose not eaten film with food, that film can be thrown and has no negative effect on the environment because of its biodegradability. It has the same biodegradability as it contains food.
- The edible film has various additives such as flavouring, colouring and sweetening which enhance the organoleptic properties of packaged food.
- Edible film and coating can be used for individual packaging of small size food such as peas, beans, nuts, etc. which is not possible in synthetic packaging.
- Incorporated between food layers – edible film can be incorporated into two-layers of food which helps to reduce deterioration of inter-components of food such as pizza, pies, etc.
- Edible film and coating have various functional properties such as antimicrobial, colour, mechanical, etc. which make it superior to synthetic packaging.
- The edible film can be used for encapsulation of flavouring and leavening agent to control its addition to a particular product [215].
- Has less gas, aroma and WV permeability [16].
- Level of carbon dioxide inside the food decreases.
- Helps to reduce emission of greenhouse gas level [30].

7.2 Disadvantages

- More expensive than synthetic packaging – the edible film has various additives and manufacturing methods which make it more expensive than synthetic packaging.
- Required secondary packaging – in some cases like an ice-cream cone, it holds the ice-cream but for packaging, it required secondary paper packaging over it.
- Unable to use in unsanitary condition – edible film can be eaten directly with food but in unsanitary condition, consumers are unable to eat such film directly [16].

- Biopolymers contain some particle of metals.
- Has low mechanical and physical property.
- Pollution in ocean cannot be fully resolved by it [30].

8. Comparison between edible films and synthetic film

Table 2 shows the comparison of WVP of edible film with synthetic packaging (PP, LDPE, etc.). Puncture strength of trilayered edible film from sodium alginate emulsion, gelatin emulsion and whey protein isolates emulsion (SAOGOWPIO) is seven times higher than that of synthetic packaging, and also edible film has more resistance to oxygen permeability than PS and PET [216]. **Table 3** shows the mechanical properties of different types of synthetic as well as edible packagings, from that we can say that in case of some mechanical properties, edible film is deprived than synthetic packaging, but it was acceptable [216].

Properties				
Packaging material	**Elongation (%)**	**Tensile strength (MPa)**	**Young's Modulus (MPa)**	**References**
Synthetic packaging				
PET	136.94	—	—	[216]
LDPE	500	9–17	—	[217]
HDPE	380	20.3	911	[218]
Edible film				
Chitosan	4.60	74.0	2451	[219]
Whey potein isolates	10.08	5.34	—	[220]
Soya protein isolates	3.95	1.93	1.19	[221]
Quinoa starch	58.14	7.56	4.59	[222]

Table 2.
Comparison of mechanical properties of edible film with synthetic packaging.

Properties		
Packaging material	**WVP [10^{-11} g.(m.s.Pa)$^{-1}$]**	**References**
Synthetic film		
Aluminium	0.0005	[223]
HDPE	0.002	[224]
LDPE	0.014	[225]
PP	0.010	[225]
PVC	0.041	[225]
Protein film		
Wheat gluten	12.97	[226]

Properties		
Packaging material	**WVP [10^{-11} g.(m.s.Pa)$^{-1}$]**	**References**
Corn zein	11.6	[227]
Soya	281.18	[228]
Lipid film		
Waxes	0.03–1.0	[224]
Carauba wax	0.114	[225]
Paraffin	0.023	[224]
Polysaccharide film		
Starch	25–78	[229]
Cellulose derivatives	9.2–11.0	[227]

Table 3.
Water vapour permeability of edible film vs synthetic film.

9. Conclusion

This review assembles the information about the need for the edible food packaging in dairy and food industry. Edible packaging can be used in two forms, i.e. film and coatings. Based on raw material, edible film and coatings are classified (polysaccharide, lipid and protein films and coating). Here, the importance of films and coating additives is also discussed, which helps to enhance the shelf life of contained food product by retarding the growth of microorganisms and bacteria, these additives also help to improve the mechanical properties such as a barrier, various strength. Different additives have their own function or effect on film and coating. The edible film has various methods of manufacturing such as casting, thermoplastic, extrusion, etc., and coating can be applied by using methods such as dipping, brushing, enrobing, panning, etc.; these methods are used to form a uniform layer on the food material and help to control the physical, mechanical and biological hazard effect on food. This review also confirms that dipping (coating method) and casting film-forming methods are the cheapest and easy methods to use at the laboratory level. Edible film and coating have various functional properties such as colour property, mechanical property, barrier property, edibility and biodegradability, etc.; this property makes the edible film and coating superior to the commercial packaging of food. Nowadays, edible film and coating have various uses/applications in the food as well dairy industry due to its positive effect on contained food. Different edible films such as starch-chitosan film contain natural antimicrobial used to improve shelf life of various cheeses. Coatings such as zein-based coating for mode cheese and nano-emulsion-based edible coating for low-fat cut cheese are used to enhance its shelf life. Edible film and coating are also used in the meat and meat product industry, deep-fat fried industry and bakery industry. From this review, it is concluded that the use of edible film and coating is suitable for dairy as well as various food industries and helps to reduce industrial waste and environmental pollution because of their edibility and biodegradability.

10. Self-citation

Self-citation is used for chapter writing.

Acknowledgements

We are grateful for the help and support provided by MAFSU, Nagpur, and Agriculture University Jodhpur, Rajasthan.

Conflict of interest

The authors declare no conflict of interest.

Author details

Bhushan Devidas Meshram[1], Vaibhav Kisanrao Lule[1], Shivani Vyawahare[1] and Rekha Rani[2]*

1 Department of Dairy Technology, College of Dairy Technology, Warud, Maharashtra, India

2 Department of Dairy Chemistry, College of Dairy and Food Technology, Agriculture University, Jodhpur, Rajasthan, India

*Address all correspondence to: verma.rekha@gmail.com

IntechOpen

© 2022 The Author(s). Licensee IntechOpen. This chapter is distributed under the terms of the Creative Commons Attribution License (http://creativecommons.org/licenses/by/3.0), which permits unrestricted use, distribution, and reproduction in any medium, provided the original work is properly cited.

References

[1] Zoe Bain. Is Edible Packaging the Future of Packaged Foods?. 2022. Available from: https://www.delish.com/food/news/a38926/edible-food-packaging/

[2] Janjarasskul T, Krochta JM. Edible packaging materials. Annual Review of Food Science and Technology. 2010;**1**:415-448

[3] Krochta JM, De Mulder-Johnston C. Edible and biodegradable polymer films: Challenges and opportunities. Food Technology. 1997;**51**:61-73

[4] Bourtoom T. Edible films and coatings: Characteristics and properties. International Food Research Journal. 2008;**15**:237-248

[5] Lakshmi S: Edible Food Packaging with Food. 15-Dec-2021. Available from: https://www.buzzonearth.com/blog/2018/06/24/edible-food-packaging/. 2018.

[6] Ribeiro P, Cerqueira MÂ. Edible packaging. Reference Module in Food Science. Boca Raton: CRC Press; 2018:173-176. DOI: 10.1201/b19468

[7] Ali A, Maqbool M, Ramachandran S, Alderson PG. Gum arabic as a novel edible coating for enhancing shelf-life and improving postharvest quality of tomato (*Solanumlycopersicum* L.) fruit. Postharvest Biology and Technology. 2010;**58**:42-47

[8] Trinetta V. Biodegradable packaging. In: Reference Module in Food Science. Amsterdam, The Netherlands: Elsevier; 2016. pp. 1-2. DOI: 10.1016/B978-0-08-100596-5.03351-5

[9] Marsh K, Bugusu B. Food packaging—Roles, materials, and environmental issues. Journal of. Food Science. 2007;**72**:39-55

[10] Qualman D. Global Plastics Production, 1917 to 2050. 2017. Available from: https://www.darrinqualman.com/global-plastics-production [Accessed: December 15, 2021]

[11] Condor: Ferries Plastic in the Ocean Statistics 2020-2021. (online). 2021. Available from: https://www.condorferries.co.uk/plastic-in-the-ocean-statistics.

[12] Pollution C, Board C. Annual report 2019-20 on implementation of solid waste management rules, 2016. Central Pollution Control Board. ;**2022**:145-287

[13] Bodamer D. 14 Charts from the EPA's latest MSW estimates. Waste 360. 2016. Available from: https://www.waste360.com/waste-reduction/14-charts-epa-s-latest-msw-estimates [Accessed: March 7, 2019]

[14] Marchese K. Edible Packaging: The Four Fab Start-ups Tackling Plastic Pollution. 2022. Available from: https://designwanted.com/edible-packaging-four-startups/.

[15] Saklani P, Siddhnath Das SK, Singh SM. A review of edible packaging for foods. International Journal of Current Microbiology and Applied Sciences. 2019;**8**(07):2885-2895 10.20546/ijcmas.2019.807.359

[16] Robertson GL. Food Packaging Principle and Practice. 2nd ed. CRC press Taylor and Francis Group; 2006. pp. 44-47

[17] Türkoğlu A, Erkmen O. Edible Food Packaging Films and Coating. Poster: GAnziantap University; 2019

[18] Kamal I. Edible films and coatings: Classification, preparation, functionality and applications—A review. Archieve of Organic Inorganic Chemical Sciences. 2019;**4**(2):501-510 10.32474/AOICS.2019.04.000184

[19] Dhaka R, Upadhyay A. Edible films and coatings: A brief overview. The Pharma Innovation Journal. 2018;*7*(7):331-333

[20] Khwaldia K, Ferez C, Banon S, Desobry S, Hardy J. Milk proteins for edible films and coatings. Critical Reviews in Food Science and Nutrition. 2004;**44**(4):239-251. DOI: 10.1080/10408690490464906

[21] Kumari M, Mahajan H, Joshi R, Gupta M. Development and structural characterization of edible films for improving fruit quality. Food Packaging and Shelf Life. 2017. DOI: 10.1016/j.fpsl.2017.02.003

[22] Zhang S, Zhao H. Preparation and properties of zein–rutin composite nanoparticle/corn starch films. Carbohydrate Polymers. 2017;**2017**(169):385-392. DOI: 10.1016/j.carbpol.2017;04.044

[23] Fernández-Pan I, Royo M, Ignacio Maté J. Antimicrobial activity of whey protein isolate edible films with essential oils against food spoilers and foodborne pathogens. Journal of Food Science. 2012;**77**:M383-M390

[24] Stuchell YM, Krochta JM. Enzymatic treatments and thermal effects on edible soy protein films. Journal of Food Science. 1994;**59**:1332-1337

[25] Sharma L, Singh C. Sesame protein based edible films: Development and aharacterization. Food Hydrocolloids. 2016;**61**:139-147. DOI: 10.1016/j.foodhyd.2016.05.007

[26] Bourtoom T. Edible protein films: Properties enhancement. International Food Re-search Journal. 2009;**16**(1):1-9

[27] Coppola D, Oliviero M, Vitale GA, Lauritano C, D'Ambra I, Iannace S, et al. Marine collagen from alternative and sustainable sources: Extraction, processing and applications. Marine drugs. 2020;**18**(4):214 . MDPI AG. DOI: 10.3390/md18040214

[28] Fadini AL, Rocha FS, Alvim ID, Sadahira MS, Queiroz MB, Alves RMV, et al. Mechanical properties and water vapour permeability of hydrolysed collagen–cocoa butter edible films plasticised with sucrose. Food Hydrocolloids. 2013;**30**(2):625-631

[29] Wolf KL, Sobral PJA, Telis VRN. Physicochemical characterization of collagen fibers and collagen powder for self-composite film production. Food Hydrocoloids. 2009;**23**(7):1886-1894

[30] Yuvaraj D, Iyyappan J, Gnanasekaran R, Ishwarya G, Harshini RP, Dhithya V, et al. Advances in bio food packaging–an overview. Heliyon. 2021;**7**(9):e07998

[31] Krochta JM, Baldwin EA, Nysperos-Carriedo M. Edible Coatings and Films to Improve Food Quality. Boca Raton: CRC Press; 1994. pp. 1-7, 25-101, 189-330

[32] Gomez-Guillen MC, Gimenez B, Lopez-Caballero ME, Montero MP. Functional and bioactive properties of collagen and gelatin from alternative sources: A review. Food Hydrocolloids. 2011;**25**(8):1813-1827

[33] Gennadios A, McHugh TH, Weller CL, Krochta JM. Edible coatings and films based on proteins. In: Krochta JM, Baldwin EA, Nisperos-Carriedo MO, editors. Edible

Coatings and Films to Improve Food Quality. Lancaster: Technomic Publishing; 1994. pp. 201-277

[34] Ge L, Zhu M, Xu Y, Li X, Li D, Mu C. Development of antimicrobial and controlled biodegradable gelatin-based edible films containing nisin and amino-functionalized montmorillonite. Food and Bioprocess Technology. 2017;**10**(9):1727-1736. DOI: 10.1007/s11947-017-1941-0

[35] Andreuccetti C, Carvalho RA, Galicia-Garcia T, Martinez-Bustos F, Grosso CRF. Effect of surfactants on the functional properties of gelatin-based edible films. Journal of Food Engineering. 2011;**103**(2):129-136

[36] Bustillos ARJ, Krocha JM. Water vapor permeability of caseinate-based films as affected by pH, calcium crosslinking and lipid content. Journal of Food Science. 1993;**58**(4):904-907

[37] Pawar PA, Purwar AH. Biodegradable polymers in food packaging. American Journal of Engineering Research. 2013;**2**(5):151-164

[38] Arrieta MP, Peltzer MA, Lopez J, Garrigos MC, Valente AJM, Jimenez A. Functional properties of sodium and calcium caseinate antimicrobial active films containing carvacrol. Journal of Food Engineering. 2013;**121**(1):94-101

[39] Chen H. Formation and properties of casein films and coatings. In: Genadios A, editor. Protein-Based Films and Coatings. Boca Raton: CRC Press; 2002. pp. 181-211

[40] Fakhouri FM, Martelli SM, Caon T, Velasco JI, Buontempo RC, Bilck AP, et al. The effect of fatty acids on the physicochemical properties of edible films composed of gelatin and gluten proteins. LWT—Food Science and Technology. 2017. DOI: 10.1016/j.lwt.2017.08.056

[41] Wieser H. Chemistry of gluten proteins. Food Microbiology. 2007;**24**(2):115-119

[42] Gontard N, Guilbent S, Cuq LL. Edible wheat gluten films: Influence of the main process variables on film properties using response surface methodology. Journal of Food Science. 1992;**57**:190

[43] Tanada-Palmu PS, Grosso CRF. Effect of edible wheat gluten-based films and coatings on refrigerated strawberry (Fragariaananassa) quality. Postharvest Biology and Technology;**36**(2):199-208

[44] Koehler P, Kieffer R, Wieser H. Effect of hydrostatic pressure and temperature on the chemical and functional properties of wheat gluten III. Studies on gluten films. Journal of Cereal Science. 2010;**51**(1):140-145

[45] Escamilla-García M, Calderón-Domínguez G, Chanona-Pérez JJ, Farrera- Rebollo RR, Andraca-Adame JA, Arzate-Vázquez I, et al. Physical and structural characterisation of zein and chitosan edible films using nanotechnology tools. International Journal of Biological Macromolecules. 2013;**61**:196-203. DOI: 10.1016/j.ijbiomac.2013.06.051

[46] Arcan I, Yemenicioglu A. Incorporating phenolic compounds opens a new perspective to use zein as flexible bioactive packaging materials. Food Research International. 2011;**44**(2):550-556

[47] Padua VG, Wang Q. Formation and properties of corn zein films and coatings. In: Genadios A, editor. Protein-Based Films and Coatings. Boca Raton: CRC Press; 2002. pp. 43-67

[48] Dangaran K, Tomasula PM, Qi P. Structure and function of protein-based

edible films and coatings. In: Embuscado ME, Huber HC, editors. Edible Films and Coatings for Food Application. New York: Springer; 2009. pp. 25-56

[49] Lungu B, Johnson MG. Potassium sorbate does not increase control of Listeria monocytogenes when added to zein coatings with nisin on the surface of full fat Turkey frankfurter pieces in a model system at 4°C. Journal of Food Science. 2005;**70**(2):95-99

[50] Soliman EA, Eldin MSM, Furuta M. Biodegradable zein-based films: Influence of γ-irradiation on structural and functional properties. Journal of Agricultural and Food Chemistry. 2009;**57**(6):2529-2535

[51] Wang, Kai, Wu K, Xiao M, Kuang Y, Corke H, et al. Structural characterization and properties of konjac glucomannan and zein. 2017.

[52] Yin YC, Yin SW, Yang XQ, Tang CH, Wen SH, Chen Z, et al. Surface modification of sodium caseinate films by zein coatings. Food Hydrocolloids. 2014;**2014**(36):1-8. DOI: 10.1016/j.foodhyd.2013.08.027

[53] Raajeswari PA, Pragatheeswari R. Edible packaging and market overview., food marketing and technology. Indian Addition. 2019 Available from: https://www.magzter.com/stories/Business/Food-Marketing-Technology-India/Edible-Packaging-And-Market-Overview

[54] Basiak E, Lenart A. Starch coatings used in food packaging industry. Food Science Technology Quality. 2013;**1**:21-31

[55] Wani AA, Singh P, Shah MA, Schweiggert-Weisz U, Gul K, Wani IA. Rice starch diversity: Effects on structural, morphological, thermal, and physicochemical properties—A review. Comprehensive Reviews in Food Science and Food Safety. 2012;**11**:417-436

[56] Kocira A, Kozłowicz K, Panasiewicz K, Staniak M, Szpunar-Krok E, Hortyńska P. Polysaccharides as edible films and coatings: Characteristics and influence on fruit and vegetable quality—A review. Agronomy. 2021;**11**(5):813

[57] Chiumarelli M, Hubinger M. Stability, solubility, mechanical and barrier properties of cassava starch – Carnauba wax edible coatings to preserve fresh-cut apples. Food Hydrocolloids. 2012;**28**(1):59-67

[58] Bilbao-Sáinz C, Avena-Bustillos RJ, Wood DF, Williams TG, McHugh TH. Composite edible films based on hydroxypropyl methylcellulose reinforced with microcrystalline cellulose nanoparticles. Journal of Agricultural and Food Chemistry. 2010;**58**(6):3753-3760

[59] Müller CMO, Borges Laurindo J, Yamashita F. Effect of cellulose fibers addition on the mechanical properties and water vapor barrier of starch-based films. Food Hydrocolloids. 2009;**23**(5):1328-1333

[60] Wittaya T. Edible films and coatings. Characteristics and Properties. 2008 Available from: https://www.researchgate.net/publication/228397700

[61] Suput D, Lazic V, Popovic S, Hromis N. Edible films and coatings: Sources, properties and application. Food and Feed Research. 2015;**42**(1):11-22. DOI: 10.5937/ffr1501011s

[62] Mamata Sahu PK, Sahoo PK. Bio polymers: Sustainable alternative for food packaging. International Journal of Engineering and Management Research. 2017:28-32

[63] Olivas GI, Barbosa-Cánovas GV. Edible Coatings for fresh-cut fruits. Critical Reviews in Food Science and Nutrition. 2005;**45**(7-8):657-670

[64] Callegarin F, Gallo JAQ, Debeaufort F, Voilley A. Lipids and bio-packaging. Journal of American Oil Chemists Society. 1997;**74**(10):1183-1192

[65] Kalkan S, Otăg MR, Engin MS. Physicochemical and bioactive properties of edible methylcellulose films containing Rheumribes L. extract. Food Chemistry. 2020;**307**:125524

[66] Abugoch LE, Tapia C, Villamán MC, Yazdani-Pedram M, Díaz-Dosque M. Characterization of quinoa protein–chitosan blend edible films. Food Hydrocolloids. 2011;**25**(5):879-886

[67] Andrady AL, Xu P. Elastic behavior of chitosan films. Journal of Polymer Science. 1997;**5**:307-521

[68] Butola BS. Recent advances in chitosan polysaccharide and its derivatives in antimicrobial modification of textile materials. Int.J. Biol. Macromol. 2019;**121**:905-912

[69] Cuero RG, Osuji G, Washington A. N-carboxymethylchitosan inhibition of aflatoxin production: Role of zinc. Biotechnology Letters. 1991;**13**:441-444

[70] Galus S, Kibar EAA, Gniewosz M, Krasniewska K. Novel materials in the preparation of edible film and coatings—A review. Coatings. 2020;**10**:1-14. DOI: 10.3390/coatings10070674

[71] Romanazzi G, Nigro F, Ippolito A, Di Venere D, Salerno M. Effects of pre- and postharvest chitosan treatments to control storage grey mold of table grapes. Journal of Food Science. 2002;**67**(5):1862-1866

[72] dos Santos NST, Aguiar AJAA, de Oliveira CEV, de Sales CV, de Melo e Silva S, da Silva RS, et al. Efficacy of the application of a coating composed of chitosan and *Origanumvulgare* L. essential oil to control Rhizopusstolonifer and Aspergillusniger in grapes (*Vitislabrusca* L.). Food Microbilogy. 2012;**32**(2):345-353

[73] Kaplan DL, Mayer JM, Ball D, McCassie J, Allen AL, Stenhouse P. Fundamental of biodegradable polymer. In: Ching C, Kaplan D, Thomas E, editors. Biodegradable Polymers and Packaging. Lancaster: Technomic Publishing; 1993, 1993. pp. 1-42

[74] Sandford PA. Chitosan: Commercial uses and potential application. In: Skjak-Braek G, Anthosen T, Sandford P, editors. Chitin and Chitosan: Source, Chemistry, Biochemistry, Physical Properties, and Application. New York: Elsevier Applied Science; 1989. pp. 51-69

[75] Kumar A, Hasan M, Mangaraj S, Pravitha M, Verma DK, Srivastav PP. Trends in edible packaging films and its prospective future in food: A review. Applied Food Research. 2022;**100118**

[76] Dhanapal A, Rajamani L, Banu M. Edible films from polysaccharides. Food Science and Quality Management. 2012;**3**(1):9-18

[77] Stabler C, Wilks K, Sambanis A, Constantinidis I. The effects of alginate composition on encapsulated_TC3 cells. Biomaterials. 2001;**22**:1301-1310

[78] Madsen M, Westh P, Khan S, Ipsen R, Almdal K, Aachmann FL, et al. Impact of alginate mannuronic-guluronic acid contents and pH on protein binding capacity and complex size. Biomacromolecules. 2021. DOI: 10.1021/acs.biomac.0c01485

[79] Blanco-Pascual N, Montero MP, Gómez-Guillén MC. Antioxidant film

development from unrefined extracts of brown seaweeds Laminaria digitate and Ascophyllum nodosum. Food Hydrocolloids. 2014;**37**:100-110

[80] Glicksman M. Utilization of seaweed hydrocolloids in the food industry. Hydrobiologia. 1987;**151-152**:31-47

[81] Hassan B, Chatha SAS, Hussain AI, Zia KM. Recent advances on polysaccharides, lipids and protein based edible films and coatings: A review. International Journal of Biological Macromolecules. 2018;**109**:1095-1107

[82] Hassan B, Chatha SAS, Hussain AI, Zia KM, Akhtar N. Re-cent advances on polysaccharides, lipids and protein based edible films and coat-ings: A review. International Journal of Biological Macromolecules. 2018;**109**:1095-1107. DOI: 10.1016/J.IJBIOMAC.2017.11.097

[83] Borchard W, Kenning A, Kapp A, Mayer C. Phase diagram of the system sodium alginate/water: A model for biofilms. International Journal of Biological Macromolecules. 2005;**35**:247-256

[84] Varela P, Fiszman SM. Hydrocolloids in fried foods. A review. Food Hydrocolloids. 2011;**25**(8):1801-1812

[85] Rhim JW, Shellhammer TH. Lipid-based edible films and coatings. In: Han JH, editor. Innovations in Food Packaging. London: Elsevier; 2005. pp. 362-383

[86] Mehyar GF, Al-Ismail K, Han JH, Chee GW. Characterization of edible coatings consisting of pea starch, whey protein isolate, and carnauba wax and their effects on oil rancidity and sensory properties of walnuts and pine nuts. Journal of Food Science. 2012;**77**(2):52-59

[87] Han JH. Innovations in Food Packaging. Academic Press, USA: Elsevier; 2014

[88] Hagenmaier RD, Baker RA. Wax microemulsions and emulsions as citrus coatings. Journal of Agricultural and Food Chemistry. 1994;**42**(4):899-902

[89] Hernandez E. Edible coating from lipids and resins. In: Krochta JM, Balwin EA, Niperos-Carriedo MO, editors. Edible Coatings and Films to Improve Food Quality. Basel, Switzerland: Technomic Publishing Company; 1994. pp. 279-303

[90] Hagenmaier RD, Baker RA. Reduction in gas exchange of citrus fruit by wax coatings. Journal of Agricultural and Food Chemistry. 1993;**41**(2):283-287

[91] Ahmad M, Khan I. Effect of waxing and cellophane lining on chemical quality indices of citrus fruit. Plant Foods Human Nutrition. 1987;**37**:47-57

[92] Cohen E, Shalom Y, Rosenberger I. Posthar vest ethanol buildup and off-flavor in 'Murcott' tangerine fruits. Journal of American Society of Horticultural and Science. 1990;**115**(5):775-778

[93] Ke D, Kader AA. Tolerance of 'Valencia' oranges to controlled atmospheres as determined by physiological responses and quality attributes. Journal of American Society of Horticultural and Science. 1990;**115**(5):770-783

[94] Petracek PD, Dou H, Malik I. A post-harvest pitting of temple oranges stimulated by high temperature storage and wax application. Florida State Horticultural Society. 1997;**110**:211-214

[95] Petracek PD, Dou H, Pao S. Influence of applied waxes on postharvest physiological behavior and pitting of white grapefruit. Post-harvest Biology and Technology. 1998;**14**:99-106

[96] Cheng E, Sun X, Karr GS. Adhesive properties of modified soybean

flour in wheat straw particleboard. Composites Part A: Applied Science and Manufacturing. 2004;**35**(3):297-302

[97] Gounga ME, Xu S, Wang Z. Whey protein isolate-based edible films as affected by protein concentration, glycerol ratio and pullulan addition in film formation. Journal of Food Engineering. 2007;**83**(4):521-530

[98] Su J, Huang Z, Yuan X, Wang X, Li M. Structure and properties of carboxymethyl cellulose/soy protein solate blend edible fimscrosslinked by Maillard reactions. Carbohydrate Polymers. 2010;**79**(1):145-153

[99] Stuchell YM, Krochta JM. Edible coatings on frozen king salmon: Effect of whey protein isolate and acetylated monoglycerides on moisture loss and lipid oxidation. Journal of Food Science. 1995;**60**:28-31

[100] Liu L, Kerry JF, Kerry JP. Application and assessment of extruded edible casings manufactured from pectin and gelatin/sodium alginate blends for use with breakfast pork sausage. Meat Science. 2007;**75**(2):196-202

[101] Zahedi Y, Ghanbarzadeh B, Sedaghat N. Physical properties of edible emulsified films based on pistachio globulin protein and fatty acids. Journal of Food Engineering. 2010;**100**(1):102-108

[102] Sothornvit R, Krochta JM. Plasticizers in edible films and coatings. In: Han JH, editor. Innovations in Food Packaging. Vol. 23. New York: Elsevier Acad; 2005. pp. 403-433

[103] Rawdkuen S. Edible films incorporated with active compounds: Their properties and application. In: Var I, Uzunlu S, editors. Active Antimicrobial Food Packaging. London, UK, Rijeka: IntechOpen; 2018. DOI: 10.5772/intechopen.80707

[104] Oussalah M, Caillet S, Salmiéri S, Saucier L, Lacroix M. Antimicrobial and antioxidant effects of milk protein-based film containing essential oils for the preservation of whole beef muscle. Journal of Agriculture Food Chemistry. 2004;**52**:5598-5605

[105] Naidu AS. Overview. In: Naidu AS, editor. Natural Food Antimicrobial Systems. London: CRC Press; 2000. pp. 1-16

[106] Rawdkuen S. Antimicrobial activity of some potential active compounds against food spoilage microorganisms. African Journal of Biotechnology. 2012;**11**(74):13914-13921. DOI: 10.5897/ajb12.1400

[107] Dehghani S, Hosseini SV, Regenstein JM. Edible films and coatings in seafood preservation: A review. Food Chemistry. 2018;**1**(240):505-513. DOI: 10.1016/j.foodchem.2017.07.034 Epub July 11, 2017. PMID: 28946304

[108] Bekhit M, Arab-Tehrany E, Kahn CJF, Cleymand F, Fleutot S, Desobry S, et al. Bioactive films containing alginate-pectin composite microbeads with Lactococcuslactis subsp. lactis: Physicochemical characterization and Antilisterial activity. International Journal of Molecular Sciences Article. 2018. DOI: 10.3390/ijms19020574

[109] Beristain-Bauza SC, Mani-López E, Palou E, López-Malo A. Antimicrobial activity and physical properties of protein films added with cell-free supernatant of lactobacillus rhamnosus. Food Control. 2016;**62**:44-51. DOI: 10.1016/J.FOODCONT.2015.10.007

[110] da Rocha M, Prietto L, de Souza MM, Furlong EB, Prentice C.

Effect of organic acids on physical-mechanical and antifungicidal properties of anchovy protein films. Journal of Aquatic Food Product Technology. 2018;**27**:316-326. DOI: 10.1080/10498850.2018.1433736

[111] Han JH. Antimicrobial packaging systems. In: Han JH, editor. Innovations in Food Packaging. London: Elsevier Science; 2005. pp. 80-101

[112] Uranga J, Puertas AI, Etxabide A, Dueñas MT, Guerrero P, de la Caba K. Citric acid-incorporated fish gelatin/chitosan composite films. Food Hydrocolloids. 2019;**86**:95-103 10.1016/j.foodhyd.2018.02.018

[113] Rocha RP, Lopes PA, De Almeida AT, Tavakoli M, Majidi C. Fabrication and characterization of bending and pressure sensors for a soft prosthetic hand. Journal of Micromechanics and Microengineering. 2018;**28**(3):034001

[114] Huber KC, Embuscado M. Edible films and Coatings for food applications. In: Edible Films and Coatings for Food Applications (Issue September 2015). 2009. Dordrecht, Heidelberg, London, New York. DOI: 10.1007/978-0-387-92824-1

[115] Bandeira SF, da Silva RDSG, de Moura JM, de Almeida Pinto LA. Modified gelatin films from croaker skins: Effects of pH, and addition of glycerol and chitosan. Journal of Food Process Engineering. 2015;**38**(6):613-620

[116] Rhim JW, Mohanty AK, Singh SP, Ng PK. Effect of the processing methods on the performance of polylactide films: Thermocompression versus solvent casting. Journal of Applied Polymer Science. 2006;**101**(6):3736-3742

[117] Jensen A, Lim LT, Barbut S, Marcone M. Development and characterization of soy protein films incorporated with cellulose fibers using a hot surface casting technique. LWT Food Science Technology. 2015;**60**(1):162-170

[118] Cha DS, Cooksey K, Chinna MS. HJ Park release of nisin from various heat-pressed and cast films. LWT—Food Science and Technology. 2003;**36**(2):209-213

[119] Sherrington DC. Introduction to physical polymer science. In: Reactive Polymers. Hoboken, New Jersey: Wiley 2003. p. 20

[120] Velaga SP, Nikjoo D, Vuddanda PR. Experimental studies and modeling of the drying kinetics of multicomponent polymer films. AAPS Pharm Science Technology. 2018:425-435

[121] Chen Q, Roether J, Boccaccini A. Chapter 6: Tissue engineering scaffolds from bioactive glass and composite materials. Topics in Tissue Engineering. In: Ashammakhi N, Reis R, Chiellini F, editors. 2008:1-27

[122] Koide Y, Ikake H, Muroga Y, Shimizu S. Effect of the cast-solvent on the morphology of cast films formed with a mixture of stereoisomeric poly (lactic acids). Polym Journal. 2013;**45**(6):645-650

[123] Asghari F, Mohammad S, Khosro A, Abolfazl A, Soodabeh D. Biodegradable and biocompatible polymers for tissue engineering application: A review. Artificial Cells, Nanomedicine and Biotechnology. 2017;**45**(2):185-192

[124] Fakhouri FM, Costa D, Yamashita F, Martelli SM, Jesus RC, Alganer K. Comparative study of processing methods for starch/gelatin films. Carbohydrate Polymers. 2013;**95**(2):681-689

[125] Sait HH, Ma HB. An experimental investigation of thin-film evaporation. Nanoscale and Microscale Thermophysical Engineering. 2009;**13**(4):218-222

[126] Gamboni JE, Colodro MV, Slavutsky AM, Bertuzzi MA. Selection of the conveyor belt material for edible film production by a continuous casting process. Brazilian Journal of Food Technology. 2021;**24**:1-13. DOI: 10.1590/1981-6723.02620

[127] Suhag R, Kumar N, Petkoska AT, Upadhyay A. Film formation and deposition methods of edible coating on food products: A review. Food Research International. 2020;**136**:109582. DOI: 10.1016/J.FOODRES.2020.109582

[128] Fishman ML, Coffin DR, Onwulata CI, Konstance RP. Extrusion of pectin and glycerol with various combinations of orange albedo and starch. Carbohydrate Polymers. 2004;**57**:401-413

[129] Hernandez-Izquierdo VM, Krochta JM. Thermoplastic processing of proteins for film formation—A review. Journal of Food Science. 2008;**73**(2):30-39

[130] Calderon-Castro A, Vega-Garcıa M, Zazueta-Morales JD, Fitch-Vargas PR, Arrillo-Lopez A, Gutierrez-Dorado R, et al. Effect of extrusion process on the functional properties of highamylose corn starch edible films and its application in mango (*Mangiferaindica* L.) cv. Tommy Atkins. Journal of Food Science and Technology. 2018;**55**(3):905-914

[131] Peressini D, Bravin B, Lapasin R, Rizzotti C, Sensidoni A. Starch-methylcellulose based edible films: Rheological properties of film-forming dispersions. Journal of Food Engineering. 2003;**59**:25-32

[132] Fitch-Vargas PR, Aguilar-Palazuelos E, Vega-García MO, Zazueta-Morales JJ, Calderón-Castro A, Montoya-Rodríguez A. Effect of a corn starch coating obtained by the combination of extrusion process and casting technique on the postharvest quality of tomato. Revista Mexicana de Ingeniera Quimica. 2019;**18**(3):789-801

[133] Otoni CG, Avena-Bustillos RJ, Azeredo HMC, Lorevice MV, Moura MR, Mattoso LHC, et al. Recent advances on edible films based on fruits and vegetables—A review. Comprehensive Reviews in Food Science and Food Safety. 2017;**16**(5):1151-1169. DOI: 10.1111/1541-4337.12281

[134] Winotapun C, Phattarateer S, Aontee A, Junsook N, Daud W, Kerddonfag N, et al. Development of multilayer films with improved aroma barrier properties for durian packaging application. Packaging Technology and Science. 2019;**32**:405-418. DOI: 10.1002/pts.2452

[135] Liu Z. Edible films and coatings from starches. In: Han JH, editor. Innovations in Food Packaging. London: Academic Press; 2005. pp. 318-337

[136] Espitia PJP, Avena-Bustillos RJ, Du W-X, Chiou B-S, Williams TG, Wood D, et al. Physical and antibacterial properties of acai edible films formulated with thyme essential oil and apple skin polyphenols. Journal of Food Science. 2014;**79**:M903-M910

[137] Andreuccetti C, Carvalho RA, Galicia-García T, Martinez-Bustos F, González-Nuñez R, Grosso CRF. Functional properties of gelatin-based films containing Yucca schidigera extract produced via casting, extrusion and

blown extrusion processes: A preliminary study. Journal of Food Engineering. 2012;**113**(1):33-40

[138] Krishna MA, Nindo CI, Min SC. Development of fish gelatin edible films using extrusion and compression molding. Journal of Food Engineering. 2012;**108**(2):337-344

[139] Raghav PK, Agarwal N, Saini M. Edible coating of fruits and vegetables: A review. International Journal of Scientific Research and Modern Education. 2016;**1**(1):2455-5630

[140] Mohammadi MA, Hosseini SM, Yousefi M. Application of electrospinning technique in development of intelligent food packaging: A short review of recent trends. Food science and Nutrition. 2020;**8**:4656-4665. Available from: https://onlinelibrary.wiley.com/doi/full/10.1002/fsn3.1781

[141] Ebrahimi S, Fathi M, Kadivar M. Production and characterization of chitosan-gelatinnanofibers by nozzlelesselectrospinning and their application to enhance edible film's properties. Food Packaging and Shelf Life. 2019;**22**:100387

[142] Janssen PJFM, Solberg RHM. Electrospinning device and method. In: Google Patents. 2019

[143] Li C, Sheng L, Sun G, Wang L. The application of ultraviolet-induced photo-crosslinking in edible film preparation and its implication in food safety. Lebensmittel-Wissenschaft & Technologie. 2020;**131**

[144] Ghorani B, Emadzadeh B, Rezaeinia H, Russell SJ. Improvements in gelatin cold water solubility after electrospinning and associated physicochemical, functional and rheological properties. Food Hydrocolloids. 2020;**104**

[145] Guan X, Li L, Li S, Liu J, Huang K. A food-grade continuous electrospunfiber of hordein/chitosan with water resistance. Food Bioscience. 2020:37

[146] Li D. Xia Y electrospinning of nanofibers: Reinventing the wheel? Advanced Materials. 2004;**16**:1151-1170

[147] Xue J, Xie J, Liu W, Xia Y. Electrospun nanofibers: New concepts, materials, and applications. Accounts of Chemical Research. 2017;**50**:1976-1987

[148] Fabra MJ, Busolo MA, Lopez-Rubio A, Lagaron JM. Nanostructured biolayers in food packaging. Trends in Food Science & Technology. 2013;**31**(1):79-87

[149] Lagarón Cabello, J. M., Martínez Sanz, M., & López-Rubio, A. (2013). Procedimiento de obtención de una película multicapa con alta barrera.

[150] Busolo MA, Torres-Giner S, Lagaron JM. Enhancing the gas barrier properties of polylactic acid by means of electrospun ultrathin zein fibers. In: ANTEC, Proceedings of the 67th Annual Technical Conference. Chicago, IL; 2009. pp. 2763-2768

[151] Frinault A, Gallant DJ, Bouchet B, Dumont JP. Preparation of casein films by a modified wet spinning process. Journal of Food Science. 1997;**62**(4):744-747

[152] Liao Y, Loh CH, Tian M, Wang R, Fane AG. Progress in electrospun polymeric nanofibrous membranes for water treatment: Fabrication, modification and applications. Progress in Polymer Science. 2018;**77**:69-94

[153] Sun B, Long YZ, Zhang HD, Li MM, Duvail JL, Jiang XY. Yin HL advances

in three-dimensional nanofibrous macrostructures via electrospinning. Progress in Polymer Science. 2014;**39**:862-890

[154] Cui H, Bai M, Rashed MMA, Lin L. The antibacterial activity of clove oil/ chitosan nanoparticles embedded gelatinnanofibers against *Escherichia coli* O157:H7 biofilms on cucumber. International Journal of Food Microbiology. 2018;**266**:69-78

[155] Wang H, Ding F, Ma L, Zhang Y. Edible films from chitosan-gelatin: Physical properties and food packaging application. In: Food Bioscience. Vol. 40. Elsevier; 2021. p. 100871. DOI: 10.1016/j.fbio.2020.100871

[156] Ghanbari A, Tabarsa T, Ashori A, Shakeri A, Mashkour M. Preparation and characterization of thermoplastic starch and cellulose nanofibers as green nanocomposites: Extrusion processing. International Journal of Biological Macromolecules. 2018;**112**:442-447

[157] Dang KM, Yoksan R. Morphological characteristics and barrier properties of thermoplastic starch/chitosan blown film. Carbohydrate Polymers. 2016;**150**:40-47

[158] Lizarraga-Laborín LL, Quiroz-Castillo JM, Encinas-Encinas JC, Castillo-Ortega MM, Burruel-Ibarra SE, Romero-García J, et al. Accelerated weathering study of extruded polyethylene/poly (lactic acid)/chitosan films. Polymer Degradation and Stability. 2018;**155**:43-51

[159] Pelissari FM, Yamashita F, Garcia MA, Martino MN, Zaritzky NE, Grossmann MVE. Constrained mixture design applied to the development of cassava starch-chitosan blown films. Journal of Food Engineering. 2012;**108**(2):262-267

[160] Nilsuwan K, Guerrero P, de la Caba K, Benjakul S, Prodpran T. Fish gelatin films laminated with emulsified gelatin film or poly (lactic) acid film: Properties and their use as bags for storage of fried salmon skin. Food Hydrocolloids. 2021;**111**:106199

[161] Huntrakul K, Yoksan R, Sane A, Harnkarnsujarit N. Effects of pea protein on properties of cassava starch edible films produced by blown-film extrusion for oil packaging. Food Packaging and Shelf Life. 2020;**24**:100480

[162] Agrawal AM, Pandey P. Scale up of pan coating process using quality by design principles. Journal of Pharmaceutical Sciences. 2015;**104**(11):3589-3611

[163] Gesford P: Challenges in panning. Manufacturing Confectioner. Global Source for Chocolate, Confectionery and Biscuit Information. Princeton, USA. 2002;82(11): 43-50.

[164] Cunning T. G. and J. M. Caulkins. Method of spray coating fruit and vegetables. US Patent 3192052 A. Filed: July 6, 1962. 1965.

[165] Chen H. Functional properties and applications of edible films made of milk proteins. Journal of Dairy Science. 1995;**78**(11):2563-2585

[166] Prateepchanachai S, Thakhiew W, Devahastin S, Soponronnarit S. Mechanical properties improvement of chitosan films via the use of plasticizer, charge modifying agent and film solution homogenization. Carbohydrate Polymers. 2017;**174**:253-261. DOI: 10.1016/J.CARBPOL.2017.06.069

[167] Kan J, Liu J, Yong H, Liu Y, Qin Y, Liu J. Development of active packaging based on chitosan-gelatin blend

films functionalized with Chinese hawthorn (*Crataegus pinnatifida*) fruit extract. International Journal of Biological Macromolecules. 2019;**2019**(140):384-392

[168] Melvin PA. The application of edible polymeric films and coatings in the food industry. Journal of Food Processing & Technology. 2012;**04**(02). DOI: 10.4172/2157-7110.1000e116

[169] Yadav S, Mehrotra GK, Bhartiya P, Singh A, Dutta PK. Preparation, physicochemical and biological evaluation of quercetin based chitosan-gelatin film for food packaging. Carbohydrate Polymers. 2020;**227**:115348

[170] Mohammadi R, Mohammadifar MA, Rouhi M, Kariminejad M, Mortazavian AM, Sadeghi E, et al. Physico-mechanical and structural properties of eggshell membrane gelatin-chitosan blend edible films. International Journal of Biological Macromolecules. 2018;**107**(Pt A):406-412

[171] NorsuzilaYa'acob M, Abdullah M, Ismail Medina M, Talarico TL, Casas IA, Chung TC, et al. We are IntechOpen, The World ' s Leading publisher of Open Access Books Built by Scientists, for Scientists TOP 1%. Vol. 32. London, UK, Rijeka: Intech; 1989. pp. 137-144

[172] Cao C, Wang Y, Zheng S, Zhang J, Li W, Li B, et al. Poly (butylene adipate-co-terephthalate)/titanium dioxide/silver composite biofilms for food packaging application. Lebensmittel-Wissenschaft & Technologie. 2020

[173] Butler BL, Vergano PJ, Testin RF, Bunn JM, Wiles JL. Mechanical and barrier properties of edible chitosan films as affected by composition and storage. Journal of Food Science. 1996;**61**

[174] Mchugh TH, Krochta JM. Permeability perperties of edible films. In: Krochta JM, Baldwin EA, Nisporos-Carredo AM, editors. Edible Coatings and Films to improve Food Quality. Lancaster, PA: Telephonic Publ. Co. Inc.; 1994. p. 139

[175] DrakE SR, Fellman JK, Nelson JW. Journal of Food Science. 1987;**52**:1283-1285

[176] McHugh TH, Aujard JF, Krochta JM. Plasticized whey protein edible films: Water vapor permeability properties. Journal of Food Science. 1994;**59**(2):416-419, 423

[177] Ramos ÓL, Fernandes JC, Silva SI, Pintado ME, Malcata FX. Edible films and coatings from whey proteins: A review on formulation, and on mechanical and bioactive properties. Critical Reviews in Food Science and Nutrition. 2012;**52**:533-552

[178] Tulamandi S et al. A biodegradable and edible packaging film based on papaya puree, gelatin, and defatted soy protein. Food Packaging and Shelf Life. 2016;**10**:60-71. DOI: 10.1016/j.fpsl.2016.10.007

[179] Amjadi S, Emaminia S, Davudian HS, Pourmohammad S, Hamishehkar H, Roufegarinejad L. Preparation and characterization of gelatin-based nanocomposite containing chitosan nanofiber and ZnO nanoparticles. Carbohydrate Polymers. 2019;**216**:376-384

[180] Moeini A, Mallardo S, Cimmino A, Dal Poggetto G, Masi M, Di Biase M, et al. Thermoplastic starch and bioactive chitosan sub-microparticlebiocomposites: Antifungal and chemico-physical properties of the films. Carbohydrate Polymers. 2020;**230**:115627

[181] Liu Y, Cai Y, Jiang X, Wu J, Le X. Molecular interactions, characterization

and antimicrobial activity of curcumin–chitosan blend films. Food Hydrocolloids. 2016;**52**:564-572. DOI: 10.1016/j.foodhyd.2015.08.005

[182] Riaz A, Lei S, Akhtar HMS, Wan P, Chen D, Jabbar S, et al. Preparation and characterization of chitosan- based antimicrobial active food packaging film incorporated with apple peel polyphenols. International Journal of Biological Macromolecules. 2018;**114**:547-555. DOI: 10.1016/j. ijbiomac.2018.03.126

[183] Kalaycıoğlu Z, Torlak E, Akın-Evingür G, Özen İ, Erim FB. Antimicrobial and physical properties of chitosan films incorporated with turmeric extract. International Journal of Biological Macromolecules. 2017;**101**:882-888. DOI: 10.1016/j. ijbiomac.2017.03.174

[184] Alfin-Slater RB, Coleman RD, Feuge RO, Altschul AM. Journal of the American Oil Chemists' Society. 1958;**35**:122-127

[185] Brusewitz GH, Singh RPJ. Food Processing and Preservation. 1985;**9**:1-9

[186] Guilbert S. In: Mathlouthi M, editor. Food Packaging and Preservation. New York: Elsevier Applied Science Publishers; 1986. pp. 371-394

[187] Krochta JM, Pavlath AE, Goodman N. International Congress on Engineering and Food. Vol. 5

[188] Raju A, Sasikala S. Natural antimicrobial edible film for preservation of paneer. Biosciences Biotechnology Research Asia. 2016;**13**(2):1083-1088. DOI: 10.13005/bbra/2136

[189] Fajardo P, Martins JT, Fuciños C, Pastrana L, Teixeira JA, Vicente AA. Evaluation of a chitosan-based edible film as carrier of natamycin to improve the storability of Saloio cheese. Journal of Food Engineering. 2010;**101**(4):349-356. DOI: 10.1016/j.jfoodeng.2010.06.029

[190] Wagh YR, Pushpadass HA, Emerald FME, Nath BS. Preparation and characterization of milk protein films and their application for packaging of Cheddar cheese. Journal of Food Science and Technology. 2014;**51**(12):3767-3775. DOI: 10.1007/s13197-012-0916-4

[191] Di Pierro P, Sorrentino A, Mariniello L, Giosafatto CVL, Porta R. Chitosan/whey protein film as active coating to extend Ricotta cheese shelf-life. LWT—Food Science and Technology. 2011;**44**:2324-2327

[192] Henriques M, Santos G, Rodrigues A, Gomes D. Replacement of conventional cheese coatings by natural whey protein edible coatings with antimicrobial activity. Journal of Hygienic Engineering and Design. 2013;**3**:34-47

[193] Artiga-Artigas M, Acevedo-Fani A, Martín-Belloso O. Improving the shelf life of low-fat cut cheese using nanoemulsion-based edible coatings containing oregano essential oil and mandarin fiber. Food Control. 2017;**76**:1-12. DOI: 10.1016/J. FOODCONT.2017.01.001

[194] Pena-Serna C, Penna ALB, LopesFilho JF. Zein-based blend coatings: Impact on the quality of a model cheese of short ripening period. Journal of Food Engineering. 2015;**171**:208-213. DOI: 10.1016/j.jfoodeng.2015.10.039

[195] Mei J, Yuan Y, Wu Y, Li Y. Characterization of edible starch-chitosan film and its application in the storage of Mongolian cheese. International Journal of Biological Macromolecules. 2013;**57**:17-21. DOI: 10.1016/j.ijbiomac.2013.03.003

[196] Lucera A, Mastromatteo M, Conte A, Zambrini AV, Faccia M, Del Nobile MA. Effect of active coating on microbiological and sensory properties of fresh mozzarella cheese. Food Packaging and Shelf Life. 2013;**1**(1):25-29. DOI: 10.1016/J.FPSL.2013.10.002

[197] Labuza TP, Hyman CR. Moisture migration and control in multi-domain foods. Trends in Food Science & Technology. 1998;**9**(2):47-55. DOI: 10.1016/S0924-2244(98)00005-3

[198] Nawab A, Alam F, Haq MA, Haider MS, Lutfi Z, Kamaluddin S, et al. Innovative edible packaging from mango kernel starch for the shelf life extension of red chili powder. International Journal of Biological Macromolecules. 2018;**114**:626-631. DOI: 10.1016/J.IJBIOMAC.2018.03.148

[199] Rangel-Marrón M, Mani-López E, Palou E, López-Malo A. Effects of alginate-glycerol-citric acid con-centrations on selected physical, mechanical, and barrier properties of papaya puree-based edible films and coatings, as evaluated by response surface methodology. LWT—Food Science and Technology. 2019;**101**:83-91. DOI: 10.1016/j.lwt.2018.11.005

[200] Milani JM, Tirgarian B. An overview of edible protein-based packaging: Main sources, advantages, drawbacks, recent progressions and food applications. Journal of Packaging Technology and Research. 2020;**4**:103-115

[201] Wang, C.; Killpatrick, A.; Humphrey, A. and Guo, M. Whey protein functional properties and applications in food formulation. In Whey Protein Production, Chemistry, Functionality, and Applications; John Wiley & Sons, Ltd.: Chichester, UK, 2019; pp. 157-204. ISBN 9781119256052

[202] Tien C, Vachon C, Mateescu M-A, Lacroix M. Milk protein coatings prevent oxidative browning of apples and potatoes. Journal of Food Science. 2001;**66**:512-516

[203] Perez-Gago MB, Serra M, del Río MA. Color change of fresh-cut apples coated with whey protein concentrate-based edible coatings. Postharvest Biology and Technology. 2006;**39**:84-92

[204] Huang L, Zhang M, Yan W, Mujumdar AS, Sun D. Effect of coating on post-drying of freeze-dried strawberry pieces. Journal of Food Engineering. 2009;**92**:107-111

[205] Ochoa E, Saucedo-Pompa S, Rojas-Molina R, Garza HDL, Charles-Rodríguez AV, Aguilar CN. Evaluation of a candelilla wax-based edible coating to prolong the shelf-life quality and safety of apples. American Journal of Agricultural and Biological Sciences. 2011, 2011;**6**(1):92-98. DOI: 10.3844/ajabssp.2011.92.98

[206] Caner C. Whey protein isolate coating and concentration effects on egg shelf life. Journal of the Science of Food and Agriculture. 2005;**85**:2143-2148 (126)

[207] Domínguez R, Barba FJ, Gómez B, Putnik P, BursaćKovačević D, Pateiro M, et al. Active packaging films with natural antioxidants to be used in meat industry: A review. Food Research International. 2018;**113**:93-101

[208] Shon J, Chin KB. Effect of whey protein coating on quality attributes of low-fat, aerobically packaged sausage during refrigerated storage. Journal of Food Science. 2008;**73**:C469-C475

[209] Cutter CN, Senevirathne RN, Chang VP, Cutaia RB, Fabrizio KA, Geiger AM. Major microbiological hazards associated with packaged fresh and processed meat and poultry. In:

Kerry JP, editor. Advances in Meat, Poultry and Seafood Packaging. Cambridge: Woodhead Publishing; 2012. pp. 3-41

[210] Mallikarjunan P, Chinnan MS, Balasubramaniam VM, Phillips RD. Edible coatings for deep-fat frying of starchy products. LWT—Food Science and Technology. 1997;**30**(7):709-714. DOI: 10.1006/FSTL.1997.0263

[211] Castro FVR, Andrade MA, Sanches Silva A, FátimaVaz M, Vilarinho F, Doutor R, et al. The contribution of a whey protein film incorpo-rated with green tea extract to minimize the lipid oxidation of salmon (*Salmosalar L.*). Mdpi.com. 2019;8:**327**(1-16). DOI: 0.3390/foods8080327

[212] Cai L, Wang Y, Cao A. The physiochemical and preservation properties of fish sarcoplasmic protein/ chitosan composite films containing ginger essential oil emulsions. Journal of Food Process Engineering. 2020;**10**:43. DOI: 10.1111/jfpe.13495

[213] Lee, S.-Y.andKrochta, J.M. Accelerated shelf life testing of whey-protein-coated peanuts analyzed by static headspace gas chromatography. Journal of Agricultural and Food Chemistry. 2002;**50**:2022-2028

[214] Maté JI, Saltveit ME, Krochta JM. Peanut and walnut rancidity: Effects of oxygen concentration and relative humidity. Journal of Food Science. 1996;**61**:465-469

[215] García MPM, Gómez-Guillén MC. In: López-CaballeroME,Barbosa-CánovasGV, editors. Edible Films and Coatings: Fundamentals and Applications. CRC Press; 2016

[216] Wang DL, Kerry DJP. Edible/ biodegradable packaging for food. New Food Magazine. 2009

[217] Hamleton A, Debeaufort F, Bonnotte A, Voilley A. Influence of alginate emulsion-based films structure on its barrier properties and on the protection of microencapsulated aroma compound. Food Hydrocolloids. 2009;**23**:2116-2124

[218] Zhao R, Torley P, Halley PJ. Emerging biodegradable materials: Starch- and protein-based bio-nanocomposites. Journal of Materials Science. 2008;**43**(9):3058-3071. DOI: 10.1007/s10853-007-2434-8

[219] Greener Donhowe IK, Fennema OR. The effects of solution composition and drying temperature on crystallinity, permeability and mechanical properties of methylcellulose films. Journal of Food Processing & Preservation. 1993;**17**:231-246

[220] Galus S, Lenart A. Optical, mechanical, and moisture sorption properties of whey protein edible films. The Journal of Food Process Engineering. 2019;**42**:1-10

[221] Galus S. Functional properties of soy protein isolate edible films as affected by rapeseed oil concentration. Food Hydrocolloids. 2018;**85**:233-241

[222] Araujo-Farro PC, Podadera G, Sobral PJA, Menegalli FC. Development of films based on quinoa (*Chenopodium quinoa*, Willdenow) starch. Carbohydrate Polymers. 2010;**81**:839-848

[223] Myers AW, Meyer JA, Rogers CE, Stannett V, Szwarc M. Studies in the gas and vapor permeability of plastic films and coated papers. Part VI. The permeation of water vapor. The Technical Association of the Pulp and Paper Industry. 1961;**44**:58-64

[224] Shellhammer TH, Krochta JM. Edible coatings and film barriers. In:

Gunstone FD, Padley FB, editors. Lipid Technologies and Applications. New-York: Marcel Dekker; 1997. pp. 453-479. Chapter 17

[225] Bourlieu C, Guillard V, Vallès-Pamiès B, Guilbert S, Gontard N. Edible moisture barriers: How to assess of their potential and limits in food products shelf-life extension? Critical Reviews in Food Science and Nutrition. 2009;**49**(5):474-499

[226] Guillard V, Broyart B, Bonazzi C, Guilbert S, Gontard N. Preventing moisture transfer in a composite food using edible films: Experimental and mathematical study. Journal of Food Science. 2003;**68**:2267-2277

[227] Park HJ, Chinan MS. Gas and water vapor barrier properties of edible films from protein and cellulosic materials. Journal of Food Engineering. 1995;**25**:479-507

[228] Rhim JW, Gennadios A, Weller CL, Hanna MA. Sodium dodecyl sulfate treatment improves properties of cast films from soy protein isolate. Industrial Crops and Products. 2002;**15**:199-205

[229] Bertuzzi MA, Castro Vidaurre EF, Armada M, Gottifredi JC. Water vapor permeability of edible starch based films. Journal of Food Engineering. 2007;**80**(3):972-978

Chapter 14

Food Preservation Packaging

Abubakar Ibrahim Garba

Abstract

The most important role of food packaging is to provide a total barrier to physical, biological and/or chemical factors that can tender the quality integrity of the packaged food, an ideal food packaging should be rigid and non-reactive to the food enclosed. That is, a packaging material should be safe by and/or for the food. Food and packaging may interact and pose effects, which may affect the quality, integrity and shelf life of the food. This chapter discusses the processes of food packaging interaction such as permeation, migration, sorption and their adverse effects on the food packaging system. Novel packaging systems such as Active packaging (packaging that preserve, communicate and protect quality integrity of the food), Intelligent packaging (packages with improved communication models that help consumers ascertain the quality and the state of the food) and Bio-active packaging (are active packaging with improve compound in them that support stability of the food) were discussed as advanced packaging systems that help in mitigating the food-package interaction as well as give consumer conveniences while extending shelf stability. Scientific models used in studying the extent of food packaging interaction are also discussed which includes the stochastic, mathematical and simulation models.

Keywords: packaging, packaging interaction, food quality, active packaging, intelligent packaging, migration, sorption

1. Introduction

Food preservation involves the art and science of extending food quality/integrity or maintaining its nature for a specified period of time [1]. The food preservation techniques includes the use of packaging enclosure to protect food from external factors which may interfere with initial nature or quality of the food. As a food preservation element, food packaging provide enclosure to food and supports the food product against external invasion. It provide an integral role of *containment, protection communication and preservation* of the inherent quality of the package food [2]. Food packaging provide several advantage in food production that includes; nutrient stability/preservation, information containment and communication including information on ingredients (type and quantity), shelf life and also quality/integrity protection. Therefore, the role of food packaging in food reservation cannot be overemphasized. Food packaging are usually in direct contact with the food, therefore there is direct association which can result to some changes. Some of these changes may render the food unsafe for consumption or may speed up chemical process which may poses organoleptic loss. The advent of food-packaging interaction depends on the nature and quality of the packaging material. Some packaging provide 100% barrier to any external factors why some may allow lite interaction with external system as such can easily pose the food to contamination and or other reactive changes, therefore a packaging selection is critical to the shelf stability of the food and needed to follow stringent process.

An important requirement in selecting packaging systems for foods is the barrier property of the packaging material. To keep a food product crisp and fresh, the package must provide a barrier to moisture. Food changes such as off-flavor and rancidity may occur if the packaging material has a poor barrier to oxygen, allow gas permeation and by allowing passage of light to the food. Original organoleptic properties of a food can be maintained by using a packaging material that offers a good barrier to moisture, gasses and aroma. Thus, properly selected packaging materials are beneficial in extending the shelf-life of foods. A food's characteristic flavor and aroma are the result of a complex construct of hundreds of individual constituent compounds interacting to produce a recognizable taste and aroma. Therefore, if one or more flavor constituents are altered or diminished, food quality may be reduced. A reduction in food quality may result from the interaction of the food component with external environment or with the internal packaging environment which hinders the ability of the packaging role. The way food package influences the integrity of the packaged food, food under packaging also affect the packaging integrity when it react or possess chemical changes to the packaging films therefore effective packaging system is the one which passes no effect to the food or to the packaging or vice-versa. Plastic and metal packaging usually react if used in direct contact with acidic/corrosive foods therefore needed to be incorporated together with second barer, transparent plastic and glass jars are not advisable to light reacting foods because light can react with food components (such as vitamin and lipid) and speed deteriorative chemical process such as oxidative rancidity. As interaction principle, permeation of volatile aromatic compound into or out of a packaging system causes unbalanced change in flavor profile which changes the sensory properties of the packaged food. Therefore effective packaging that can extend shelf life of a flavored food should provide a total barrier to volatile aromatic compound permeation. Packaging such as aluminum foil are used which provide inert barrier than other polymers/plastic [3]. Plastic packaging such as low-density polyethylene (LDPE), polypropylene (PP), polycarbonate (PC) and polyethylene terephthalate (PET) provide permeation flavor and oxygen as such cannot be used on highly flavored compound or products that are sensitive to oxygen such as fat and oils. The characteristic permeation of oxygen and flavor in LDPE and PP increase linearly while flavor permeation increase with increase in oxygen permeation in PC with increase in flavor absorption in PET with increase in storage duration [4]. Migration on the other hand, causes sensory changes at smaller rate and affect consumer health at a higher rate of occurrence. The effect of migration as interaction may be seen on food as color change, taints and off-flavors [5]. This chapter explored the interaction principles in a food packaging system as well as the preservation effect of the interaction and ways to estimate/ascertain the extent of the interaction, the chapter covers the interaction methods, the effects and included the methodical approach to ascertain the effect of food-packaging interaction.

1.1 Food-packaging interaction

When food products are packaged, the food is in direct contact with the inside surface of the packaging. It is possible for interaction between the food and the packaging to occur and for components of the packaging to be absorbed by, or react with, the food. Food packaging interaction involves mass transfer through which packaging material migrate to food or external environmental factors (such as odor, light or gas) permeate into the packaged food or process that involves sorption through adsorption (by packaging) or absorption (by the food) or combination of all which at end affect the initial quality/integrity of the packaging or the food product. The mass transfer occurs as a result of difference in concentration between the two medium (High and low) which causes migration at a rate equal to the driving force of the specific food component. Different packaging materials allows different mass transfer rate example; metal or glass package approve very low mass transfer if compared with package made up of plastic. The mass movement allows permeation to small molecules of water vapor, volatile gasses, organic vapors, flavors, aroma and additives from food. These movement causes change in the component's concentration gradient as such more molecule migrated with extended storage which causes quality reduction. Exchange of materials in packaging system offers both positive and negative advantage. Depending on packaging type selected, active package react or changes the packaging environment by interacting with the food enclosed (packaged) thereby extending the food shelf life. The molecular exchange in food packaging can either be through; (i) permeation (ii) migration (iii) sorption/scalping and begin from the moment the package contacts the food during production, and extends throughout the package shelf life and can produce adverse effects on the food, and/or the package. As shown in the **Figure 1** below, food-package interaction follows different pattern (permeation, migration, sorption/scalping) and causes varying degree in quality effect on the food (drying, aroma loss, moisture loss/absorption etc.). Different food components follows different exchange pattern with volatile compound mostly adopting to permeation and very low molecule motile compound (such as water molecule) following a migration and permeation sequence. The whole food-packaging interaction methods involves three elements/environment viz. (i) External environment (ii) Internal environment (food) (iii) Packaging. In Permeation pattern molecule of moisture, CO_2, O_2 and other volatile compounds move either from the

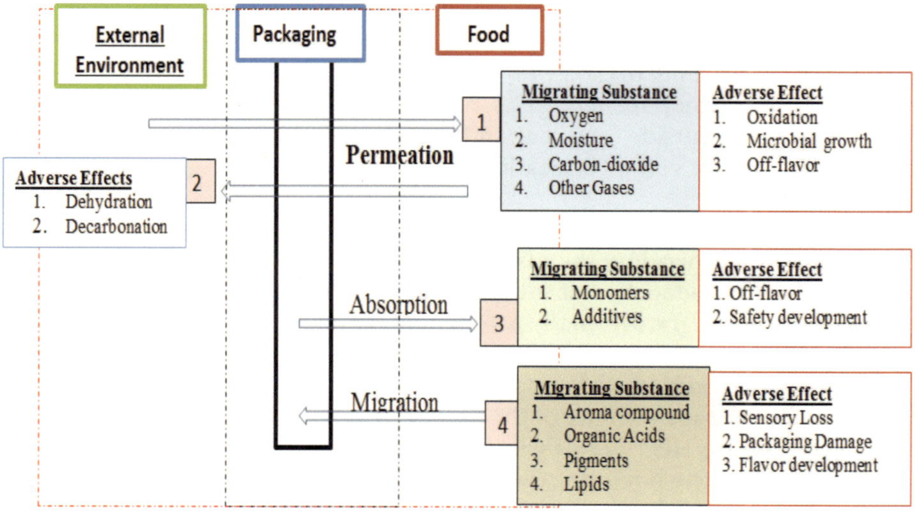

Figure 1.
Food package interaction. Adapted with modification from [6].

food or from external environment to the internal packaging environment or to the external surrounding environment through the packaging material [6] this may be due to porosity or less intactness of the packaging film. Example, some nylon packaging can allow moisture to pass thereby causing dissolution of the food packaged. In a case of sachet packs, poor sealing may allow rapid permeation (leakage) which may give rise to moisture loss/absorption, drying (dehydration), de-carbonation, off-flavors and microbial growth (as in **Figure 1** below). Leakages on sachets foods can be detected by using vacuum leak test apparatus (VLTA). VLTA machines are made in a way that they absorbed all available gasses around the packaging environment thereby given chance to only permeation of gas from the internal packaging environment that can be physically detected through gas bubble formation (when they escape from the internal packaging environment due to pressure surrounding the external environment formed by vacuum creation) as leakages from the sachets. In the testing process, the sachets are first immersed in a water in the vacuum creation section of the machine, pressure level are then set from the program which will be used to create the vacuum. While permeation covers both external and internal environment of the packaging system, migration on the other hands involves movement of packaging materials from the packaging film to the food. Migration reduce the compactness of the packaging film and can give rise to permeation when the surface packaging materials erodes and migrate to the food which then increase the characteristics porosity of the packaging film as such increase rate of components migration, migration causes sensory quality loss, packaging damage and flavor development. Except the food packaging allows, food-package interaction though inevitable but causes very limited effect to the quality of the food enclosed.

1.1.1 Permeation

Caner [6] defines permeation as the movement of gases, vapors or liquids through homogenous packaging materials, and excludes the passage of materials through perforation, cracks, or other related defects. In his own definition, Caner [6] excluded any internal or external defects that allow passage of molecule through packaging as permeation as such defines it as the characteristics of the package that allow such movement therefore; permeation to his own definition is inherent to the packaging materials used. According to Caner [6] permeation involves molecular movement of volatiles and aroma component of gases, moisture and other low molecular weight substances from the outside (external) environment into the food through the packaging materials or vice versa. Permeation process through a packaging film occurs in a three steps:

1. **Solution or Absorption:** the adsorption involves the dissolution of the food materials on to the surface of the packaging materials. In adsorption process the penetrant disintegrate from the parent food mass and the absorbed to the surface of the packaging. Adsorption process depends on the: (i) particulate nature of the food (ii) condition of the packaging environment e.g. pH, Moisture content, Acidity (iii) the absorptivity of the packaging materials to penetrant. Adsorption principle follows Henry's rule which states that the amount of gas absorbed by a given volume of a liquid at a given temperature is directly proportional to the partial pressure of the gas therefore the absorptivity of the food depends on the pressure of the internal packaging system which depends on the seallability of the package and the relative condition of the food prior to the initial packaging.

2. **Diffusion:** Through diffusion, the penetrant passes through the packaging film and are then transferred in or out of the package. Molecules in a diffusion process move from a region/place of high concentration to a region/place of a lower concentration through a permeable membrane (here the membrane is the packaging film). Diffusion principle follows Fick's law which stated that the quantity of diffusing gas is proportional to concentration and time and inversely proportional to the thickness of the substrate through which it is diffusing.

3. **Emergence/desorption**: this involves escape of penetrant from opposite surface of the packaging (either from external system or from the internal). Graham's Law best explain desorption process and states that the velocity of diffusion of a gas is inversely proportional to the square root of the density. Absorption and desorption depend on the solubility of the permeant, and solubility is greatest when penetrant and material have similar properties.

Permeation process as explained in the **Figure 2** below follows a concentration gradient because molecules are exchanged from the region where their concentration is high to a region where their concentration is low. Therefore, generally permeation process is affected by the (i) Concentration (ii) Density (iii) Solubility (iii) Internal and External Pressure/Temperature (iv) Permeability of the packaging film (v) Time (vi) Thickness of the packaging film and (vii) Relative humidity. Permeation impact the shelf life of foods, since they gain or lose components, or undergo unwanted chemical reactions with the permeating substances, therefore offers both detrimental and positive advantage. Permeation causes unbalanced flavor profile in a flavored foods, leading to change in sensory properties of the food product therefore packaging selection should ensure packing of food products with packaging material of effective barrier properties that can protect the foods packed in them for longer time not only provide enclosure.

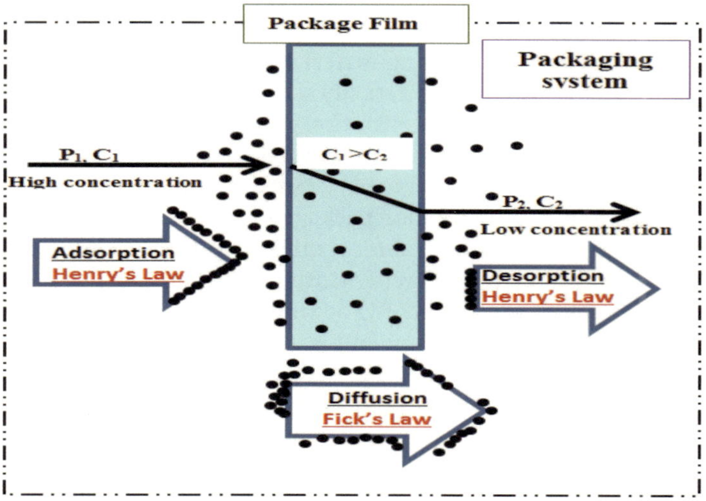

Figure 2.
Permeation principle in a food packaging system. Adapted from [6] with modification.

1.1.2 Migration

There are more than 85,000 chemicals used on consumer products [7] and more than 6000 chemicals are not to be used in food packaging that have direct contact with food [8], and all these chemical compound are prone to cause toxicity hazard if consumed when they migrated into food. Migration or otherwise called diffusion in food packaging is the movement of food substance from a region of higher concentration to a region of a lower concentration through a permeable membrane (packaging film). In food packaging, migration process involves the transfer of packaging materials from the packaging surface to the food product which causes a relative change in the original integrity/quality of the food. According to Abbes et al. [9] migration can either be global (when it involves movement from the packaging itself to the food) or specific (when only specific material moves to the food surface). Ferrara et al. [10] explain migration process in a packaging system in four steps:

1. Diffusion: Movement of the food substance through the packaging film.

2. 1st Desorption: movement of packaging material from the surface of the packaging film.

3. Sorption: movement at the food-packaging interface.

4. 2nd Desorption: movement from the packaging material to the food.

Migration in food packaging is affected by (i) physico-chemical characteristics of the food (ii) Storage time (iii) Temperature (iv) packaging size (v) type of packaging material/coating (vi) type of contact (vii) mobility of the packaging migrant. Depending on the packaging materials, diverse chemical compound migrated from the surface of food packaging and affect the shelf stability of the food. An ideal packaging should be inert (non-reactive) to the food. For inert packaging materials such as stainless steel, ceramics and glass only chemicals from the interior surface of the package migrated to the food and this causes tearing/wearing of the packaging as such affect its strength and its permeation inertness. Migration occurs due to the concentration gradient of the packaging materials and or the food. Migration follows simple Fick's law which states that; as a steady state, the rate of movement of diffusing compound is proportional to the concentration gradient [9, 11].

However, to determine the extent of migration in food packaging, food simulant are used. The simulant are formulated to have exact physicochemical characteristics of the food to be packaged, they are then packaged into the packaging materials as substitute for the food analyzed for chemical migration after a stipulated length of storage. The simulant differs in types representing different types of foods; hydrophilic (water based), lipophilic (fat based) or amphiphillic foods (food of varying properties). Example, vegetable oil simulant is used to measure migration into oily foods, 10% ethanol or 3% acetic acid are used for a water based drinks, 50% ethanol solution is used as simulant for butter and other amphiphillic foods. Using simulant for the estimation of migration gives only probable estimated values which are close to the actual values. Other processes such as Chemical Risk Assessment (CRA) are conducted to determine the risk of toxicity due to chemical migration into foods; Migration Models (such as Stochastic, Probabilis tic, and Empirical) are also employed to determine the risk and extent of packaging migration.

1.2 Sorption

Sorption or otherwise called scalping involves the mass adsorption or absorption of integral sensory quality components of food such as Favor, Aromas, Lipids and Moisture to the or by the packaging material [12] resulting in the reduction of quality of the packaged products. Sorption in packaging system occurs through both Adsorption and Absorption procedure. Adsorption involves mass transfer from the food to the surface of the packaging material, this resulting to an increased in the concentration of the food components at the interface. Sorption of aromatic components results in degradation of quality. This phenomenon of loss in quality of food product by absorption of flavor from food by polymer or vice versa is known as 'scalping' [13]. For scalping to occur, a thermodynamically favorable condition must exist [14]. But the major concern of flavor sorption is that loss in very small amount of flavor has significant effect on quality of the stored food product depending on the component sorbed [14]. A study on packing of orange juice in glass bottles and polyethylene- laminated cardboard packages found that after 24 weeks of storage at 4°C, up to 50% of d-limonene and little aldehydes and alcohols were absorbed into polyethylene-laminated cardboard packages [15]. But, this had little/no effect on the sensory properties of the orange juice [15]. At times, sorption can result in swelling of the packaging material, leading to increased migration and permeation. Further, it can reduce the mechanical properties of the polymer [16–18]. For instance, the absorption of limonene on LDPE increased the oxygen permeability of the polymer. Flavor absorption depends on characteristics of polymer (polarity, crystallinity, chain stiffness), flavor compounds (concentration, chemical composition, polarity) and environmental and external factors like temperature, relative humidity, duration of storage and composition of food matrix [6]. For liquid foods, sorption is mainly affected by partition coefficient of flavor components, whereas for solid foods, the sorption is affected by solubility and vapor pressure of components in the polymer [19].

Similar to migration, sorption is also a diffusion process. Hence, Fick's law governs sorption as well [14]. Further, sorption process is also influenced by temperature. In general, sorption and temperature are positively related. Exceptionally, sorption and temperature are negatively correlated in few cases. For instance, sorption of vinyl chloride by dry casein particles was found to reduce with increasing temperature [17, 18]. Pressure on the other hand increases sorption process the way temperature do, the internal pressure of the packaging system reduce the mechanical quality of the packaging at a higher level, high swelling pressure of the packaging film may cause mechanical disintegration of the components of the packaging film thereby making them more available to be sorbed as such increases sorption process. Pressure level with prolonged storage condition increase sorption that is why it is generally required that a packaging material should be inert to external factors. The nature of the sorbing material also affect the rate of sorption process, with chemical (such as coding inks and surface monomers) of high reaction affinity been sorbed faster. In sorption process, the quantity of volatile component sorbed by the polymer can be measured by the parameter 'solubility' [20].

1.3 Effect of food-packaging interaction

The interaction of food and packaging possess negative and positive effect to both the food and the packaging materials. When packaging components migrate from the surface of the parent package it causes withering which reduces the compact characteristics of the package. When external environmental factors migrated into the internal packaging system, it causes physical, chemical and microbiological changes to the food. Flavors and other aromatic volatile compound easily migrate out of the package from the surface of food to the external environment through a withered package. The mass movement of vapor from surface of aqueous food causes drying which affect the organoleptic and physical characteristics of the packaged food. Movement of fillers, colorant, inks and other additive from the surface of plastic films causes dangerous toxicity effect. As described in **Figure 1** above, food-package interaction possess different adverse effect ranging from:

1. Physical changes (color change, dehydration/drying and packaging damage)

2. Microbial growth (contamination gasses/chemical possess serious health effect)

3. Chemical changes (oxidation of fats, browning, off-favor, decarbonation)

4. Sensory changes (crunches, softness, off-flavor/odor)

Food-packaging interaction as well as Food-packaging and environment interaction plays a major role on quality of the product as well as integrity of the package. These effects may have direct or indirect effect on the sensory attributes of the food product, consumer health and shelf stability of the food thereby causing a direct impact on the market value and the overall acceptability of the product as such affect business good-will. Moreover, Food and Drugs Administration of the United States (FDA) and National Agency of Food and Drugs Administration (NAFDAC, 2022) of Nigeria sets a strict legislations for a 'zero tolerance' of carcinogenic migrants, and also considered all migration compounds in food as indirect food additive. Hence, knowledge on factors affecting interaction phenomena and its effect are of high importance. In this section, factors affecting the food-packaging interaction and their possible effects on food quality are discussed.

1.4 Factors affecting packaging interaction

The shelf life of packaged food is dependent on numerous factors such as the intrinsic nature of the food, e.g. acidity (pH), water activity (aw), nutrient content, occurrence of antimicrobial compounds, redox potential, respiration rate and biological structure, and extrinsic factors, e.g. temperature, relative humidity (RH) and the surrounding gaseous composition. These factors will directly influence the chemical, biochemical, physical and microbiological spoilage mechanisms of individual food products and their achievable shelf lives. By carefully considering all of these factors, it is possible to evaluate existing and developing active packaging technologies and apply them for maintaining the quality and extending the shelf life of different food products.

1.4.1 Food composition

Food is a complex compound containing varying amount of components composing of volatile and non-volatile substance is primarily of protein, lipids, carbohydrate and water [21, 22]. These components of food interact with each other and with the wall of packaging materials [23]. These particulate component of food are held together by a chemical bonding such as (i) hydrogen bond (ii) covalent bond (iii) hydrophobic bond/vander waals (iv) physical binding [24] and they disintegrate/ separated at a different petition rate depending on the amount of interaction with micro-molecules in the food matrix. The composition containing fat/oil had major effect on flavor interaction, followed by proteins and polysaccharides and then by disaccharides [4]. Example lipids (fats and oils) determine the physical and sensory characteristics of non-water soluble foods as well as determine their ability to absorbed or loose flavor compounds to or from packaging films. Aqueous and high water foods tend to interact with packaging more than solid or powdered foods therefore very prone to packaging interaction. Increase in food pH also increase the rate of package migration [6], nature and concentration of migrating compound in the food also influences the rate of packaging interaction, example presence of same flavor compound of food on the packaging materials will speed of the rate of scalping process [12]. In nutshell, the selection of packaging for foods should consider the nature of the foods such as; (i) acidity (carbon chain and structure (iii) viscosity (iv) concentration (v) flavor and aroma compound presence (vi) molecular weight of the food (vii) carbon functional group and polarity.

Depending on the food nature, interaction of food with packaging is greatly affected by the food composition, the components of food can either be interactive/migrating (occurs through leaching or volatile system) or non-interactive/ non-migration. Migrating compound in food are usually very reactive, loose and less chemically bonded to the food therefore can easily interact with the packaging film. Migrating volatile aromatic compound in dried foods can easily be loss through either diffusion or evaporation, desorption from product or adsorption on to the product without direct contact with the packaging film. For a leaching migrating food system, components of food interact when they are in direct contact with the packaging film, that is the food (such as fat) components diffuse from the food surface, dissolved on the surface of the packaging and the adsorbed by the packaging film, or by its diffusion from the packaging materials, followed by dissolution and dispersion into the food. High fatty foods leach faster than high fiber foods due to their higher affinity and reduced viscosity.

1.4.2 Nature of the packaging material

For an ideal packaging, the packaging material should be; (i) non-reactive to the food (ii) non corrosive (iii) impermeable (iv) sealable (v). Packaging materials such as glass and metal provide total barrier to external factors therefore migration occurs only form the internal contact surfaces. Semipermeable packaging material such as plastics offers limited resistance to permeation and migration therefore used only on properly selected food materials. Porous materials such as paper and paper boards facilitate rapid migration [25]. Interaction rate depends on factors such as (i) molecular weight (ii) density (additive present (iii) temperature (iv) crystallability of the packaging material use. With the increase in use of polymer based packaging materials, over 42% of a the polymers produced are used as packaging materials, and about half are used for food packaging applications [26, 27]. They are usually preferred in food packaging due to their flexibility, low density, strength, ease to mold, cost, controlled hydrophobicity and moldability into different sizes and shape. Several researches shows that there is increase in interaction of food if package in polymer based packaging which can induce undesirable quality changes in packed food. For instance, degree of browning and ascorbic acid degradation of orange and grapefruit juice was found to be high when packed in polyethylene-laminated cartons than in glass [28]. Bott et al. [29] reported that the rate of interaction of LDPE and polystearine decrease by 10 fold with the increase in the molecular weight of the packaging material. Migration rate in polypropylene increases with decrease in the crystallability [30] rate of migration in polyprpyene packaging increase with decrease in crystallability of the polymer.

Interaction of packaging material with the food and the environment plays a major role on quality of the product as well as integrity of the package. These effects have impact on market value, goodwill of the manufacturer, sensory attributes of products and health effect of consumers. Moreover, strict legislations are set for 'zero tolerance' of carcinogenic migrants, and also the compounds that migrate into food are considered as indirect food additive by FDA. Hence, knowledge on factors affecting interaction phenomena and its effect are of high importance. The ability of the packaging to absorbed or release light should also be considered in selecting package for oily/fatty foods, example, polyolefins are not advisable for the packaging of lipophilic food [30]. Other than the polymer itself, the nature of additives like colorants, plasticizers, stabilizers, fillers, blowing agents, antioxidants, antimicrobials, slip compounds and printing ink also migrates into the food material, and their characteristics impact the rate of migration [31] which may be toxic if consumed with food. Example, residual ethylene oxide on metal cans is highly toxic [32], concentration of tin lacquer at about 500mgkg^{-1} can cause gastrointestinal ailments [33] and usually attain acute threshold at about >730 mgkg^{-1}. Lacquers are usually added to reduce interaction of food and packaging as well as with external environment by reducing oxygen scavenging [34] lead coating on beverage containers can cause damage to central nervous system and has negative growth impact, it could also result to mental retardation [35] chemical components of paper packaging such as dioxins, nitrosamines, chlorophenols, chloroanisoles and benzophenone are toxic if found in food [35–37]. 0.06–1.0% of acetaldehyde was detected in different beverages packaged on PET plastic containers by Lau & Wong [38], acetaldehyde usually impact odors on food especially in cola-type beverages [39] therefore its presence in food is of great significance.

Health related risk assessment from chemical and materials to be used in producing food packaging should be considered and thoroughly studied before used, to prevent contact and potential migration of these chemical/materials into the food which may be carcinogens or caused other ailments. Trace of metal, additives, inks and plastics from the packaging or from environment due to failure of packaging should be eliminated to prevent the food and the consumer health. Hazard related to presence of metals contaminants in foods raised serious health concerns. Acute and chronic symptoms such as dizziness, nausea, diarrhea, loss of appetite, disorders, vomiting and reduced contraception rate has been associated with metal toxicity, and these symptom may results to a serious cardiovascular diseases, suppressed growth, impaired fertility, immune disorders/failure and or neurological ailments which may lead to death [39].

1.4.3 Environmental and storage factors

Environmental factors such as (i) temperature (ii) relative humidity (iii) storage time (iv) moisture content (v) oxygen and other gaseous content. Change in environmental temperature affect the stability of the packaging-food interaction rate by increasing the mobility of interacting material, free volume and the swelling capacity of the packaging film. Storage time also increases the rate of food-packaging interaction with heat processed cans suffering more migration effect than non-heat processed materials [40]. Different packaging material offers different migration rate in relation to change in the environmental relative humidity. Permeation through EVOH packs increase with increase in environmental humidity while there is decrease in permeation in PET and nylon with increase in relative humidity of storage environment.

2. Novel food packaging technology

The world population is on the increase. Currently, there are over 7.8 Billion people living on this planet, this number is expected to reach 9 billion by 2050. With this rise in population and pressure on food, from ancient system of gathering food to modern day food, food production value chains fears that global food production could not meet up with the population growth of animals on earth. This leads to the need to improve people's access to sufficient, nutritious and value adding food while handling the climate impact on the environment. And as the demand for food increases, peoples also require food access that will meet their personal convenience/need. Novel arose as a result of consumer's desire for convenient, ready to eat, tasty and mild processed food products with extended shelf life and maintained quality. Recent trend of lifestyle changes with less time for consumers to prepare foods posed a great challenge toward food packaging sector for the evolution of novel and innovative food packaging techniques. The novel food packaging techniques includes; (1) Active packaging (AP) (2) Intelligent packaging (IP) and (3) Bio-active packaging which involve intentional interaction with the food or its surroundings and influence on consumer's health have been the major innovations in the field of packaging technology. These novel techniques act by prolonging the shelf life, enhancing or maintaining the quality, providing indication and to regulate freshness of food product. The advancement in novel food packaging technologies involves retardation in oxidation, hindered respiratory process, prevention of microbial attack, prevention of moisture infusion, use of CO_2 scavengers/emitters, ethylene scavengers, aroma emitters, time–temperature sensors, ripeness indicators, biosensors and sustained release of antioxidants during storage. The novel food packaging technologies besides the basic function of containment increase the margin of food quality and safety. The novel food packaging techniques thus help in fulfilling the demands throughout the food supply chain by gearing up toward persons own lifestyle (**Figure 3**).

Figure 3.
Model for packaging function [41].

2.1 Active packaging

Active packaging refers to the integration of certain additives into packaging film or within packaging containers with the aim of preserving and prolonging shelf life [30]. Packaging may be designated active when it accomplishes some desired role in food preservation other than providing barrier to external environments [42, 43]. The primary aim of food packaging is to protect and extend the shelf-life of the packaged foods thereby making it available for later use or extend transport advantage. Active packaging provides such advantage without endangering the quality and safety integrity of the food. By definition, active packaging technology are those packaging technology that provide protection, safety, store as well as maintain food quality integrity without impacting to its original sensory attributes, this is achieved through modification of the food pack condition. It is also a packaging system which provide enclosure to the product and the environment interaction to prolong shelf-life or to enhance safety and/or quality of the food. Active packaging provide packaging protection to all agent of food deterioration; such as Chemical activities (such as lipid oxidation and volatile chemical loss), Physical impact (including protection against impact and compression factors), Moisture invasion (such as sorption and drying), Physiological processes (Such as respiration in root and tubers, Microbiological effect (including spoilage microorganism attack) and Insect/small animal infestation. Active packaging includes additives or

Active packaging system	Mechanisms	Food application
Preservatives releasers (Usually added on the surface of a packaging that inhibit or slow down some deteriorating activities in or on the package food)	1. Extract (from spices, herbs & Animal). 2. Antioxidants (e.g. Vitamin E and BHA/BHT) 3. Ethanol capsule 4. Organic Acids 5. Silver zeolite	Meat, Fish, Cheese, Baked products (e.g. bread and snacks), cereal food products, fruits and vegetables
Carbon dioxide Scavenger/ Emitters (they are used to increase or reduce the CO2 content in a package thereby inhibiting the surface growth of microorganism [44])	1. Activated charcoal 2. Iron oxide/hydroxide of calcium 3. Ascorbate/sodium bicarbonate 4. Ferrous carbonate/metal halide	Snack foods and cakes, coffee and tea, nuts, fish-meat and their products.
Ethylene Scavengers	1. Activated carbon 2. Activated clays/zeolite 3. Potassium permanganate	Fruit and vegetables
Oxygen Scavengers (Used to remove or reduced oxygen in a package of food thereby extending shelf life usually in a Modified atmosphere packaging [45])	1. Iron based 2. Enzyme based 3. Metal catalyst 4. Metal acids 5. Metal salts	Baked foods (biscuit, breads and cakes), cheese, meat and fish, dried foods, beverage, pasta and cured meat.
Ethanol emitters	1. Alcohol sprays 2. Encapsulated ethanol	Baked foods (bread, biscuit, etc.), pizza
Moisture Absorbers	1. PVA blanket 2. Silica gels 3. Activated catalyst	Fish, meat, cereal, dried foods, fruit and vegetables, sandwiches and snacks.

Active packaging system	Mechanisms	Food application
Flavor/ordor absorbers	1. Cellulose triacetate 2. Acetylated paper 3. Citric acid 4. Ferrous salt/ascorbate 5. Activated carbon/clays/zeolite	Poultry, dairy foods, fruit and fruit juices, fried snacks and cereals
Temperature control package	1. Non-woven plastics 2. Double walled containers 3. Lime/lime water 4. Ammonium nitrate/water 5. Hydroflourocarbon gas	Meat and meat products, fish, poultry and beverages

Table 1.
Active packaging system and their food application.

freshness enhancers that are capable of scavenging oxygen; adsorbing carbon dioxide, moisture, ethylene and/or flavor/odor taints; releasing ethanol, sorbates, antioxidants and/or other preservatives; and/or preserving temperature control. **Table 1** contain list of active packaging and their food application [30, 43].

2.2 Intelligent packaging

Intelligent packaging is an emerging and existing area of food technology that can provide better food preservation and extra convenience benefits for consumers [46]. Intelligent packaging provides information about the food product enclosed in them prior to provision of the natural packaging properties. They are integrated with a target-specific sensor (TSS), which can store information about the history of a quality attribute, such as freshness, gas leakage, microbial contamination, product demography and footprints, etc., and deliver this information to a consumer. Intelligent or smart packaging refers to a packaging system that senses and informs [30]. They are integrated in a packaging material with a sensing devices that are capable of sensing and providing information about the function and properties of food and assurances of pack integrity, tempering evidence, product safety and quality, and also utilized in authenticating, anti-theft and product traceability [30]. Sensing device in Intelligent packaging system include (i) Time–temperature indicators (TTI), (ii) Gas sensing dyes, (iii) Microbial growth indicators, (iv) Physical shock indicators, and (v) Tracing device such as tamper proof, anti-counterfeiting and anti-theft technologies [30]. Intelligent packaging also provide detailed information about shelf life and quality state of the food in order to ensure consumer satisfaction and safety while potentially improving logistics and minimizing the losses. Other sensing devices in Intelligent Packaging includes biosensors, Radio frequency identification (RFID) tags or electronic tracking tags that stores and wirelessly transmit information about the food packaged and ease traceability.

1. **Time–temperature Indicators:** In Time–temperatures Indicators, the shelf life of food products or their freshness is determined based on selected indicators, such as vitamins, color or flavor change, enzyme activity, etc. through predictive models which are developed based on accelerated shelf-life tests, usually on meat and meat product, fish and sea foods. TTIs sensors can either be

(i) diffusion-based (diffusion of colored esters forms color contrast alongside a reference scale), (ii) Enzymatic (certain enzymatic reaction that can change the environmental conditions, such as pH, which then causes color change), or (iii) Polymer-based (polymerization reaction forms a color contrast with a reference scale) depending on their material of formation. All TTIs commonly require an activation step to start the sensing process at the same time when the product enters the package.

2. **Gas Sensors:** In Gas sensors packaging, sensitivity to change in gas levels is used as a quality marker inside the package and is used to detect any gas leakage in a Modified Atmosphere Packaging Systems (MAPS). This gas leakage can either be due to microbial contamination, fruit ripening by the formation of aromatic volatiles and degree of fermentation or by the formation of organic acids. Gas sensors works with binding reactions such as redox reactions, pH change, or luminescent dyes to produce color (for easy to interpretation) controlled by the change in the target gas concentrations.

3. **Biosensors** are used to provide online about microbial contamination, growth, and related biological reactions. Biosensors accomplish their usefulness by detecting and following the development of secondary metabolites, such as volatile nitrogen compounds, sulfide indicators, ethanol, organic acids, etc., due to biological activities (e.g., respiration and fermentation) (**Table 2**).

2.3 Bio-active packaging

Bio-active packaging is those packaging materials that employ the use of active biological materials to support/improve shelf stability of package foods. They are novel concept of technologies intended to help in the production of functional foods, whose bioactive principles and actuators are devised to be contained within packaging or coating materials [47]. The bioactive compound is extra nutritional

Indicator	Principle	Information	Application
Microbial growth Indicators (use both in internal and external package surface).	1. pH dyes. 2. Metabolite reacting dyes.	Microbial quality of the packaged food.	Perishable goods such as tomato, meat, fish and poultry.
Time–temperature indicators (TTI) (placed externally).	1. Mechanical. 2. Chemical. 3. Enzymatic.	Storage condition.	For food stored under very low temperature condition (chilled and Frozen).
Carbon dioxide indicators (used internally).	1. Chemical	1. Storage condition. 2. Leakage on packs.	Modified and controlled atmosphere packaging.
Pathogen Indicators (internally).	Chemicals that react with microbial released toxins.	Used for pathogen specific bacteria such as *E. coli*.	For perishable food such as meat, fish and poultry.

Table 2.
Intelligent packaging indicators and their application.

components present in a small amount in foods such as fruits, vegetables, legumes, herbs and spices. They are usually extracted from these foods as probiotics, polyphenos, essential oils, fatty acid or vitamins or as bacteriocin extracted from bacteria.

3. Food-packaging interaction testing

The most important quality of packaging materials is to provide protection to the food by providing enclosure which provide total barrier to external impact. As seen above, food-package interaction follows different pattern and offered several detrimental effect to the food quality/integrity. Packaging testing is necessary step in package selection to ensure that applicable packaging material is used. Quality testing as well as testing to measure the amount of interacting substance in a packaging system is very important for the manufacturer to ensure that standard packaged are employed for enclosure of the food and also to make sure that packaging material has negligible effect on the final product getting the consumer and it does not cause any negative impact on health after consumption of the packed food material. For the detection and analysis of the interacting components in a packaging system the following system are employed (i) simulants (ii) mathematical models (iii) predictive models and (iv) analytical/chromatographic techniques are used.

3.1 Simulation model

Food simulant are compound developed to mimic the natural characteristics of food to be packaged. Food simulant are developed to have the exact physicochemical properties of the food under study. After the development, the simulant are then packaged into the packaging film kept under controlled/studied condition for a stated time to determine the interacting characteristics of the food. Example, vegetable oil simulant is used to measure migration into oily foods, 10% ethanol or 3% acetic acid are used for a water based drinks, 50% ethanol solution is used as simulant for butter and other amphiphillic foods. Using simulant for the estimation of migration gives only probable estimated values which are close to the actual values. The migration quantity is determined by evaporating the simulant and then calculating the weight of the remaining residue. Simulant procedure is limited especially on fatty foods due to their difficulty in vaporization. ASTM standard are put in place for the estimation of odor and taste transfer from a packaging material, the standard employed 0.9 M^2 test material kept in a required environment for a minimum of 20 hours (**Table 3**) [49].

Solvent	Simulant	Type of food
Distilled water	Simulant A	Aqueous food (pH > 4.5)
Aqueous acetic acid (3%w/v)	Simulant B	Acidic foods such as fruit juices (pH <4.5)
Aqueous ethanol (15% -50% w/v)	Simulant C	Diary foods and emulsions
Vegetable oils e.g. Sunflower/olive oil	Simulant D	Oily foods, high fat content foods

Table 3.
List of some common simulant used for food-package interaction testing [3, 48].

3.2 Mathematical and predictive models

Mathematical models are predictive equation developed based on a simulation principle to predict the migration in food packaging. The models were developed to determine the diffusion and partition coefficient and explain the interacting substance concentration [50] since it is only when there is partition and when diffusion occurs that packaging materials are said to be loss. Permeation in packaging can also be determine in term of gaseous and moisture transfer rate. Since material permeability depends of the type and characteristics of the packaging film (such as thickness) as well as the condition of the surrounding environment. This is achieved by placing the packaging material in both low and high pressure environment followed by the measurement of pressure difference (both pressure and volume determined) usually expressed as the amount of gas permeated per unit time. Moisture permeation is determined by placing the package in between environment of different humidity rate and estimated by determining the vapor pressure difference in M^2day^{-1} [51].

3.3 Analytical/chromatographic method

Chromatographic techniques were employed to determine trace of metal compound that migrate to food surface. Though there is no standard or direct analytical method for the estimation of packaging migrant in a packaging system, different chromatographic method such as GC–MS and LC-UV. Begley [52] uses LC–MS method to determine interacting component in a packaging system. Although not all migrating compound in a packaging system can be detected using conventional method but the application of chromatographic method prove to be effective.

3.4 Stochastic/predictive models

Stochastic models or predictive models are probabilistic mathematical function that provide a prediction of certain level of migration (food-packaging interaction) of packaging materials in a packaging system [53]. Stochastic models takes account the variability and uncertainty as well as the probability of the occurrence of food-packaging interaction. Latin Hypercube Sampling (LHS) and Monte Carlo Models (MCM) are example of stochastic models that gives numerical values of food migration based on numerical distribution data build through simulation method. Stochastic thermodynamics models measure the thermodynamic changes due to food packaging interaction such as temperature, pressure and moisture changes while molecular models of a stochastic methods measure the rate of molecular changes/interaction of gaseous or chemical molecules arises due to disintegration of either packaging or food components and subsequent molecular interaction [8, 54]. Stochastic models such as Life Cycle Assessment (CA) and High-Throughput Risk-based Screening (HTRS) tools are used to determine extent of migration in plastic packaging while empirical Weibull Model has been used to determine the chemical migration curves in paper packaging [55], Mechanistic Models are used to predict the migration of toxic metals into acidic food [56] while diffusion models are used for the determination of migration in ceramic packaging as well as packaging with low migration risk or plastic additives with low diffusion potentials.

Predictive models are used to generate data and also been associated with statistic functions to develop software programs that simplify the determination methods for food-packaging interaction. The European Flavors, Additives and food Contact materials Exposure Task (FACET) developed a software for the prediction of probabilistic exposure to chemicals from food or packaging contact materials. The software measure the migration, permeation as well as sorption of flavors, moisture, food additives and packaging surface materials in a packaging system using existing probabilistic data in its database [8], the software consist of data that will be able to study over 6000 substances covering metal coatings, paper and paper boards, inks and adhesive, plastic and plastic coverings as well as components of food flavors and additives [8] with over 600 statistics functions all links to packaging use, composition, application, pack size, storage, environmental conditions and food composition/nature. Measurement is achieved through clustering based on chemical, physical properties (polarity and diffusion) [8] and physic-chemical parameters [57] other modeling software includes; MIGRATEST lite 2000/2001 [53], AKTS-SM by Advanced Kinetics Technology Solutions AG Switzerland, SMEWISE (Simulation of Migration Experiments with Swelling Effect), MULTITEMP, MULTIWISE, and SFPP3 by National Institute for Agricultural Research, and FMECAengine (Failure Mode Effect and Critically Analysis) [53, 57]. With the continues rise of artificial intelligence, the future of predictive models is bright and look at possibility of developing robotic system that will simplify the study of food-packaging interaction and its advent.

4. Conclusion

Food-packaging interaction is a systematic procedure that may take long to proliferate, but has direct solid effect to the food integrity, it directly either render the food or the packaging helpless to either external or internal factors which may speed changes that will affect the shelf stability of the food. If the interaction is on the negative side, the food and the packaging losses their initial state of intactness and therefore losses their natural role in packaging system. As evolved from this module, different packaging materials react differently to adverse chemical, physical and environmental factors and that can facilitate to improve or mitigate interaction with food. The compounding effect of food-packaging interaction is critical to selection and even for the preparation of food therefore, critical attention should be made on studying the nature and type of the packaging to be used as well as to the studying the length of storage, condition, the logistic requirement to delivering the product to consumer. Packaging is not only the enclosure of food but part of the food system that can either be detriment or of advantage to the food depending on the type of interaction it has with the food. Novel packaging technologies as new trends brought about solution to reduction of the adverse effect of food-packaging interaction which improve on and reduce to cost of production, shelf stability of food products as well as the environmental impact of packaging. Because it is inevitable, food-packaging interaction can be determine through several methods that are evolving with the continuous evolution of informative computer management system and the artificial intelligence. More work need to be done on the development of better predictive tools for the determination of migration/interaction in paper packaging. Since the science of food is analytical, effort should be made also on the development of fast and precise analytical methods for the determination of food-packaging interaction.

Conflict of interest

The author declares zero conflict of interest and no funding from any person or organization is received in an aim to deliver this chapter. All citation and reference are duly acknowledged.

Notes

I hereby declared that this work was carried out by me, unless where reference is made. I thank my Mom for her immeasurable support throughout the course of delivering this content.

Author details

Abubakar Ibrahim Garba
Federal University Dutsin-Ma, Katsina, Nigeria

*Address all correspondence to: abubakargzango@gmail.com

IntechOpen

© 2023 The Author(s). Licensee IntechOpen. This chapter is distributed under the terms of the Creative Commons Attribution License (http://creativecommons.org/licenses/by/3.0), which permits unrestricted use, distribution, and reproduction in any medium, provided the original work is properly cited. (cc) BY

References

[1] Muhammed Shafiur Rahman Food Preservation: Overview in Handbook of Food Preservation, Second Edition. CRC Press. 2007. p. 3-19

[2] Marshall MR, Adams JP, Williams JW. Flavor absorption by aseptic packaging materials. In: Aseptipak: Proceedings of the 3rd International Conference and Exhibition on Aseptic Packaging. Princeton, NJ: Scotland Business Research. 1985;**02**:299-312

[3] Lamberti M, Escher F. Aluminium foil as a food packaging material in comparison with other materials. Food Review International. 2007;**23**:407-433

[4] van Willige RW, Linssen JP, Voragen AG. Influence of food matrix on absorption of flavour compounds by linear low-density polyethylene: Proteins and carbohydrates. Journal of the Science of Food and Agriculture. 2000;**80**:1779-1789

[5] Risch SJ, Hotchkiss JH. Food and Packaging Interaction II. Washington DC: American Chemical Society; 1991

[6] Caner C. Fundamentals of the sorption (scalping) phenomena in packaged foods. Reference Module in Food Science. 2017:1-12. DOI: 10.1016/B978-0-08-100596-5-22346-9

[7] Goldman LR, Koduru S. Chemicals in the environment and developmental toxicity to children: A public health and policy persfective. Environmental Health Persfective. 2000;**108**(3):443-448

[8] Oldring PI et al. Development of new modeling tool (FACET) to assess exposure to chemical migrant from food packaging. Food Additives & Contaminants. Part A, Chemistry, Analysis, Control, Exposure & Risk Assessment. 2013;**31**:444-465

[9] Abbes B, Abbes F, Guo Y-Q. Interaction phenomena between packaging and product. In: Alavi S, Thomas S, Sandeep K, et al., editors. Polymers for Packaging Applications. 1st ed. New Jersey: Apple Academic Press; 2015. pp. 39-70

[10] Ferrara G, Bertoldo M, Scoponi M, Ciardelli F. Diffusion coefficient and activation energy of Irganox 1010 in poly(propylene-co-ethylene) copolymers. Polymer Degradation and Stability. 2001;**73**:411-416

[11] Bhunia K, Sablani SS, Tang J, Rasco B. Migration of chemical compounds from packaging polymers during microwave, conventional heat treatment, and storage. Comprehensive Reviews in Food Science and Food Safety. 2013;**12**:523-545

[12] Caner C. Sorption phenomena in packaged foods: Factors affecting sorption processes in package–product systems. Packaging Technology and Science. 2011;**24**:259-270

[13] Sajilata MG, Savitha K, Singhal RS, Kanetkar VR. Scalping of flavors in packaged foods. Comprehensive Reviews in Food Science and Food Safety. 2007;**6**:17

[14] Gnanasekharan V, Floros JD. Migration and sorption phenomena in packaged foods migration and sorption phenomena in packaged foods. Critical Reviews in Food Science and Nutrition. 1997;**37**:519-559

[15] Pieper G, Borgudd L, Ackermann P, Fellers P. Absorption of aroma volatiles of orange juice into laminated carton packages did not affect sensory

quality. Journal of Food Science. 1992;**57**:1408-1411

[16] Piringer O. Mathematical modelling of chemical migration from food contact materials. In: Barnes KA, Sinclair CR, Watson DH, editors. Chemical Engineering Journal. 1st ed. New York: CRC Press; 2007. pp. 180-202

[17] Helmroth E, Rijk R, Dekker M, Jongen W. Predictive modelling of migration from packaging materials into food products for regulatory purposes. Trends in Food Science and Technology. 2002;**13**:102-109

[18] Biran D, Giacin JR, Hayakawa K, Gilbert SG. Vinylchloride sorption by dry casein particles: Mechanistic considerations. Journal of Food Science. 1979;44:59-61

[19] Lebossé R, Ducruet V, Feigenbaum A. Interactions between reactive aroma compounds from model citrus juice with polypropylene packaging fi lm. Journal of Agricultural and Food Chemistry. 1997;**45**:2836-2842

[20] Siracusa V. Food packaging permeability behavior: A report. International Journal of Polymer Science. 2012;**2012**:302029. DOI: 10.1155/2012/302029

[21] Dury-Brun C, Chalier P, Desobry S, Voilley A. Multiple mass transfers of small volatile molecules through flexible food packaging. Food Review International. 2007;**23**:199-255

[22] Guth H, Rusu M. Food matrices - of odorant partition coefficients and application of models for their prediction. Food Chemistry. 2008;**108**:1208-1216

[23] Giroux HJ, Perreault V, Britten M. Characterization of hydrophobic flavor release profile in oil-in-water emulsions. Journal of Food Science. 2007;**72**:125-129

[24] Naknean P, Meenune M. Factors affecting retention and release of amour compounds in food carbohydrates. International Food Research Journal. 2010;**17**:23-34

[25] Castle L. Chemical migration into food: An overview. In: Barnes KA, Sinclair CR, Watson DH, editors. Chemical Migration and Food Contact Materials. 1st ed. New York: CRC Press; 2007. pp. 1-13

[26] Silvestre C, Duraccio D, Cimmino S. Food packaging based on polymer nanomaterials. Progress in Polymer Science. 2011;**36**:1766-1782. DOI: 10.1016/j.progpolymsci

[27] Rhim J, Park H, Ha C. Bio-nanocomposites for food packaging applications. Progress in Polymer Science. 2013;**38**:1629-1652. DOI: 10.1016/j.progpolymsci

[28] Mannheim CH, Miltz J, Letzter A. Interaction between polyethylene laminated cartons and aseptically packed citrus juices. Journal of Food Science. 1987;**52**:737-740

[29] Bott J, Störmer A, Franz R. Migration of nanoparticles from plastic packaging materials containing carbon black into foodstuffs. Food Additives and Contaminants. 2014;**31**:1769-1782

[30] Day BPF. Underlying principles of active packaging technology. Food, Cosmetics and Drug Packaging. 2001;**23**(7):134-139

[31] Guart A, Bono-Blay F, Borrell A, Lacorte S. Migration of plasticizers, phthalates, bisphenol a and alkylphenols from plastic containers and evaluation of risk. Food Additives & Contaminants. Part A, Chemistry, Analysis, Control, Exposure & Risk Assessment. 2011;**28**:676-685. DOI: 10.1080/19440049.2011.555845

[32] Food Standards. Survey of Chemical Migration from Food-Contact Packaging Materials in Australian Food. Food Standards Australia New Zealand; 2010

[33] Omori Y, Takanaka A, Tanaka S, Ikeda Y. Experimental studies on toxicity of tin in canned orange juice. Journal of the Food Hygienic Society of Japan. 1973;**14**(1):69-74

[34] Oldering PKT. Exposure-the missing element for assessing the safety of migrant from food packaging materials. In: Barnes KA, Sinclair R, Watson D, editors. Chemical Migration and Food Contact Materials. Cambridge, UK: Woodhead Publishing; 2007. pp. 122-157

[35] Robertson G. Safety and legislative aspect of packaging. Food Packaging Principle and Practice. 2006;**3**:473-502

[36] Ackermann PW, Herrmann T, Stehr C, Ball M. Status of the PCDD and PCDF contamination of commercial milk caused by milk cartons. Chemosphere. 2006;**63**:670-675

[37] Kirwan M, Brown H, Wiliams J. Packaged product quality and shelf lif. In: Coles R, Kirwan M, editors. Food and Beverage Packaging Technology. Second ed. London: Willey-Blackwell; 2011. pp. 59-83

[38] Lau O, Wong S. Contamination in food packaging material. Journal of Chromatography. A. 2000;**888**:255-270

[39] Claudio L. Packaging and public health. Environmental Health Perspectives. 2012;**120**(6):A232-A237

[40] Munguia-Lopez EM, Gerardo-Lugo S, Peralta E, et al. Migration of bisphenol a (BPA) from can coatings into a fatty-food simulant and tuna fish. Food Additives and Contaminants. 2005;**22**:892-898

[41] Yam KL, Takhistov PT, Miltz J. Intelligent packaging: Concepts and applications. Journal of Food Science. 2005;**70**:R1-R10

[42] Kwapong OY, Hotchkiss JH. Comparative sorption of aroma compounds by polyethylene and lonomer food-contact plastics sensory evaluation. Journal of Food Science. 1987;**52**:761-763

[43] Rooney ML. Active Food Packaging. Blackie Academic and Professional. Glasgow, UK: Chapman & Hall; 1995

[44] Lee DS. Carbon dioxide absorbers in food packaging applications. Trends in Food Science and Technology. 2016;**57**:146-155

[45] Miltz J, Perry M. Evaluation of the performance of iron-based oxygen scavengers with comments on their optimal application. Packaging Technology and Science. 2005;**18**:21-27

[46] Jung H. Han Innovations in food packaging: In Food Science and Technology, Academic Press 2005. P. 14

[47] Lapez-Rubio Bioactive Packaging Strategies. Woodhead publishing. 2011. p. 460-482

[48] Ajaj A, Jbari S, Ononogbo A, Buonpcore F, Mayes A, Morgan A. An insight into the growing concerns of styrene monomer and poly(styrene) fragment migration into food and drink simulant from poly(styrene) packaging. Food. 2021;**10**:1136

[49] Tice P. Packaging materials as a source of taints. In: Saxby MJ, editor. Food Taints and off-Flavors. Berlin: Springer Science and Businees Media; 1996. pp. 226-263

[50] Samsudin H, Auras R, Mishra D, et al. Migration of antioxidants from polylactic acid films: A parameter

estimation approach and an overview of the current mass transfer models. Food Research International. 2018;**103**:515-528

[51] Huang J, Qian X. Comparison of test methods for measuring water vapor permeability of fabrics. Textile Research Journal. 2008;**78**:342-352

[52] Begley TH, White K, Honigfort P, et al. Perfluorochemicals: Potential sources of and migration from food packaging. Food Additives and Contaminants. 2005;**22**:1023-1031

[53] Pocas MF et al. A critical survey of predictive mathematical models for migration from packaging. Critical Reviews in Food Science and Nutrition. 2008;**46**(10):913-928

[54] Nguyen PM et al. Moecular thermodynamics for food science and engineering. Food Research International. 2016;**88**(A):91-104

[55] Cai H et al. Migration kinetics of four photoinitiators from paper food packaging to solid food simulants. Food Additives & Contaminants. Part A, Chemistry, Analysis, Control, Exposure & Risk Assessment. 2017;**34**(9):1632-1642

[56] Dong Z et al. Migration model of toxic metals from ceramic food contact materials into acidic food. Food Packaging Technology and Science. 2015;**28**:545-556

[57] Seiler et al. Correlation of food stuffs with ethanol-water mixture with regards to solubility of migrants from food contact materials. Food Additives & Contaminants. Part A, Chemistry, Analysis, Control, Exposure & Risk Assessment. 2013;**31**:498-511